T0229135

Three-Dimensional Object Recognition Systems

ADVANCES IN IMAGE COMMUNICATION 1

Three-Dimensional Object Recognition Systems

Edited by

Anil K. Jain
Department of Computer Science
Michigan State University
East Lansing, Michigan
USA

Patrick J. Flynn
School of Electrical Engineering and Computer Science
Washington State University
Pullman, Washington
USA

ELSEVIER
Amsterdam – London – New York – Tokyo   1993

ELSEVIER SCIENCE PUBLISHERS B.V.
Sara Burgerhartstraat 25
P.O. Box 211, 1000 AE Amsterdam, The Netherlands

ISBN: 0 444 89797 6

© 1993 Elsevier Science Publishers B.V. All rights reserved.

No part of this publication may be reproduced, stored in a retrieval system or transmitted in any form or by any means, electronic, mechanical, photocopying, recording or otherwise, without the prior written permission of the publisher, Elsevier Science Publishers B.V., Copyright & Permissions Department, P.O. Box 521, 1000 AM Amsterdam, The Netherlands.

Special regulations for readers in the U.S.A. – This publication has been registered with the Copyright Clearance Center Inc. (CCC), Salem, Massachusetts. Information can be obtained from the CCC about conditions under which photocopies of parts of this publication may be made in the U.S.A. All other copyright questions, including photocopying outside of the U.S.A., should be referred to the copyright owner, Elsevier Science Publishers B.V., unless otherwise specified.

No responsibility is assumed by the publisher for any injury and/or damage to persons or property as a matter of products liability, negligence or otherwise, or from any use or operation of any methods, products, instructions or ideas contained in the material herein.

This book is printed on acid-free paper.

Printed and bound by CPI Antony Rowe, Eastbourne

Preface

1. Introduction

The design and construction of three-dimensional object recognition systems has long occupied the attention of many computer vision researchers. The variety of systems that have been developed for this challenging and difficult task is evidence both of its strong appeal to researchers and its applicability to modern manufacturing, industrial, military, and consumer environments. Three-dimensional object recognition is a problem of interest to researchers due to both a desire to endow computers with robust visual capabilities, and the variety of applications which would benefit from mature and robust systems. However, 3D object recognition is also a very difficult problem, and few systems have been developed for production use; most existing systems were developed for experimental use by researchers. Part of the reason for the lack of penetration into the market by 3D recognition systems is undoubtedly the high computational requirement of current recognition methods.

Our goal in organizing this edited collection is twofold. We first want to summarize the state of the art in three-dimensional object recognition using examples of systems developed by leading researchers in the field. We also hope to give practical advice to those implementing model-based 3D object recognition systems. To achieve these goals, we felt that an edited collection of fairly large chapters, describing in most cases complete systems, was preferable to a research monograph.

It is appropriate to give pointers to surveys of this popular sub-area of computer vision research. Comprehensive surveys of the state of the art (*circa* 1985) in 3D object representation and recognition appear in Besl and Jain [2] and Chin and Dyer [3]. Suetens *et al.* [6] is a broad survey article, not limited to 3D recognition. Arman and Aggarwal [1] and Flynn and Jain [4] are very recent surveys of progress in object recognition from range data. The topic of 3D object recognition was the subject of an IEEE Computer Society Workshop on the Interpretation of 3D Scenes held in November 1989 in Austin, Texas. More recently, the October 1991 and February 1992 issues of the *IEEE Transactions on Pattern Analysis and Machine Intelligence* were devoted to 3D scene interpretation. Of related interest is the area of *CAD-Based Vision*, which was the focus of an IEEE Computer Society Workshop in June, 1991 held in Maui, Hawaii. Expanded versions of several papers presented at that workshop appear in the February, 1992 issue of *Computer Vision, Graphics, and Image Processing: Image Understanding*. Broader comments about the productivity and scope of computer vision research appear in [5].

2. The Object Recognition Problem

A three-dimensional object recognition system makes decisions about the content of a scene imaged by a sensor. These decisions can include *identification* of one or more objects in the scene using a previously-constructed database of models, *localization* of those models (estimation of a rigid transformation which aligns the object model with its image), and *learning* of new object models from the image data. Applications of interest to researchers in this area include inspection, sorting, and navigation.

Object recognition systems are frequently decomposed into modules devoted to *sensing, early processing, feature extraction, modeling,* and *matching.* Most current systems employ either range or intensity sensors. Early processing tasks include image filtering for noise removal, edge pixel identification, and image-structured feature estimation (*e.g.,* surface curvature estimation for range data). In the feature extraction stage, the primitive curves, surfaces, or volumes to be used in matching are identified and described. The modeling system contains a representation of each object in terms of features comparable to those extracted from the input image, and should include facilities for addition of new objects as well as refinement of existing models. The matching module builds correspondences between image and model features.

The primary differences between current techniques relate to the types of features used (*i.e.,* the modeling strategy) and the technique for building feature correspondences. Popular choices for object models employ contours (edges), surfaces (planar, quadratic, or free-form), or volumes (superquadric primitives). An emerging theme in the literature is the use of CAD models for vision tasks. Popular matching strategies include constrained search of the tree of interpretations (sets of bindings between scene and model features), geometric hashing (exploiting invariant image and model features to build correspondences), and automated programming (generation of an object recognition program directly from an object model).

While 3D object recognition research has shown significant progress over the past several years, some barriers to continued advancement must be overcome before these systems receive significant use in applications. One critical problem is the *scaling behavior* of the system; as the size of object model databases increase, techniques for *indexing* the model database to quickly reject models from consideration are beginning to receive serious attention. Another barrier is representational in nature; current modeling strategies can competently represent rigid polyhedral or piecewise-quadric objects, but few systems can accommodate objects with free-form ('sculpted') surfaces or articulations. Relaxing these restrictions will allow the object recognition systems of the future to handle a much wider class of objects. We also believe that standardized image and software databases have an important role in current research. Sharing of data and software among research groups is important to research productivity, and such sharing is already taking place (albeit on an informal basis).

3. In This Collection

Each chapter in this volume was written by a separate researcher or research group. Most of the chapters relate to object recognition systems but we also invited chapters on biological vision, sensing, and early processing because appreciation for the problems involved in the latter can and should influence design decisions made in the former. Opinions expressed in these chapters reflect the opinions of the authors alone, and should not be taken as representative of the opinions or policies of the agencies funding the research, the other authors, or the Editors.

3.1. Biological Vision, Sensing, and Early Processing

Caelli *et al.* compare and contrast computational and biological 3D object recognition systems and theories, emphasizing both shared themes and lessons already learned in one domain but not yet learned in the other. Jarvis reports on the state of the art in range sensing, organizing current sensing technology along the orthogonal axes of active/passive, image-based/direct, and monocular/multiview. Ferrie *et al.* present a comprehensive treatment of the estimation of differential-geometric features from dense range data; such features (being view-invariant) are widely used in the higher-level processing stages of an object recognition system. Stevenson and Delp describe an approach for reconstruction of piecewise-smooth 3D surfaces from discrete samples.

3.2. Recognition and Pose Estimation

Eleven chapters address specific systems for recognition and/or localization of 3D objects. Arman and Aggarwal describe a strategy tree-based recognition system employing range data and CAD models developed on a commercial system and described using the IGES standard. Bolle *et al.* encode object models in active networks with the explicit goal of rapid indexing in the model database. Camps *et al.* describe the PREMIO object recognition system which exploits knowledge about lighting and viewing (more specifically, their effects on feature detectability) to optimize a constrained correspondence search. Dhome *et al.* address the problem of 3D object localization from several different combinations of intensity image contours. Dickinson employs volumetric primitives (superquadrics) to efficiently index an object model database for recognition using 2D features in intensity imagery. Gremban and Ikeuchi describe the appearance-based vision paradigm and the use of vision algorithm compilers for recognition with explicit sensor and object models; two example systems are presented. Grimson *et al.* summarize the constrained-search approach to object recognition and describe a practical system for recognition and localization of piping. Gupta and Bajcsy address recognition of models described as superquadric aggregates in range data, including explicit treatment of the extraction of 3D primitives from surface features. Huttenlocher proposes a closed-form method for calculation of an alignment transformation from corresponding features and describes a recognition system based on this idea. Ponce *et al.* summarize their work in curved object recognition and localization using algebraic surface representations and intersection curves. Stein and Medioni describe the TOSS object recognition system, which uses encoded image contours and invariant local features to rapidly index a model database of objects not necessarily described in terms of simple surface types.

3.3. Emerging Topics

Four additional chapters address ideas that we feel will assume greater importance as 3D vision research matures. The chapter by Asada addresses a data fusion problem of great interest to researchers building autonomous systems; his system integrates range views into an evolving world model incorporating information about (for example) obstacles. DeCurtins *et al.* incorporate sensor specifications, range data, and physics to characterize the spatial arrangement of sets of objects; this information will prove useful when manipulation of objects in the arrangement is required. Pentland *et al.* motivate modal object descriptions for different applications; such models are based on the finite element method and offer both increased geometric descriptiveness and modeling of object dynamics. Finally, Sutton *et al.* describe the GRUFF object recognition system which uses object form and function (rather than strictly geometric features) in modeling and matching.

4. Conclusion

We wish to thank the chapter authors for their timely and valuable contributions to this volume. We also thank Cathy Davison for her help with contractual matters. We are grateful to Ir. Hans van der Nat and Ir. Mark Eligh of Elsevier, and Prof. Jon Biemond of the Delft Technical University (Editor of the series *Advances in Image Communication*) for their interest in this project. To the reader: we hope you find this work useful in your application area.

East Lansing, Michigan   Anil K. Jain
Pullman, Washington   Patrick J. Flynn
December 1, 1992

REFERENCES

1 Farshid Arman and J.K. Aggarwal. Model-based Object Recognition in Dense Range Images–A Review. *ACM Computing Surveys*, in press.
2 Paul J. Besl and Ramesh C. Jain. Three-Dimensional Object Recognition. *ACM Computing Surveys*, 17(1):75–145, March 1985.
3 Roland T. Chin and Charles R. Dyer. Model-Based Recognition in Robot Vision. *ACM Computing Surveys*, 18(1):67–108, March 1986.
4 Patrick J. Flynn and Anil K. Jain. Three-Dimensional Object Recognition. In Tzay Y. Young, editor, *Handbook of Pattern Recognition and Image Processing: Computer Vision*, volume 2. Academic Press, in press.
5 Shahriar Negahdaripour and Anil K. Jain. Challenges in Computer Vision: Future Research Directions. In *Proc. IEEE Computer Society Conference on Computer Vision and Pattern Recognition*, pages 189–199, June 1992.
6 Paul Suetens, Pascal Fua, and Andrew J. Hanson. Computational Strategies for Object Recognition. *ACM Computing Surveys*, 24(1):5–61, March 1992.

Table of Contents

Contributors

J.K. Aggarwal, University of Texas at Austin
Farshid Arman, Siemens Corporate Research
Minoru Asada, Osaka University
Ruzena Bajcsy, University of Pennsylvania
Ruud Bolle, IBM T.J. Watson Research Center
Kevin Bowyer, University of South Florida
Terry Caelli, University of Melbourne
Andrea Califano, IBM T.J. Watson Research Center
Octavia Camps, Penn State University
Cregg Cowan, SRI International
Jeff DeCurtins, SRI International
Edward Delp, Purdue University
Michel Dhome, Blaise Pascal University of Clermont-Ferrand
Sven Dickinson, University of Toronto
Irfan Essa, Massachusetts Institute of Technology
Frank Ferrie, McGill University
Keith Gremban, Carnegie Mellon University
W. Eric L. Grimson, Massachusetts Institute of Technology
Alok Gupta, Siemens Corporate Research
Robert Haralick, University of Washington
Bradley Horowitz, Massachusetts Institute of Technology
Daniel Huttenlocher, Cornell University
Katsushi Ikeuchi, Carnegie Mellon University
Ray Jarvis, Monash University
Michael Johnston, University of Melbourne
Rick Kjeldsen, IBM T.J. Watson Research Center
David Kriegman, Yale University
Jean-Thierry Lapresté, Blaise Pascal University of Clermont-Ferrand
Tomás Lozano-Pérez, Massachusetts Institute of Technology
Shailendra Mathur, McGill University
Gérard Medioni, University of Southern California
Rakesh Mohan, IBM T.J. Watson Research Center
Prasanna Mulgaonkar, SRI International
Norman Noble, SOCS Research
Alex Pentland, Massachusetts Institute of Technology
Sylvain Petitjean, University of Illinois
Jean Ponce, University of Illinois
Marc Richetin, Blaise Pascal University of Clermont-Ferrand
Gérard Rives, Blaise Pascal University of Clermont-Ferrand
Terry Robison, University of Melbourne
Stan Sclaroff, Massachusetts Institute of Technology
Linda Shapiro, University of Washington
Gerard Soucy, McGill University
Louise Stark, University of South Florida
Fridtjof Stein, University of Southern California
Robert Stevenson, University of Notre Dame
Steven Sullivan, University of Illinois
Melanie Sutton, University of South Florida
Gabriel Taubin, IBM T.J. Watson Research Center
B. Vijayakumar, Yale University
Steven White, Massachusetts Institute of Technology

Three-Dimensional Object Recognition Systems
A.K. Jain and P.J. Flynn (Editors)
© 1993 Elsevier Science Publishers B.V. All rights reserved.

3D Object Recognition: Inspirations and Lessons from Biological Vision

Terry Caelli, Michael Johnston and Terry Robison[a]

[a] Department of Computer Science
The University of Melbourne
Parkville, Vic. 3052
Australia

1. Introduction

Visual object recognition may be defined as the process by which image data is referenced to known descriptions of objects. Consequently, models of both machine and human object recognition involve the extraction and interpretation of image features which can index 3D world structures from what is sensed. However, "visual systems" are restricted to the sensing and processing of information which can be displayed as 2D projections, possibly varying over time, and also restricted by specific constraints such as the lack of transparency. Such restrictions on acceptable representations for vision systems prohibit, for example, higher dimensional analysis, seeing "inside" without a view, and specific topologies. It does not restrict vision to the study of images produced by passive means nor exclude images generated by artificial or active sensors (for example, ultra-sound, infra-red or range sensors). Any object recognition system (ORS) which claims either to describe human object recognition, or to be useful in machine object recognition applications must include the following properties:

- It must be possible to recognize objects with some degree of view invariance.

- The data structure for the stored description of objects should be such that it is possible to access it through partial information as, for example, with single views, degraded, or partially occluded object data.

- A description must be general enough to include any known stimulus within the object class to which it refers, and specific enough to exclude other stimuli.

Theories of machine object recognition are inherently algorithmic while physiological and psychophysical descriptions are typically more qualitative, often due to practical difficulties in the observation of brain structure and function but, sometimes, also due to what the biological vision (BV) scientific community is prepared to accept as "theory". That is, most theories in biological vision do not consist of *algorithms* which actually predict behavior in explicit detail but, rather, are *descriptions* of encoding and information processing.

Machine vision (MV) provides representations and algorithms for solving 3D vision problems (e.g. Jain and Hoffman [1], Fan, Medioni and Nevatia [2]). However, these theories are not aimed at satisfying biological constraints. Rather, they are typically developed as a result of current applications of engineering technologies and preconceived notions of just what constitutes an 'efficient' solution. Many ORS's developed in the MV tradition have not been adequately evaluated and often lack the generality of a biologically-based ORS. Further, there is no standards for quantifying the performance machine ORS's - at least up to recently.

Experimental research on human object recognition has focussed mainly on sub-processes and general qualitative descriptions rather than providing complete computational models. Those "models" which do exist are typically undetailed and lack algorithmic description (e.g. Triesman [3], Biederman [4]). Because of the complexity and abstractness of the psychological approach to object recognition, it has been easier to investigate it in a piecemeal fashion. To preempt the conclusion of this chapter, the MV conceptual analyses of how we *understand* the processes of recognition in machines, may provide BV with questions and tools which will allow for a truly explicit computational theory of BV to evolve. From a BV perspective, MV theories may provide a theoretical outline which must be translated into the language of experimental psychology and specified in ways which allow for psychophysical experimentation. The thesis of this chapter is that BV and MV can mutually learn from each other and, in particular, MV offers the theory necessary for BV to ask appropriate questions of biological information processing.

Indeed, one can well argue that in many areas of biological inquiry, successful languages and models are borrowed from more formal areas of science: Mathematics, Physics, Engineering and Artificial Intelligence. Perhaps the most relevant example of this, in BV, is that of the concept of a *receptive field* (RF). This is the notion that we can describe the information processing characteristics of a nerve cell in terms of the stimulus which best activates it. RF's were first delimited in the mammalian retina by Ratliff and Hartline [5] during a period when the formulation of filter theory, adaptive filters, and their underlying control system (network) formalization were taking place. That is, the formal definition and analyses of such structures gave BV scientists the "tools" for representing what **may** be processed at a given level of function. This notion of filtering became very popular in the 1960's to the present in most areas of BV since it fit well with the notion that the activity of individual neurons is sufficient to describe visual processing - the "Neuron Doctrine" [6]. The notion that BV systems do filtering has, in turn, motivated many engineers and computer scientists to construct pre-processing procedures which incorporate such mechanisms. However, the latter is done less critically than one would expect and so there are many misconceptions in the MV community about what we really know about BV. For example, it is by no means proven that there are only four, six or even a fixed number of frequency bands of the signal that the visual system is selectively sensitivity to. It is certainly not the case that the system is insensitive to phase nor even that the global Fourier transform is computed by the biological vision system [7,8].

In a similar vein, a number of BV scientists have borrowed representations and procedures from the MV literature without due critical analysis of the status of such models. For example, the work of Marr [9] has been taken too literally and definitively representative of developments in MV by the BV community and, here, we wish to clarify these misconceptions and demonstrate the both BV and MV have much to offer each other so long as each area understands the status of theory and results in the other. In particular, MV scientists should remember that BV is less formal and less is known objectively about how the system works: BV is driven by observation whereas MV is typically part of Information Science.

This paper is focused on one central problem in vision: Object Recognition Systems (ORS) – the ability of systems to store and/or learn the descriptions of 3D objects and recognize them in 2D data structures invariant to their rigid motion, viewer position and, to some extent, environmental complexity. Common to all approaches to ORS, in man or machines, are the following two major questions:

- How do we generate/store 3D models including their explicit feature-based representation?

- How do we match 2D data to such models?

Associated with both questions are specific questions related to:

- How do we sense depth or shape from image data?

- How do we encode basic surface information?

- How do we segment models and/or data?

- What features do we use for model/data representations?

- How do we actually do 3D-ORS in BV, or, efficiently in machines?

In the following sections we briefly review each such process and, where possible, compare ideas and results from BV and MV.

2. Theory in Biological and Machine Vision

In MV, the term "theory" typically refers to a description of a process by algorithms which are implementable on digital computers. In the psychological literature the term is used more loosely to refer to a description of a process which is testable with psychophysical experimentation. To illustrate this difference, two psychological theories are examined in brief. These "theories" are the Recognition By Components model of Biederman [4], and the Feature Integration Theory of Treisman [3].

The Recognition by Components(RBC) model provides a non-algorithmic and sketchy account of "primal access: The first contact of a perceptual input from an isolated, unanticipated object, to a representation in memory" (Biederman [4], p. 32). Briefly, according to this theory, an image is segmented into regions and combined with information about non-accidental properties in order to describe the 3D components in the scene – which are known as "geons". Geons have the highly advantageous property of being finite in number; they are defined in terms of non-accidental properties of generalized cones. In fact, apart from lacking explanatory validity, in that it includes no algorithms, this model is not even descriptively complete. No account is given of how geons and their spatial relationships may be encoded. However, there has been some psychological research following up the RBC theory but, as can be seen from the following examples, because its processes are not described in detail, any number of models, other than RBC, could explain the findings. For example, Biederman and Ginny [10] examined surface and edge information in recognition. They found that subjects recognized full color photographs and line drawings with the same degree of speed and accuracy. They concluded that line descriptions of visual images are what are used for object recognition, and that surface information such as color and texture gradient may be useful in edge extraction, but is not directly useful in object recognition tasks. This conclusion is claimed to be supportive of the RBC model, though, from the authors' perspective, it is difficult to imagine any theory that it wouldn't support!

On the other hand, Price and Humphreys [11] found that both color and surface detail information can enhance object recognition. They suggested that the degree to which surface detail is useful is proportional to the amount of within category discrimination required: features are relative to the classification problem – a characteristic of human pattern recognition well-documented in 2D vision (see, for example, Caelli, Bischof and Liu [12]). Other research, in the geon tradition, includes that of Biederman and Cooper [13], who primed subjects on images with either half the features (lines and vertices), or half the geon components removed. Priming facilitation was tested, using identical, complementary, and different class samples. For feature primed subjects, there was no difference in priming between identical and complimentary images, meaning the actual features present in the original image had nothing to do with the priming. Different exemplars showed somewhat less priming. For component primed subjects, the identical images showed more priming than the complementary ones: priming took place at the component level, rather than the feature level. This suggests that early preprocessing in biological ORS involves some form of segmentation and it was claimed that these results support RBC. The difficulty with this is that RBC is not well defined and, again, the results are so predictable from most ORS theories that one can hardly use them to strongly support RBC, per se.

Physiological evidence shows that in early stages of visual processing, color, orientation, motion, etc. are analyzed by separate but related channels (Livingstone and Hubel [14]). According the feature integration theory (FIT) of Treisman and Galade [15], a visual scene is processed first along these lines of separate dimension. Where attention is focussed, these features are combined into "unitary objects". They are then encoded in memory in this unitary form. However, with memory decay, the features may once again become dissociated. Without attention, these features cannot become associated in the first place. This model is interesting in that it brings visual attention into play, an aspect which is neglected in most models of ORS. However, it suffers from the same problems as Biederman's RBC model, in that it is descriptive, but not explanatory, in terms of algorithms. Further, it should be noted that such features are not even conceptually independent in so far as color and motion must be indexed by space, pattern and form – unless they are derived from some perceptual "ether." Indeed, we have recently shown how this notion of modularity actually misses rich properties of the signal, at least with respect to spatio-chromatic information processing (Caelli and Reye [16]).

3. Sensing Objects

During a time when most vision research was related to understanding how to encode and process images as signals, per se – in man and machines – one scientist made a stand which was to become the basis for a change in vision research: the "Gibsonian" position (see Gibson [17]). Paraphrasing and consolidating what was really a very qualitative and introspective inquiry into visual perception, we can simply point out that Gibson's essential "insight" was that the various algorithms which constitute the "act of perception" must have evolved to reference **world** structures and not just image features, per se. This perspective on vision was taken up by the Artificial Intelligence (AI) community while both Image Processing and BV scientists were still mainly focussed on how the image is encoded, etc. However, the early AI approaches to such a view were mainly symbolic, as in "Block World" interpreters [18] and not until the late 1970's was there any formal attempt to integrate known image processing technologies of visual systems with properties of objects and structures within the world around us. This is not to say that BV avoided such issues. Indeed, "ecological optics" (a term used by Gibson to describe this perspective) was of central interest to many BV researchers from motion, spatial encoding, through to color vision research programmes. The main difference, however, was that experiments were conducted which really did not enable us to unambiguously infer how given processing systems actually encode the signal to infer world structures. This is particularly true of most of the work done on the threshold detection of grating patterns which not only have not provided of consistent theory of threshold spatial vision but has not been able to predict how humans process above threshold images, in general, nor how such information is used to solve problems of inferring depth or shape. On the other hand, exceptions to this situation are the studies of Reichardt [19] on understanding the encoding of motion by flies in flight, and the experimental work by Johannson [20] on how observers solve motion correspondence problems with the motion of realistic shapes. However, fundamental research into the variety of passive sensing sources and just how they are used in inferring shape from image intensity information is still evolving in the BV literature.

Though most recent techniques for ORS in MV use range data, there is still need for intensity-based ORS not only because of its relevance to BV but also because of the robustness of passive sensing for industrial and other applications. Hence there is still strong interest (see, for example, Seibert and Waxman [21]) in both MV and BV circles in problems of inferring depth from passive sensing resources such as focus, stereo, motion, perspective/texture. What follows is a brief resume of the known relationships between MV solutions to Shape-from-X and the associated results, where possible, from BV.

- **Shape-from-Shading.** Here, with appropriate assumptions and one image, or, with photometric stereo, etc., it is possible to generate shape and depth from intensity information with knowledge of the rendering model. The algorithms are generally of the relaxation type, propagating depth from boundaries - as has been also shown to be the case with human vision [22,23].

- **Shape-from-Focus.** Here, at least 2 images are required and the degree of blur about a pixel (measured, say, by variance) is used to fit a blur [24] or even point spread function [25] from which depth is determined. Since the human eye, under photopic vision, functions with a small aperture, focus is not a strong cue, per se. However, accommodation of the lens is, implicitly, a response and/or source for depth though the exact degree of afferent-to-efferent processing with this is still yet to be fully determined. That is, the visual system does adjust the power of the lenses with respect to distance but the degree to which this can serve as a depth cue is not resolved.

- **Shape-from-Stereo.** Whether it be with two or multiple images, the use of image disparity information to infer depth is well established in photogram-metry or stereo procedures. However, the problems with this resource's insensitivity to small disparities and the "correspondence problem" with large ones, still leaves this as a source needing additional analyses [26]. However, human vision does use binocular disparity to infer depth and cells have been found within the vertebrate visual cortex which directly encode this and so produce a signal which can be be used by higher-order neurons to infer depth [27].

- **Shape-from-Motion.** Motion flow has been studied in both areas as a strong resource for depth and, again, without large disparities the process is quite insensitive. With large displacements however, there is, again, a "correspon-dence problem". Recent neurophysiological data shows that vertebrates also have neurons selectively sensitive to various types of relative motions including looming and linear motion components [28]. Also, recent evidence suggests that the same type of constraint satisfaction algorithm used to solve motion correspondence problems in MV may well explain and predict similar problems in BV [29].

- **Shape-from-Perspective/Texture.** This resource for MV is not that popular due to the fact that many imaging devices have little perspective. In the human visual system, depth may be inferred from edges, which shows that the preprocessed image, in terms of lines, bars and edges, described by Marr [9] as the "primal sketch", for example, is enough to provide sparse depth information. The psychological literature has many reports of how observers use this cue for depth – so long as these types of features are present [17].

However, for passive sensing there is one major limitation: no depth or shape can be inferred from pixels whose neighborhood variances are zero: where there is no variation in light or "features". This shows that the inference of full range from intensity information cannot be obtained without prior or "top-down" knowledge or constraints. Indeed, recent solutions to this "interpolation" problem [30,31] demonstrate how this knowledge can be introduced either locally and algorithmically, or, globally via knowledge of the actual models involved.

To this stage, however, only broad relations between the known physiology of intensity processing and the extraction of depth are available in the sense that it is well established that the visual system is capable of differentiating the intensity image via the hierarchy of orientation-specific receptive fields. However, at this stage the only depth resources with direct physiological substrate are stereo and motion.

4. Parts and Features

It is widely accepted in the both the psychological (e.g. Biederman [4], Hoffman and Richards [32], Braunstein, Hoffman and Saidpour [33]), and in the machine literature (e.g. Jain and Hoffman [1], Fan, Medioni and Nevatia [2]), that some form of image segmentation must occur before recognition can take place.

Current practice in the MV literature is to segment surface data into parts as a function of various degrees of prior knowledge or constraints. The purpose of segmentation in this literature is to reduce the matching problem and obtain descriptions which are more robust and less dependent of specific pixel-based matching criteria. However, the segmentation problem is underconstrained without additional knowledge and, equally, segmentation does not necessarily imply that the complete surface needs to be partitioned into surface "parts". That is, one form of segmentation is the location of surface "edges", corners, etc., without determining what is non-edge, etc.

The issue of segmentation for ORS's, and for range data, specifically, has received a good deal of attention in recent years [35,36]. Common to most approaches is the development of surface part clustering in terms of similarities in surface point position, normals, or curvature information or surface curve fitting parameters [35,36]. Actual techniques vary from simply grouping via curvature sign (-,0,+) values or by complex clustering algorithms with hybrid constraints [36] to actually merging and splitting initial clustered regions to be consistent with known part properties in the model database [1].

Incorporating the notion of image segmentation into a model of human ORS is a very good way of providing the flexibility required for recognition from novel views, and recognition from partially occluded or degraded images. Segmentation reduces geometric information about an object into discrete, manageable chunks. In this way, parts may be recognized in isolation, making it unnecessary for all object parts to be visible for object recognition to take place. It is necessary, however, to include information not only about the segmented parts themselves, but also about the spatial relationships between these parts. One common view of the segmentation problem is that of defining where the boundaries between parts should occur. Herein lies the strong contrast between segmentation in MV and BV.

Hoffman and Richards proposed an algorithm for how humans segment surfaces that many others have already used in MV – in one form or another [2], and they termed it the minima rule: "Divide a surface into parts at loci of negative minima of each principal curvature, along its associated lines of curvature" (Hoffman and Richards [32], p.275). In intuitive terms, this means that part boundaries will occur at regions of concavity on the surface of the object, so that parts would consist of protruding surfaces. Segmentation using the minima rule has two obvious benefits: It yields a segmentation schema which is view invariant, and it requires no knowledge about the object itself; it is dependent only on the geometry of the object surface. Braunstein, Hoffman and Saidpour [33] found that when presented with object parts partitioned either at negative minima, or at positive maxima, and asked to choose a part that matched an object, subjects usually chose one partitioned at negative minima. Furthermore, when asked to mark part boundaries of novel objects manually, all subjects partitioned the objects at negative minima. This research shows evidence for the minima rule, although it does have some problems. The stimuli used were random dot patterns, which were rotationally symmetrical in the vertical plane. It is also clear that simple curvature extrema are not sufficient for all types of segmentations - as is clear from the MV segmentation algorithms where, for example, it is important to identify different surface types, etc. [2]. Another important consideration for segmentation is the role of jump boundaries, which commonly mark regions at which adjacent objects meet. If a model of object segmentation is to have high ecological validity, it should account for these additional requirements for part type descriptions.

For MV, surface features are usually of three generic forms [1]. One, Morphological(M): features derived from the complete object or model; two, unary(U): features extracted from individual parts, and three, binary(B): features derived from part relationships. Unary features can refer to typical surface patch pixels (local, as in curvatures), global patch properties (such as areas) and can also refer to patch boundary properties (such as perimeter). Binary features typically capture part relationships such as distances, angles, and also include boundary relationships. Those used in the literature are shown in Table 1(right column). These features, again, fall into seven types: Morphological (M) unary curvatures(U.C), unary distance (U.D), unary boundary (U.B) and binary boundary (B.B), binary distance (B.D), and binary angles (B.A) [37]. Those computational forms, as shown in the right column of Table 1, correspond to ways of "measuring" such features as real-valued functions whose values, wherever possible are invariant to rigid motions.

5. Encoding Objects

MV model representations are typically surface-based, either obtained via active sensing, CAD, or the use of Shape-from-X methods. Given the constraints of surface "visibility," most approaches represent models via "aspect graphs" where the fully view-independent representation is defined by having adequate numbers of views. This, we believe is consistent with what we understand as 3D model knowledge in so far as it is presumably impossible to have fully view-independent models

Table 1
Typical Surface Features used in Machine ORS's

Feature	ORS1-measure
Morphological	M.1 Perimeter M.2 Number of Parts M.3 Total Area M.4 Genus (Mean K)
Sense	U.C.1 Mean H U.C.2 Mean K
Size	U.D.1 Area U.D.2 3D Spanning Distance (max)
Boundary Features	U.B.1 Perimeter U.B.2 Mean Curvature U.B.3 Mean Torsion
Boundary Type	B.B.1 Length of Jumps B.B.2 Length of Creases
Part Distance Relations	B.D.1 Bounding Distance B.D.2 Centroid Distance B.D.3 Max Distance
Part Angle Relations	B.A.1 Normal Angle Differences (average) B.A.2 Bounding Angle between surfaces (average) B.A.3 Normal Angle Differences (average)

without having viewed all surface points or without having equivalent alternative knowledge. This is not to be confused with the issue of how such data is symbolically encoded. For example, one can have enough data to have a full description of a given 3D object but if surface descriptors are used which are not invariant to rigid motions then the model cannot be said to be "view-independent". For this reason, most MV models consist of range surface mean and Gaussian curvature descriptors which are invariant to rigid motions whereas their original (x, y, z) surface coordinates are not. As will be seen in the following section, for MV, objects are fundamentally represented via "shape descriptors" which are "positionless" in the sense that the curvatures are fully independent of the 3D location of the object. As seen in Table 1, such descriptors are then used to segment and, in general, extract model features which can be found in data – invariant to position and pose.

The internal representation of objects in humans is a dynamic memory process, in that it involves a model of objects which may be both accessed and modified by perceptual information. Current models of internal representation are rooted in early work on 2D shape recognition. For example, the work of Deutsch [38] and Sutherland [39] found that shapes could be recognized independently of location, size and brightness, but not orientation. Theories of internal representation in human object recognition may be grouped into three divisions (see Pinker [40] for a review): view-independent versus object centered models, single-view-plus-transformation models, and multi-view models.

According to the view-independent models, objects are represented as a collection of spatially independent features, such as intersections, angles, curves and surfaces. View-independent theories assign an object a representation that is the same regardless of its orientation, location, or size. Opposing this are object-centered theories, under which an object representation may be described as a data-base consisting of a store of descriptions, from multiple view-points, with which an image may be directly compared. Rock, DiVita, and Barbeito [41] found that novel wire objects shown in one position, are not often recognized when they were rotated about a vertical axis and presented later. Rock, DiVita, and Barbeito [42] also found that mirror images and left-right reversals are difficult to discriminate. Rock, Wheeler, and Tudor [43] asked subjects to imagine how these 3D wire objects would appear from positions other than the one they were in. They found that subjects were unable to perform this task unless they made use of strategies that circumvent the process of visualisation. This suggests that objects are not simply rotated about in space as seen on a CAD computer screen, but, in fact, draw on several of the available heuristics (as, for example, 2D projected feature similarities) available to humans when performing such a task.

According to single-view-plus-transformation models, object recognition is achieved via transformation, typically mental rotation, of an input stimulus into either a perspective, or an orthogonal (canonical orientation) view. In these models, mental representations are defined in terms of the end products of the transformation process. Shepard and Metzler [44] found that reaction time in a "same-different" (binary classification) matching task was a linear function of the angular difference between two geometrically identical figures. This was true for rotations in the plane and for rotations in depth. Countering this research, experiments investigating orientation-independence have provided arguments against the notion of reaction time being a linear function of the angular difference in rotation. Corballis and Nagourney [45] suggested that the time required to name normal versions of letters and digits was largely independent of the orientation of the characters. Orientation-independence in recognition time seems to occur only for highly familiar combinations of shapes and orientations; when unfamiliar stimuli must be recognized, orientation effects reminiscent of mental rotation appear. This suggests that humans may sometimes use mental rotation to recognize unfamiliar shapes or examples of shapes. However, the actual computational procedure including internal data structures and matching criteria have not usually been specified.

Because there is no theory of form perception that explains the necessity of mental rotation, Takano [46] became interested in why mental rotation appears to happen only in certain instances but not in others. If mental rotation does not occur, then simple template matching theories of form recognition would have difficulty coping with the storage and computational requirements in performing their task adequately. To deal with this problem, feature extraction theories have been proposed. These theories suggest that recognition is based on those features that are not affected by rotation (see, for example, Sutherland [39]). In instances where these features are unavailable, or not relevant, objects may need to be aligned by mental rotation. Takano used a mental rotation paradigm to see whether the distinction between orientation-free information and orientation-bound information played a significant role in human form perception. The conclusions drawn from this experiment suggest that mental rotation is unnecessary if the forms differ in either type of orientation-free information, provided that the difference is actually encoded as such.

Multi-view theories are hybrids of object centered and transformation models. Representations consist of pools of object views in familiar orientations. Recognition will occur rapidly when an image is oriented according to a stored, familiar view. When an image does not match one of these views, transformation is necessary. Jolicoeur and Kosslyn [47] investigated long term memory representation of three dimensional shapes. Their study provides converging evidence that people can store three dimensional shapes in long term memory using both object-centered and viewer-centered recognition, and that these dimensions were very stable appearing in every subjects data for every family of stimuli. A second experiment showed that the introduction of memory requirements did not seem to mitigate the subjects tendency to use both sorts of coordinate systems to compare the stimuli within each family. This contradicts Marr and Nishihara's [48] claim that recognition proceeds solely through the use of object-centered representations. Tarr and Pinker [49] presented participants with several objects each at a single rotation. They were given extensive practice at naming and classifying them as normal or mirrored- reversed at various orientations. Their preliminary findings were consistent with the early 3D mental rotation studies in that response times increased with departure from the study orientation. This suggests that subjects mentally transform the orientation of the input shape to one they had initially study or familiar with. An interesting finding from this research was that whenever mirror images of trained shapes were presented for naming, subjects required the same amount of time at all orientations. This finding suggests that mental rotation transformations of orientation can take the shortest route of rotation that allows the alignment of the input shape with its memorized counterpart.

6. Recognition Processes

One of the most difficult problems in ORS's is that of grouping object (model) parts and feature states into a form which can optimize recognition. This is because different objects share similar feature values on different parts of their surfaces. That is, given adequate features and the situation where, *from any view*, object part features are clearly differentiated in feature space, then well-known optimal classification procedures can be employed - precisely the scenario which does not occur in generic ORS environments. This situation is similar to problems in concept learning where different concepts share common states and the learning procedure, whether in modeling human function or machine applications, has to be capable of forming a structure which captures common properties within class (object) examples and highlight features which differentiate between classes. Techniques in MV which attain these goals are of two basic types: Feature Indexing (FI) and Evidenced-Based Systems (EBS's). In the former case object parts, and their feature states are ordered according to their discrimination power – along the lines of being representative of given classes of objects and discriminating others. The search procedure usually is in the form of a decision/strategy tree or some form of graph matching (where binary constraints prune the state space) initiated from more critical parts - indexed for their complexity (see, for example, McLean [50], Grimson [51] and Lowe [52]).

Evidence-Based Systems (EBS) solutions to this problem revolve around the generation of clusters of different object samples in feature space which, to various degrees, "evidence" different objects [1]. A feature space is simply an n-dimensional Euclidean space, on which each dimensional axis corresponds to some property of the data. For example, object part features such as perimeter and average Gaussian curvature would constitute two such feature dimensions. These predicates are normally either unary (describing properties of single features), or binary (describing relations between pairs of features). Evidence-based systems use bounded regions of feature space as probabilistic evidence rules for the occurrence of objects which have features represented within that region. An object in an evidence-based system is represented in terms of a series of these rules, which are triggered by the occurrence of the features of that object. Once triggered, a rule provides a certain amount of evidence for each object in the database, according to the likelihood of any given object having features which fall within the bounded region defining that rule. An object is recognized on the basis of accumulated evidence over all triggered rules.

The role of world knowledge and semantic association is an important issue for human object recognition, which has not been well addressed in the literature. Most researchers who put forward models of object recognition tend to ignore this issue. For example, Biederman [4], in proposing the recognition by components model defined object recognition in terms of matching geons to non-accidental image features without specifying the search process. Without fear of over-generalizing it appears that there is no search model for human object recognition – certainly one which makes specific predictions about the complexity of the matching process for given classification problems.

Human ORS must make sense of complex scenes containing multiple objects, some of which may be occluded to a very high degree. The evidence-based approach is able to account for both perceptual and semantic considerations in object recognition, with explanatory efficiency. When evidence rules (which may be abstractly described as bounded regions of feature space) are triggered, they provide a level of activation to objects represented within that region of feature space, proportional to the amount of evidence provided by that rule for each object. Activation may be mediated by either perceptual or a semantic processes. Perceptual processes build up the visual percept of a scene. Semantic processes use perceptual information to invest perceptual entities, including objects in the scene, with meaning, including names and properties. These semantic processes feed back into the perceptual channel to provide contextual information as an aid to the visual organization of the scene. Whenever an evidence rule is triggered by perceptual information, all representations which are semantically associated with each of the objects represented on the area of feature space designated by the triggered rule are also activated. The degree of activation is proportional to the strength of memory association, as well as the amount of evidence for any particular object, provided by the triggered rule. This allows for rapid recognition of objects in familiar environments, as well as recognition of objects which are highly occluded. For example, in an office scene, a telephone may be entirely occluded by papers, apart from the cord. As an isolated cue, a cord alone would probably provide insufficient evidence for the recognition of the telephone. Nevertheless, recognition of the cord will provide considerable semantic activation for the telephone representation, which together with semantic activation due to the triggering of evidence rules for various other items of office paraphernalia, may well be enough to facilitate the recognition of the telephone. This concept of semantic facilitation is drawn from an extensive literature in psycholinguistics (e.g. Meyer and Schvanaveldt [53]).

Modeling human ORS in terms of the evidence based system accounts well for the issues of view-independence, partial occlusions, variation between object within object classes, and novel exemplars of object classes. View independence is achieved simply by including all the segmented features of an object in the database. No matter which direction an object is viewed from, it will trigger rules which give evidence for the presence of that object. Preferred views are also accounted for, in terms of those views which provide the most evidence for the object in question. Partial occlusion is accounted for in much the same way. Provided at least some of the features of an object are displayed, some rules will be triggered, giving evidence for the object. Variation within object classes, and novel exemplars of object classes are also accounted for, because each rule is a bounded region in feature space, and each feature is simply a point. Therefore, any given feature predicate may vary in its precise position within a bounded region, and still provide the same evidence for its object class. As long as an object has enough similarity to the other objects in its class, that it triggers approximately the same set of evidence rules, it will be recognized as a member of that object class.

A final point should be made about the underlying physiology of object recognition. First, it should be noted that vision, as a sense, is passive and so any theory of biological object recognition that does not address the problem of inferring depth or shape from intensity – as an integral part of the processing system – is not complete. For these reasons it is difficult to evaluate results from neurophysiological studies showing certain sensitivities to photographs of faces, etc.(see Perrett [54], as the results confound the perception of *patterns* (as 2-D structures) in comparison to *objects* (as 3-D structures). Unfortunately, to this stage, we know very little about how the visual system solves, and uses solutions to, Shape-from-X in the *act of object recognition* though experiments are underway to investigate such issues. The evidence-based paradigm should prove useful in the study of these processes as it lays out a blueprint for how BV may learn relationships between surface types and expected image intensities, or, vice-versa.

REFERENCES

1   Jain, A., and Hoffman, R. (1988). Evidence-Based Recognition of Objects. *IEEE:PAMI*, 10, 6, 783-802, 1988.
2   Fan, T., Medioni, G., and Nevatia, R. (1989) Recognizing 3-D Objects Using Surface Descriptions *IEEE:PAMI*, 11, 11, 1140-1157, 1989.
3   Treisman, A. (1986) Features and objects in visual processing. *Scientific American*, 255(5),114-126.
4   Biederman, I. (1985). Human image understanding: recent research and a theory *Computer Vision, Graphics, and Image Processing*, 32, 29-73.
5   Ratliff, F., and Hartline, H. (1959) The response of *Limulus* optic nerve fibres to patterns of illumination on the retina mosaic *Journal of General Physiology*, 42, 1241-1255.
6   Barlow, H. (1972) Single units and sensation: A neuron doctrine for perceptual psychology? *Perception*, 1, 371, 394.
7   Caelli, T., and Oguztoreli, N. (1988) Some task and signal dependent rules for spatial vision *Spatial Vision*, 2, 4, 295-315.
8   Caelli, T., and Moraglia, G. (1987). Is pattern matching predicted by the cross-correlation between signal and mask? *Vision Research*, 27(8), 1319-1326.
9   Marr, D. (1982) *Vision*. San Francisco: Freeman.
10 Biederman, I., and Ginny, J. (1988). Surface versus edge-based determinants of visual recognition *Cognitive Psychology*,20, 38-64.
11 Price, C., and Humphreys, G. (1989). The Effects of Surface Detail on Object Categorization and Naming. *Journal of Experimental Psychology: Learning, Memory and Cognition*, 15, 797-825.
12 Caelli, T., Bischof, W. and Liu, Z. (1988) Filter-based models for pattern classification *Pattern Recognition*, 21, 6, 639-650.
13 Biederman, I., and Cooper, E. E. (1991). Priming contour deleted images: evidence for intermediate representation in visual object recognition Cognitive Psychology, 23, 393-41.
14 Livingstone, M., and Hubel, D. (1988). Segregation of form, color, movement and depth: anatomy, physiology and perception *Science*, 240, 740-750.
15 Triesman, A., and Gelade, G. (1980). A feature integration theory of attention *Cognitive Psychology*, 12, 97-136.
16 Caelli, T., and Reye, D. (1992) Classification of Images by Color, Texture and Shape *Pattern Recognition* (in press).
17 Gibson, J. (1966). *The Senses Considered as Perceptual Systems*. Boston: Houghton-Mifflin.
18 Waltz, D. Understanding Line Drawings of Scenes with Shadows. In Patrick Winston (ed.) *The Psychology of Computer Vision*, McGraw-Hill, New York.
19 Reichardt, W. (1987) Evaluation of optical flow information by movement detectors *Journal of Comparative Physiology A*, 161, 533-547.

20 Johansson, G. (1974) Vector analysis in visual perception of rolling motion *Psychological Forschung*, 36, 311-319.

21 Seibert, M., amd Waxman, A. (1992) Adaptive 3-D Object Recognition from Multiple Views *IEEE:PAMI*, 14, 2, 107-124.

22 Bulthoff, H., and Mallot, H. (1988) Integration of depth modules: stereo and shading *Journal of the Optical Society of America*, 5, 10, 1749-1758.

23 Todd, J., and Mingolla, E. (1983) Perception of surface curvature and direction of illumination from patterns of shading *Journal of Experimental Psychology: Human Perception and Performance*, 9, 4, 583-595.

24 Caelli, T., and Xu, S. (1990) A new method for Shape-from-Focus. In C P. Tsang(Ed.) *AI'90*, Singapore: World Scientific.

25 Pentland, A. (1987) A new sense of depth of field. *IEEE:PAMI*, 9, 4, 523-531.

26 Horn, B. (1986) *Robot Vision*. Cambridge, Mass.: MIT Press.

27 Pettigrew, J. (1973) Binocular neurons which signal change of disparity in area 18 of cat visual cortex *Nature(London)*, 241, 123-124.

28 Frost, B., and Nakayama, K. (1983) Single visual neurons code opposing motion independent of direction *Science*, 220, 744-745.

29 Caelli, T., Manning, M. and Finlay, D. (1992) A general correspondence approach to apparent motion *Perception* (in press).

30 Dillon, C., and Caelli, T. (1992) Inferring Shape from Multiple Images using Focus and Correspondence Measures(In Submission).

31 Dillon, C., and Caelli, T. (1992) Generating Complete Depth Maps in Passive Vision Systems *IAPR-92 Proceedings, Hague, September.*

32 Hoffman, D., and Richards, W. (1986) Parts of Recognition. In A. Pentland(Ed) *From Pixels to Predicates*. New Jersey: Ablex, 268-294.

33 Braunstein, M., Hoffman, D., and Saidpour, A. (1989) Parts of Visual Objects: an Experimental test of the Minima Rule *Perception*, 18, 817-826.

34 Besl, P., and Jain, R. (1988) Segmentation through variable-order surface fitting *IEEE:PAMI*, 10, 2, 167-192.

35 Fan, T., Medioni, G., and Nevatia, R. (1987) Segmented Descriptions of 3-D Surfaces *IEEE Journal of Robotics and Automation*, vol RA-3, 6, pp527-538.

36 Yokoya, N., and Levine, M. (1989) Range Image Segmentation Based on Differential Geometry: A Hybrid Approach *IEEE:PAMI*, 11, 6, 643-649.

37 Caelli, T., and Dreier, A. (1992) Some New Techniques for Evidenced-Based Object Recognition *IAPR-92 Proceedings, Hague, September.*

38 Deutsch, J. (1955) A theory of shape recognition. *British Journal of Psychology*, 46, 30-37.

39 Sutherland, N. (1968) Outliners of a theory of visual pattern recognition in animals and man *Proceedings of the Royal Society of London(Series B)*, 171, 297-317.

40 Pinker, S. (1984) Visual cognition: An introduction. *Cognition*, 18, 1-63.

41 Rock, I., DiVita, J., and Barbeito, R. (1981) The effect on form perception of change of orientation in the third dimension *Journal of Experimental Psychology: Human Perception and Performance*, 7, 719-732.

42 Rock, I., DiVita, J. (1987) A case of viewer-centered object perception *Cognitive*

Psychology, 19, 280-293.

43 Rock, I., Wheeler, D., and Tutor, L. (1989) Can we imagine how objects look from other viewpoints? *Cognitive Psychology*, 21 185-210.

44 Shepard, R., and Metzler, J.(1971) Mental rotation of three-dimensional objects *Science*, 171(3972), 701-703.

45 Corballis, M., and Nagourney, B. (1978) Latency to categorize disoriented alphanumeric characters as letters or digits *Canadian Journal of Psychology*, 32, 186-188.

46 Takano, T. (1989) Perception of rotated forms: A theory of information types *Cognitive Psychology*, 21, 1-59.

47 Jolicoeur, P., and Kosslyn, S. (1983) Coordinate systems in the long-term memory representation of three-dimensional shapes *Cognitive Psychology*, 15, 301-345.

48 Marr, D., and Nishihara, H. (1978) representation and recognition of the spatial organization of three-dimensional shapes *Proc. Royal Soc. (Lond)*, 200, 269-294.

49 Tarr, M., and Pinker, S. (1989) Mental rotation and orientation-dependence in shape recognition *Cognitive Psychology*, 21, 233-282.

50 McLean, S., Horan, P., and Caelli, T. (1992) A Data-Driven Indexing Mechanism for the Recognition of Polyhedral Objects *SPIE Proceedings: Advances in Intelligent Robotic Systems*, 1609.

51 Grimson, W., and Lozano-Pérez, T. (1985) Recognition and Localization of overlapping parts from sparse data *MIT A.I. Memo No. 841*.

52 Lowe, D. (1987) Three-dimensional object recognition from single two-dimensional images *Artificial Intelligence*, 355-395.

53 Meyer, D., and Schvanaveldt, R. (1971) Facilitation in recognizing pairs of words: evidence of a dependence between retrieval operations. *Journal of Experimental Psychology*, 90, 2, 227-234.

54 Perrett, D., Mistlin, A., and Chitty, A. (1987) Visual neurons responsive to faces *Trends in Neuroscience*, 10, 9, 358-363.

Three-Dimensional Object Recognition Systems
A.K. Jain and P.J. Flynn (Editors)
© 1993 Elsevier Science Publishers B.V. All rights reserved.

Range Sensing for Computer Vision

Ray Jarvis[a]

[a] Intelligent Robotics Research Centre, Monash University, Clayton, Victoria 3168
Australia

Abstract

The overall effectiveness of a Computer Vision System depends critically on both the quality of the computational algorithms devised (speed, capability, robustness, etc.) and that of the acquired sensory data. Whether "semantic free" or domain specific methodologies are being pursued (or better still, if an appropriate balance has been struck), the ideal of high quality raw data, acquired from appropriate transduction devices, is still very much sought after. Real vision problems are usually so difficult that few would choose to ignore the support provided by good raw data. Many modern vision systems focus attention on the acquisition and processing of 3D data. Range sensing is a natural first requirement for such systems.

There are very many range sensing methodologies, each with its own strengths and weaknesses in terms of constraints of applicability, robustness, compactness, accuracy, human correlates, quality, cost and safety. Three dichotomies help to organise this plethora of methodologies: whether the method is active or passive, whether image based or direct, and whether monocular or multiview. A number of strands of early and continuing work are maturing into solid ropes of hope for a future where inexpensive, off-the-shelf ranging sensors with a wide range of applicability, will become available to Computer Vision researchers and the relevant industrial users.

This chapter will review the nature of range sensing for Computer Vision, give varied examples of approach methodologies together with their strengths and weaknesses and introduce some early processing ideas with respect to dense 3D data which strongly support robust subsequent higher level processing including shape and pose determination, grip site selection, segmentation, 3D recognition and, thus, complete scene analysis.

1. Introduction

There are a variety of reasons why range data is acquired as well as a very many ways of extracting it [1,2]. To better appreciate why so many and varied attempts have been made to obtain what in some sense is a trivially defined data set, merely a list of cartesian triples (x, y, z) on surfaces and edges of objects, it is of value to understand both the motivations and the methodologies involved. There are at least three fairly distinct basic motivations for gathering range data:

1. To better understand biological vision systems, especially that of humans, both for slaking our curiosity (of the purest kind) and also to be able to exploit that knowledge for building robust artificial vision systems to be used as part of a variety of automation processes in industry and elsewhere.

2. To model 3D environments in the sense of determining space occupancy, shape, observer position, egomotion, volume, saliencies and compliance with standards, perhaps in relation to collision-free planning operations, grip-site detection, costing for transport, industrial quality control, vehicle tracking, solid modelling, aerodynamic studies, etc.

3. To describe scenes in terms of the identity of the objects within them, their positions, poses, sizes and inter-relationships. It is within this domain that 3D Object Recognition Systems belong, including segmentation, parameter extraction and classification.

In this chapter we will be dealing with the second and third of these motivational arenas, both being firmly part of 3D Computer Vision, but the last most clearly of direct relevance to 3D Object Recognition Systems.

Two other broad, motivation related, classification partitions are also helpful:

Firstly, in a robotics context, the third motivational arena, described above, is more closely linked with artificial hand/eye coordination systems involving robotic manipulator arms and bench top or manipulator mounted (eye-in-hand) vision sensors, whilst the second one is of more relevance to autonomous mobile robots and other tracking and control scenarios, usually of a medium to large scale, both indoor and out, underwater and in mines and space.

Secondly, the particular application requirements may dictate whether sparse, medium or dense ranging is required; the corresponding instrumentation of such specifications may vary markedly in terms of dynamic range, accuracy, speed, cost, safety, power consumption, compactness and ruggedness. For example, it may be sufficient to make a small number of particular range measurements in an industrial metrology application for process control; the accuracy and ruggedness requirements might be extreme but the cost not particularly critical. On the other hand, more dense but less accurate range data may be needed for mobile robot environmental mapping for path planning and obstacle avoidance; in this case cost, power consumption, speed and weight may be critical. A third example may be where very dense and accurate range measurements are required for modelling to input an object into a CAD database, perhaps for subsequent modification and reforming; in this case time, cost, weight and power consumption are not likely to be critical.

Range sensing methodology can be neatly classified within the octants defined by three sets of dichotomies [see Figure 1]:

I Passive versus Active Passive methods rely on ambient lighting conditions and, in particular, do not impose any intrusive energy sources upon the environment. Typical methodologies in this category are lateral (multiple camera) and temporal (moving camera) stereopsis, range from focus and range from texture. These methodologies usually have strong correlates with human visual functionality. Active methods, by contrast, impose structured energy sources (light, near light, ultrasonic, x-ray or microwave) on the environment, only ultrasonics haveing any biological correlates (bat navigation). Typical methodologies include ultrasonic, microwave and laser time-of-flight (and phase detection) and structured lighting. The scope of application must permit the use of such structured energy sources; in many cases a restriction to indoor and short range usage is implied.

II Image Based versus Direct Image based methods use the analysis of single or multiple images as a basis for range extraction, whether or not structured energy sources are applied. Direct methods obviate the need for image analysis, as in the case of time-of-flight laser ranging.

III Monocular versus Multiple View Monocular methods extract range data from a single viewpoint, as in the case of range from focus, brightness, attenuation or texture. Multiple view methods are essentially triangulation based and rely on identifying areas of interest in more than one image and determining range related disparities by correspondence matching; two intrinsic difficulties partially frustrate this methodology - the missing parts/obscuring edge problem and diminishing accuracy with actual range.

In what follows, a number of range finding techniques will be described in terms of how they work, their methodology class, scope of applicability (motivation-related) and relative cost, safety, accuracy, robustness (reliability and ruggedness), weight and power consumption.

2. Range Sensing – The Universal Ideal

In an earlier publication [3] this author described a "magic powder" ranging dream:

> *It would be ideal if the exposed surfaces of objects could be sprayed with a magic powder which, when gathered up, could, particle by particle, be interrogated as to where it had been in space and what the properties of the surface it touched were.*

This dream has not yet been realised, nor was it meant to be; it was essentially a way of stating simply what was wanted. Four aspects of the problem are revealed:

1. The exact 3D location of points on the surfaces of objects were sought

2. Such locations were to be densely packed over all surfaces, even those hidden from external view due to the complex shapes of the objects and their juxtapositions and obscurances

3. Local surface properties such as colour, texture, surface normal, curvature, hardness, temperature, perhaps even thermal and electrical conductivity were to be associated with each surface location

20

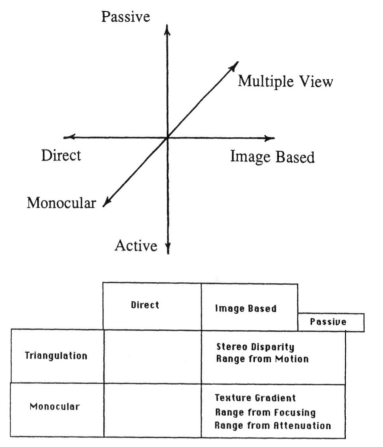

Figure 1. Range Finding Methodology Classification

4. The above data was to be acquired remotely, i.e. no mechanical contact sensing was necessary - each particle of powder would come away from the surface, perhaps by being vacuumed up, and bring with it position and surface property data. (Of course the "magic powder" touches the surfaces, but it is only a metaphor for some type of probing electromagnetic energy source).

It was argued that if the type of data proposed above were available it would be a rich source of input to subsequent stages of processing leading up to complete scene analysis which would answer all the "what, where and how" questions about the scene necessary to support the intended application, whether robotics-related or not. The philosophical (and somewhat pragmatic, also) stance was taken that "semantic-free" early processing with late binding of domain specific or model based knowledge was a viable and more general purpose alternative to "purposive vision" which is driven by goals in the "active vision" sense, where specific sensor information is retrieved only when required to disambiguate competing model matching hypotheses. Of course, these seemingly opposed approaches are extremes; most robust systems will mix them in proportions dictated by the need to maintain some degree of generality whilst fulfilling specific goals in an efficient and cost effective way. This is a debate which is not to be explicitly taken up in any detail in this chapter; however, it will emerge now and then in fragments in what follows. Just one question for now: if, in the near future, a small, accurate, affordable, fast, robust and widely applicable dense ranging device were to appear on the market and be as accessible as the common video camera (may be even built into it and producing an image registered rangepic), how would the "semantic-free" versus "purposive" paradigm mix shift?

The examples used to represent various general strategies for range sensing described in the following section are only representative of many instances developed in laboratories all over the world. In many cases experimental equipment for which some photographic and diagramatic illustrations were readily available, was favoured; each such individual piece of equipment may not represent the best example of its class but was easier to use to illustrate a method for the purpose of this chapter, which is largely tutorial in nature, in support of the central theme of this text.

3. Passive Range Sensing

3.1. Range from Focus

There are a number of versions of this approach to ranging. All exploit the optical physics of imaging through a lens onto a plane which permits distances from the optical centre of the lens to visually distinctive local patches on the 3D scene to be related to the degree of sharpness or blurring of the corresponding components of the image. The direct method of exploiting this relationship [4,5,6] is to collect a series of images for a set of controlled lens focus positions and to associate, with each visually distinctive component of the image, the range related lens position index corresponding to when they are "sharpest" amongst the set of images. This can be carried out in parallel for all the image components. Calibration of the system allows the lens position index to be directly associated with a range value through a look-up table. Adaptive sub-ranging of lens positions can be used to refine ranging in selected range slices of the scene. Large aperture settings sensitise the search for best focus position for each image component at the cost of some distortion, unless high quality lenses are used. The zooming effect caused by moving only the lens can be compensated for by adjusting the position of the camera or simply moving the image plane instead of the lens. A recent re-look at the direct method [6] suggests that it should now be possible to build a fast range scanner using this method by using specialised electronics to search the video signals for best focus positions on the fly. The proposed device would use two video cameras, one rotated 90° about the optical axis, set up in an optically coaxial arrangement using one half silvered mirror and one standard mirror. Video rate electronics would evaluate the absolute intensity difference between the adjacent pixels in each scan line, compare this with the largest value found so far for this position, and retain the larger value in one frame sized memory array, keeping its focus position index in another. At the end of a full readout of a field from each camera, two sets of maximum absolute differences and corresponding focus index positions will be available for comparison. Thus there will be two entries for each pixel position. For simplicity, the lens position index which produced the higher maximum absolute difference is used as the range index for that pixel. If the index values are very similar this suggests that confidence can be placed in the result. However, for vertical or horizontal lines one would not expect this to occur. If neither maximum absolute difference is sufficiently high that pixel can be marked as unranged to; this will occur over homogeneous intensity areas of the images. Some interference effects in busy high contrast edge areas also need attention. Blur spread from some edges can cause false best focus position determinations in neighbouring areas. A simple, fairly effective remedy is to use a non-maximum suppression method on difference data in the eight neighbourhood region around each pixel, but this is a relatively costly exercise, computationally. Some examples of range sensing using the direct focus method (but not with specialised electronics) are shown in Figure 2. A number of measures of sharpness are analysed in [5] and one of these was used for the results shown in [6]. Using specialised video cameras which were directly readable (instead of through a serial video signal) and flexible lenses whose focal length could be controlled rapidly, a fast, robust direct focus method range sensor could be realised.

An alternative approach described by Pentland [7] introduces the notion of "blur gradient" which, in the simplest cases, can be estimated at visually distinct patches by comparing only two images, one with infinite depth of focus (ideally, pin hole lens) and one with a large aperture lens. If the large aperture lens is set, say, to focus the most distant objects of interest, the blur of all visually discernable patches increases inversely with range. In [7] various blur measures are analysed in detail. An experiment inspired by the blur gradient idea is described in [8]. In this work the blur measure used was the ratio of slope at the zero crossings of the $\nabla^2 G$ [9] filtered images for a all in focus image (small aperture) and a large

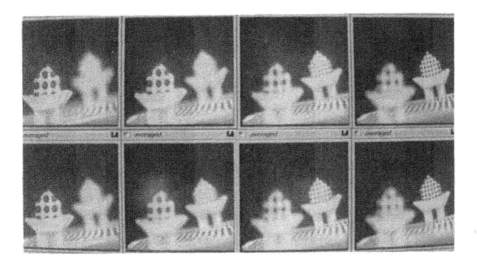

Figure 2. Direct Range from Focus Example

aperture image, similarly filtered and attenuated in intensity by using a neutral density filter. Typical results are shown in Figure 3. For both approaches to range from focus, the range to homogeneous intensity areas cannot be directly extracted. However, the monocular approach is not subject to hidden parts and occluding edge problems and can be used both indoor and out and even at night in an active mode by using infra red illumination with the appropriate camera (many standard monochrome CCD cameras are quite sensitive in the infra-red region).

3.2. Range from Attenuation

Where image brightness variation caused by transmission through an attenuating medium (e.g. fog, smoke in air, underwater, etc.) is unavoidable, that very attenuation effect can be exploited for range sensing [10,11,12]. In the simplest but unlikely situation of all surfaces of the scene being Lambertian, equally illuminated and of homogeneous albedo, neither the cosine nor the illumination variation effects apply and image brightness variation is due only to attenuation through the medium. An interesting experiment is being carried out [11] to see if attenuation ranging through water deliberately dyed can be used to simulate an ideal mobile robot range sensor supporting higher level navigation studies. All obstacles, including the walls of a water tank, could be back lit to closely approximate uniform illuminations of the surfaces, but this elaborate process may not be necessary if colour images are used. The idea (yet to be tested) is quite simple: if green dye is used, the red and blue components should be selectively attenuated relative to the green component. Non attenuation related effects such as variation in illumination, albedo and surface normal direction (cosine law) can, ideally, be factored out using two colour components of the image (say red and green since blue is often produced indirectly for inexpensive colour cameras). Figure 4 shows what the head of David looks like in a fish tank filled with blue liquid and the corresponding range result. It is clear that attenuation can be exploited for ranging. The nice thing about this approach is that high density ranging to all visible surfaces can be obtained at high speed with little computation; the silly side is the need for immersion (which may not be acceptable for many applications). The author believes this idea has an important future. In air, an artificial mist like that produced by an ultrasonic humidifier may possibly be used to exploit this methodology.

3.3. Range from Texture

Visual texture provides many natural surfaces with a regular structure not entirely unlike some projected image patterns one might use for active ranging. If the texture of a surface can be considered homogeneous despite local randomness (and this is what sometimes makes it hard to define except in probablistic terms), then, from a fixed viewpoint, it will seem to grow finer with its distance from the observer. That is, the 2D image will exhibit a texture fineness gradient which can be used as a depth cue; such a measure has strong human vision associations. Gibson [13] considered this effect to be a strong depth cue for humans, particularly with respect to the ground plane. Bajcsy and Lieberman [14] proposed a Fourier descriptor based measure of texture gradients in natural outdoor scenes. Further details are to be found in [1,14]. If a natural scene contains a number of differently

Figure 3. Range from blur gradient example results.

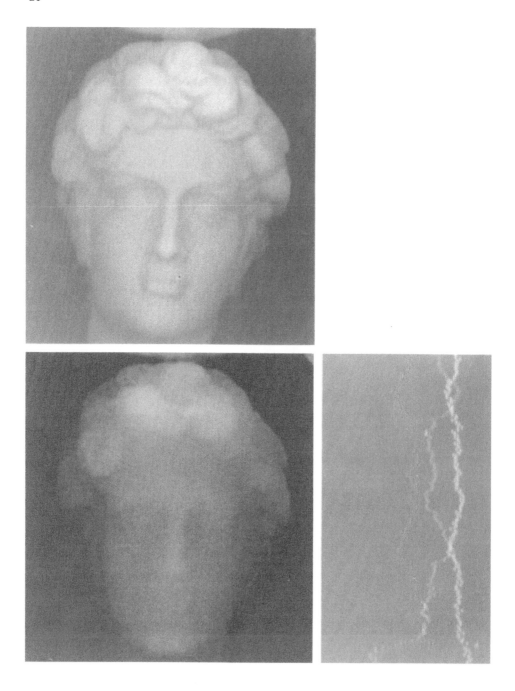

Figure 4. Coloured Liquid Range from Attenuation Example.

textured regions, each should be considered separately and homogeneity assumptions must hold for each. The method is particularly useful if used in conjunction with other depth cues. Since natural visual texture can only be defined in statistical terms, the window size used for measurements of texture coarseness must be chosen to provide sufficient reliability of measure without degrading resolution beyond necessity. Other texture measures have also been proposed [15,16] and could be used in ranging evaluations. The depths determined from texture gradient evaluations are only relative; absolute depth can be derived through calibration based on knowledge of texture element size.

The requirement of prior scene segmentation into areas of homogeneous texture is a serious short-coming of the method for complex natural scenes. Computational cost can also be relatively high.

3.4. Lateral Stereopsis Range Sensing

Stereo disparity refers to the shift in the image position of a 3D localised entity in the scene when the camera is moved [See Figure 5]. The type of entities used and the geometries of the multiple camera positions determine the style of disparity based ranging being implemented. The entities can be entire objects (which would need to be segmented out and labelled), tokens (visually discernable events which hopefully have some degree of uniqueness), points, lines, corners or simply window patches. The more complex the entity in terms of disernability, structure and uniqueness, the more reliable the likely match but more expensive the pre-match preparation. Lateral stereopsis refers to disparity based ranging where two or more cameras are placed in a plane with their optical axes approximately normal to it. That this geometry may be achieved by repositioning a smaller number of physical cameras is of no real significance except in terms of speed of image capture and real time cycles of measurement. Vergence refers to inward turning so that the intersection of camera optical axes occurs at points closer than infinity (where parallel lines meet). Increasing the vergence of a set of cameras can improve disparity accuracy at the cost of reducing the size of the range field measurable. For example, if the optical axes converge at a point one metre away, disparity directionalities will be reversed for points further than one metre and ambiguities would result for such cases. In practical terms, no part of the scene of interest should be beyond the vergence point. The simplest case is to use two cameras with parallel optical axes. The line joining the optical centres of the cameras is known as the base line. Since disparity ranging is essentially triangulation, the accuracy of ranging is proportional to base line length. Range accuracy is inversely proportional to actual range and the hidden parts/obscuring edge problems increases with base line length. Some compromise needs to be struck. Camera placement geometry also determines the linear direction for the correspondence matching search. For any pair of cameras, search directionality is parallel to the base line; this is often referred to as the "epipolar constraint". The "aperture problem" frustrates matching of target entities extended along the search direction; for example, horizontal line segments cannot be adequately matched with cameras placed along a horizontal base line.

For n cameras $(n > 2)$, $\frac{n(n-1)}{2}$ pairs of target matches are possible. The known

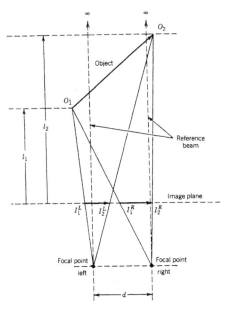

Figure 5. Lateral Stereo Matching Geometry (Copyright © 1983 IEEE)

geometry of camera placement can be used to improve both the accuracy and reliability of match search. In an early study Moravec [17] racked a camera to 9
positions along a linear rail and refined the matching process using the 36 pairs
of matches possible. A recent in-a-line multi-camera fast stereopsis matching system is described in [18]. This paper also describes what is now referred to as the
"Moravec interest operator" which is essentially a corner detector. Corners (with
orthoganal fragments) are ideal targets since they do not fall prey to the "aperture
problem".

Baker [19] uses edge data for matching targets with a horizontally displaced
pair of cameras. At the lowest level matches along horizontal lines establish correspondence candidates. At a higher level, edge continuity constraints (termed
"cooperative continuity enforcement") are used to confirm or reject these edge pairings. This is an excellent example of using global constraints to preserve local
consistency by a filtering process to improve reliability.

Marr et al [20,21,22] proposed stereo matching methods with strong biological
vision correlates based on $\nabla^2 G$ correlation mask zero crossing edge finding methodology with varying mask sizes to support coarse to fine hierarchical search guidance.
The $\nabla^2 G$ operator has the form

$$\left(2 - \frac{r^2}{\sigma^2}\right) e^{-\frac{r^2}{2\sigma^2}},$$

where r is the radial distance out from the centre of the mask and σ controls the spread of the masks. It is derived by combining the 2D Gaussian distribution with the 2D Laplacian operator

$$\nabla^2 = \left(\frac{\partial^2}{\partial x^2} + \frac{\partial^2}{\partial y^2} \right)$$

The convolution theorem permits the rewriting of $\nabla^2(G * I)$ as $(\nabla^2 G) * I$, where I is the image and "$*$" the convolution operator. Thus, the process of applying the Laplacian operator over images smoothed by a Gaussian low pass filter can be carried out in one convolution pass over the image. The zero crossings (usually at high differentials) of the resulting "image" are considered to be edge points of significance. Large mask sizes produce coarse edges and small mask sizes produce fine edges. Since correspondence matching can be frustrated by edge complexity (density), the relative paucity of coarse edges allow reliable matching at this coarse level and these matches can guide searches at a finer edge level and so on. These methods tend to be computationally expensive; also, large mask size zero crossings are often displaced from the positions of true edges in the image due to interference effects. This fraility can be compensated for, usually by looking for consistent edges through multi-level (various mask sizes used) reinforcement.

A fast hardware implementation of the Marr-Poggio stereo matcher is described in [23]; prior to the development of those specialised systems, 512×512 resolution images took in the order of 30 minutes to compute on a DEC mainframe of the day.

Using more than two cameras in an other than all-in-a-line configuration has advantages over and above simply that of more image pairs to test for matches. A number of triopsis systems [24,25] with the third camera not on the same base line extension as the other two have been developed. The image from any of the three cameras can be used to confirm matching hypotheses developed between the images of the other two. Also, some hidden parts difficulties can be resolved; nor is the "aperature effect" so seriously felt since lines in any direction can be handled.

A four camera system with parallel optical axes on the vetices of a square is described in [26,27] for small scale indoor work and large scale outdoor work, respectively. The operational characteristics are shown in Figure 6 and a range extraction example in Figure 7. The simple idea being exploited here is that horizontal and vertical disparities for valid matches should be identical and a diagonal match should be exactly times this value. With the camera scan lines properly aligned, the image arrays collected can be edge (Sobel) searched in a synchronised step mode, testing for vertical, horizontal and diagonal matches simultaneously and accumulating 0, 1, 2 or 3 votes for every tested step. Reliability generally increases with vote but some lower votes may indicate partial hidden parts obscuring eye problems and some higher votes may reflect partial "aperture" problems. Nevertheless, neither of these two problems seriously confound the approach because of the camera placement geometry.

The computation involved is trivial, hence fast (4 seconds for 128×128 images). It is also intended to use intensity, colour and continuity constraints to prune away any false matches which get through the voting process. A hardware version of this system could probably be easily built. The large version [27] of the "Quadvision"

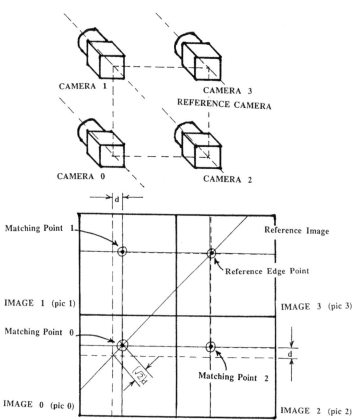

Figure 6. Quadvision Configuration

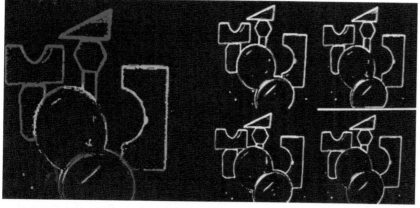

Figure 7. Quadvision Result Example.

system [See Figure 8] is built on a cruciform system of linear slides and a chain system to permit base line size variation whilst maintaining the strict square placement camera geometry. The base line length variation facility can be used for two distinct purposes. The first is to compromise between a high proportion of common views low incidence of obscuring edge problems for small base line length and the higher accuracy with large baseline length over the different scales of ranging required for outdoor application (e.g. 10 metres to 100 metres). In fact, a small baseline length scan can be used to approximate range to evaluate the need for a larger baseline length. Alternatively, or in conjunction with the above use, a set of image data collected over the four cameras for a number of equispaced baseline length positions can emulate collection from multiple sets of four cameras each, this data can be used to improve matching accuracy and robustness along the lines of the Moravec scheme [17].

Earlier devised methods using correspondance matching by area [28,29] correlation rather than edge or token matching have recently been resurrected as special purpose hardware [30,31] systems overcome the serious computational complexity impediments to this approach. The particular advantage of this approach is to range to local regions rather than tokens, points or edges; there must be, of course, intensity variations in these regions but they need not be sharp. A particular application area is in support of autonomous mobile robot navigation on Mars, whose surface is strewn with boulders.

A two camera stereo matching system using corner detection [32,33] and a "disparity gradient limit" [34,35] hypothesised match rejection filter is described in [36]. Fairly sparse but reliable ranging is obtained rapidly using special hardware [PARADOX parallel processor [37]]. Figure 9(a) and (b) show left and right images of a laboratory scene with detected corner points superimposed; Figure 9(c) shows the range related disparity map using the right image as reference; Figure 9(d) shows a Cartesian grid cast onto the disparity computed scene [38].

3.5. Range from Motion and Target Tracking

When a video camera moves at constant velocity in a straight line along its optical axis, all visual distinct local features of stationary objects move radially out from a point on the image known as the "focus of expansion". The 2D image velocities measured in the image plane are inversely proportional to depth. This effect can be exploited for ranging and is called "temporal stereopsis", a minimum of two camera positions being required. As image components radiate from the focus of expansion off the limits of the image plane they obviously cannot be ranged to. Hidden parts/obscuring boundary problems also exist, as with lateral stereopsis. Searching for correspondence matches can be constrained along radial lines out from the focus of expansion. Aperture problems occur for edges along a radial expansion line. As one would expect, using more than two positions allows more accurate and reliable results to be obtained. When the focus of expansion is not known before hand it can be calculated [39]. Polar coordinate $((r, \theta)$ or $(\log r, \theta))$ mapping using the focus of expansion as the origin can simplify the problem of detecting moving objects, since their edges would not normally be observed as

Figure 8. Large scale Quadvision system for outdoor use.

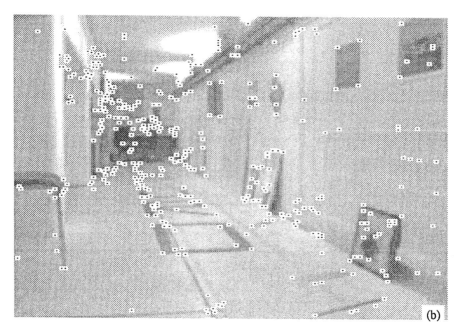

Figure 9. Corner Matching Stereopis. (a): Left image showing corners. (b): Right image showing corners. (c): Disparity map. (d): Cartesian grid.

(c)

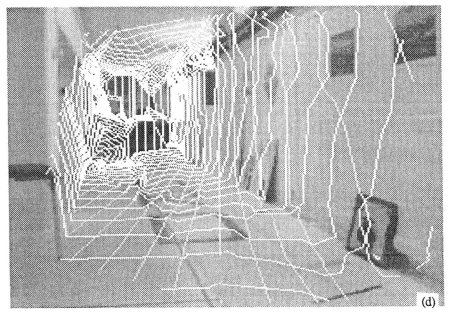

(d)

moving out from the focus of expansion along radial lines [40] and consequently are able to be detected as components not moving perpendicularly to the θ axis (constant θ lines).

The application of specialised area correlation hardware [30,31] permits a speed solution to three interrelated problems, target tracking, velocity field calculations and stereo ranging. In the work reported in [31], a commercially available electronic chip (with its initial commercial target in image bandwidth compression applications) is used to return motion vectors computed over various sized targets tested for motion by area correlation simultaneously carried out for up to 256 shifted positions. For a 16x16 target size, the motion vector is returned in 1.15 m sec. Obviously, a number of targets can be tracked simultaneously at video frame rate. If a target is selected from one image and a match found in a second image (from a separate camera with a known camera geometry relative to the first), stereo ranging is available for the target, even if it is in motion, since matches to detect motion in images from one camera and those which detect disparity with respect to an image of a second camera can both be carried out at high speed. Details of velocity field calculations are to be found in [31].

3.6. Camera in the Hand Range Sensing

If a camera is held in the "hand" of a robotic manipulator, the controlled mobility of the camera allows a number of ranging methods to be realised. For example, moving the camera along its optical axis into a scene allows temporal stereo or structure from motion range recovery; alternatively, moving the camera to effect lateral stereo with two or more positions in a plane perpendicular to the range axis is also easily realised. If, for the latter mode, the motion is small and known, a simple range recovery procedure inspired by [41] can be implemented [42].

4. ACTIVE RANGE SENSING

Once regarded by Computer Vision purists as "cheating", active range sensing is now fairly universally accepted, particularly for industrial applications. It has been admitted that the problem of remotely determining the 3D shape of an object is of sufficient difficulty that to despise active methods as "impure" is merely "old fashioned". Thus, in the last ten years or so, the first few brave active sensing probes of the previous decade have given way to a plethora of shameless attempts to extract shape from a distance without any great need for "biological correlate" considerations to "bless" the methods used.

4.1. Ultrasonic Range Sensing

Transit times of ultrasound pulses over fractions to tens of metres in air or water are easily measured electronically. It is therefore not surprising that active range sensing using such measures should be fairly widely used, particularly in the context of indoor autonomous mobile robot navigation. Ultrasonic transducers are relatively inexpensive and robust and systems using them for rangefinding were popularised by their appearing in camera systems (mostly Polaroid); the methodology has become very accessible for less than expert users and even undergraduate

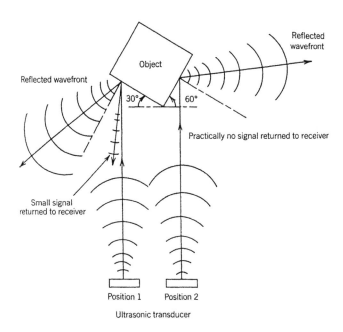

Figure 10. Ultrasonic Ranging Signal Reflections. (Copyright © 1983 IEEE)

students.

The most common use of ultrasonics for range sensing is in the form of measuring transit times of "chirps", often a pulse envelope of a number of ultrasound frequencies, to and from solid surfaces in man made environments. Two basic problems frustrate the use of this otherwise very attractive methodology. Firstly, simple ultrasonic transducers tend to have relatively large solid angle beam spreads in the main energy lobe direction, in the vicinity of 30°. Thus high spatial resolution (dense ranging) is difficult to achieve except by either attenuating the beam in directions other than within a narrower beam angle (by using ultrasonic energy absorbing tubes, for example) or by array beam focussing methods. Secondly, fairly normal man made surfaces cause specular reflection of ultrasound energy beams (surfaces with undulations of less amplitude than the energy wave length, according to Huygen's principle). If the main lobe centre is directed at more than approximately 40° to the surface normal, a major component of the beam energy will be specularly reflected, only a small fraction returning towards the source [See Figure 10]. At best, no range will be recorded for this situation; at worst secondary reflections off nearby surfaces may return a false reading which is a severe range overestimate and may not trigger diversionary actions of a mobile robot in an imminent collision situation. For all intents and purposes, an ultrasonic ranging system is like a torch in a mirrored room in terms of promoting ambiguity unless some special safeguards are put in place. For example, Elfes [43,44] introduces "sonar barrier" tests and various probabilistic mapping techniques to clarify results.

4.2. Radar Range Sensors

Skolnik [45] provides details on radar systems, including aspects of frequency, beam widths and specularity, which are of importance in range sensing. The summary given here comes from [46]. Again, as with ultrasound, the granulatity of surfaces in relation to wavelength determines the relevance of specularity to the analysis. Thus, surfaces which are specular for a 4GHz device may be diffuse for a 100GHz device. Low frequency radars may be used for over-the-horizon measurements but high frequency radars are essentially line-of-sight. Higher frequency radiation attenuate more severely, limiting range. For a given beam-width (and hence angular resolution), the higher the frequency the smaller the transmission and receiver devices (antenna).

Pulse (envelope of a mixture of frequencies) time-of-flight is the most commonly used ranging technique. The frequency of interrogation is limited by the wait time for the most distant target to return a pulse, high power is needed for the duration of the pulse and range accuracy is limited due to the short time periods being measured. An alternative scheme is to use frequency modulated continuous wave where the frequency over a period is swept in a precisely known way and the frequency shift of the returned signal used to estimate range. These measurements are more easily made than time-of-flight and, consequently, more accurate results can be obtained. The power requirements are also continuous so that the high peak power requirements of time-of-flight pulse methods are avoided.

A Marconi Defence Systems frequency modulated continuous wave system operating at 94.3GHz with a frequency sweep of 500MHz is described in [46]. Range accuracy is 30 cm over ranges up to 1 km. For an antenna size of 10cm, beam-width is approximately 1°. Scanning rate is 1HZ. Proliferation of metal objects in the application environment can distort and absorb electromagnetic radiation but this effect is serious only for large wavelength signals ($\gg$ 1 metre). At smaller wavelengths, specularity dominates over the distortion effects caused by metal objects. "Cooperative targets" can be used to resolve specularity ambiguities; this is the main operational mode used for the device described in [46], thus limiting its use to spot distance checking to specific targets rather than modelling the environment.

4.3. Laser Time-of-Flight Range Sensors

Most man made surfaces except mirrors and polished metals or shiny plastics scatter incident light energy in a Lamberian manner (surface undulations greater than wavelength) with equal flux density over a hemisphere at each point of the surface. A collimated beam from a laser range sensing device is detected back at the source after reflection as a much reduced energy beam. It suffers dimunilion by partial absorption by the surface it reflects off, according to its albedo (reflected/incident energy ratio), through inverse square law reduction on the return path and by some scattering by particles in that path. The large dynamic range of the energy which must be measured is a major challenge for the instrumentation used. The larger the receiving aperture the more energy captured, because of the Lambertian scattering at the surface. Two basic types of time-of- flight laser range finders have been developed, those which measure the phase shift of a continuous wave-modulated beam between outgoing and return components, and those which measure directly the time-of-flight of a pulse of laser light in transit from the source and back again after reflection off the target. In both cases highly collimated beams are used, as this insures high spatial resolution. Outward and return beams are arranged to be approximately coaxial. For the second type, time resolutions in the vicinity of 50ps must be provided by the instrumentation to achieve an accuracy of $\pm 1/4$in since light travels approximately 1ft/ns. In both types it is possible to measure the intensity of the return signal to evaluate the albedo of the target surface after inverse square law factors are factored out using the phase shift or direct timing information. This was done for the continuous wave instrument described in [47,48] leading to completely registered image intensity/ranged readings. The beam can be directed over an area of the scene simply by using mirror galvanometer defection system. The block diagrams of both types of instrument are shown in Figure 11. A result using the pulse direct time-of-flight instrument described in [49] is shown in Figure 12, which took approximately 40 seconds to collect a 64x64 range pic with ten samplings of each range to improve the signal/noise ratio. The earlier continuous wave instrument [47,48] took 2 hours to collect a 128x128 range pic but with 7-8 bits accuracy in the range of 3-16 ft. Another pulse timed instrument is described in [50]. A number of commercially available laser range scanners have appeared on the market [51,52,53] but cost in the vicinity of $100,000 each.

One clear advantage of laser time-of-flight (both phase shift and direct) range sensors is that they are capable of collecting dense range data without suffering any hidden parts problems, if the outward and return beams are truly coaxial. However, high dynamic range considerations and the need to use powerful lasers (which are both expensive and perhaps hazardous for humans) to extend the reach has restricted their use, in general, to no more than up to 10-15 metre ranging. Two exceptions come to mind. For surveying, point at a time range sensing instruments capable of much greater (perhaps up to 10 km) exist, but they use a "cooperative target" in the form of a retro reflective prism which can return strong signals back in the source direction. Military range sensors used for gun sight purposes cannot, obviously, rely on cooperative targets but can range to 10 km or so. However, a range accuracy of ± 10m is often sufficient for adjusting a weapon system, since the trajectory of the missile is usually not straight down on the target and its spread of destruction may be quite large.

Laser time-of-flight range sensors are limited in accuracy by the need to measure very small time intervals and hence, to date, tend not to be used on small volume scenes requiring high accuracy ranging; however, technological improvements in optronics over the last few years may well soon remove this limitation.

4.4. Shape from Brightness

Horn [54] first (1970) raised issues concerning the recovery of surface shape from image shading data. There has been a renewal of interest in this approach in more recent times [55]. The work by Ikeuchi and Horn [56] is representative of the approach to surface orientation recovery. Orientation data permits range recovery over continuous surfaces by integration; discontinuties frustrate absolute range

(a)

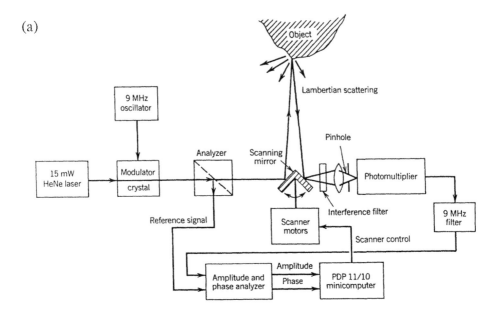

(b)

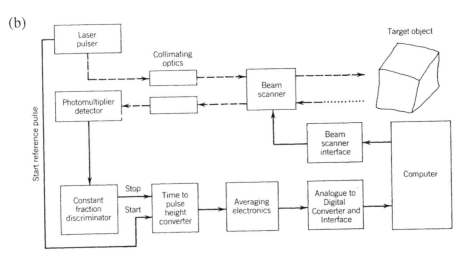

Figure 11. Laser Range Sensors. (a): Continuous wave phase shift detection system. (b): Direct pulse time-of-flight system. (Copyright © 1983 IEEE)

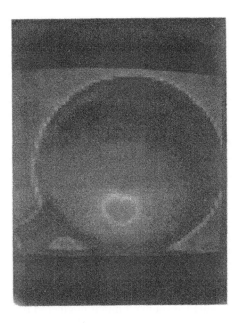

Figure 12. Laser direct pulse time-of-flight example results. (Copyright © 1983 IEEE)

determination over the entire scene of interest. A "reflectance map" which relates surface orientation to image intensity (shading) for a known albido-uniform surface is developed.

Denoting p, q as the surface slope components in a 3D cartesian (x, y, z) system where z is depth,

$$p = \frac{\partial z}{\partial x}$$
$$q = \frac{\partial z}{\partial y}$$

$R(p, q)$ is the reflectance map which associates brightness with surface orientation; this function must be evaluated for each known surface material to be analysed (one at a time). Varying surfaces or objects cannot be handled until prior segmentation and identification has been carried out.

The image brightness at a point (x, y) is related to the reflectance map as

$$E_p(x, y) = R(p, q, ps, qs),$$

where $E_p(x, y)$ is image irradiance at the image plane at (x, y), (orthographic projection assumed), (ps, qs) are the light source direction components, and $R(p, q, ps, qs)$ is the reflection map function defined over surface orientation and light source direction.

$R(p, q, ps, qs)$ is calculated off line for a particular surface material (assumed uniform over the scene, in the simple case). A set of $E_p(x, y) = R(p, q, ps, qs)$ equations with different light source positions (but camera and scene stationary), are solved to derive the surface slope components (p, q) at each point (x, y) of the image. Materials with highly specular reflective properties can be dealt with this way, but the method is not restricted to only these type of materials. Off-line computation of the reflectance map may be expensive, but on-line scene analysis is largely by rapid table look-up.

Multiple reflection illumination effects cannot be reliably analysed this way. Application in industrial vision systems would seem very promising since the capability to handle specularity, the high likelihood of occurrence of known materials and the rapid on-line computational properties are all of particular relevance here. The intrusion of special lighting conditions is also acceptable.

This approach to shape recovery is often referred to as "photometric stereo" since lighting variations are used; however only one image position is involved and the method is strictly monocular with all its advantages.

4.5. Range from Brightness

The use of the simple inverse square law of perceived brightness with point light source distance from a surface seems a much neglected but promising approach to range sensing. Neither range from texture gradient nor range from focus methods, amongst monosular approaches, are able to range to non-busy visual parts of a scene. Thus, many smoothly shaped uniform surfaced objects cannot be readily dealt with by such methods. For Lambertian scattering surfaces, whether varying in albedo or visual texture, the following approach yields satisfying results, with trivial computational requirements (which are also parallisable), provided that secondary reflection illumination effects are not severe [57].

Three images are acquired - one under ambient lighting conditions, one with a single point light source some distance from the scene and the third with the same light source (or equivalent) closer to the scene, the point light sources being at known positions approximately coaxial with the camera principal axis. If the near light source is not too close to the scene, the cosine component of observed brightness in the image can be neglected. For each pixel position, let I0, I1 and I2 refer to the image intensities for the three images taken in the order they are described above.

Denoting the point light source separation d, the range l to the point on the scene corresponding to each pixel can be calculated simply. With R defined as $\frac{(I2-I0)}{(I1-I0)}$, which is always grater than 1.0 since $I2 > I1 > I0$,

$$l = \frac{d(1 - \sqrt{R})}{(R - 1)}.$$

Provided that the camera returns values which are linear with intensity or a suitable calibration look up table is used, albedo and visual texture variation will not effect the result and ranging to all surfaces viewable from the camera position can be achieved. For both monochrome and colour cameras, range and intensity will be registered by construction, another bonus of using this approach. Whilst spatial distortaion in the camera is not critical to the method, a high signal/noise ratio camera should be used (or frame averaging used to same effect). Figure 13 shows a typical rangepic result.

4.6. Beam Scanning Range Sensing

The simplest beam scanning rangefinder, used in some early auto-focus camera systems [58], consists of a sweeping collimated pencil beam of light energy (infra-red or visible, the former have the advantage of being detected against ambient lighting conditions) and a laterally displaced detector, in the plane of the swept beam, with a small aperture. The swept source angle at which the detector "sees" a lit spot of light on the scene is simply related to range, by $l = b \tan \theta$, where b is the base line distance [see Figure 14]. To use such a point-at-a-time device for dense ranging in an array mode, the device must be moved in a 2D scanning pattern over the field of view. The detector and the source can be swapped with the detector swept and the source fixed. A swept detector may be replaced by a stationary linear array of detectors or by a linear lateral effect diode (position sensing diode) which returns an analog position of the centre of area of a detected target. A two dimensional deflection system used with a 2D lateral effect diode can be used for 2D range scanning [59].

Extension of the above arrangement to sheet beam scanning is the natural way to capture a whole column of range readings for each beam position. In the arrangement shown in Figure 15 a vertical slit projector sweeping a sheet of light beam across the scene viewed by a video camera laterally displaced with the base line perpendicular to the sheet of light plane. Points along a single, perhaps broken, stripe in the camera's image will be displaced to the right by an amount related to the depth of the corresponding patch on the scene, according to the simple geometric relationship established earlier. The processing of a single image is sufficient to recover one whole column of range data. The more usual alternative to using a slit camera is to use a laser or light emmitting diode and a cylindrical lens to produce a sheet of light.

It is tempting to use more than one stripe at a time in the above scheme but unless they are sufficiently widely separated in the image or otherwise identifiable they can be confused in the image at obscuring boundaries in the scene. Stripe identification can be by colour [60], intensity [61], width, pattern [62] or binary coding amongst a smaller number of images than stripes [63].

In the active binary coded stereopsis scheme described in [63], $n = \log_2 N + 1$ images are collected, where N is the total number (power of 2) of stripe positions (thus, for $N=64$, $n=7$). A liquid crystal stripe light valve (the size of a 35 mm film slide) with N stripes (e.g. 64 or 128) where each stripe can be made opaque or transparent under computer control is used to create patterns which can identify the stripes using a binary code distributed amongst the acquired images. First an image with all stripes transparent is acquired to establish the positions (but not identity) of all stripes. Then one image is acquired with every second stripe transparent, the other opaque. A third image is acquired with a pattern of two

Figure 13. Range from brightness example result. (Copyright © 1983 IEEE)

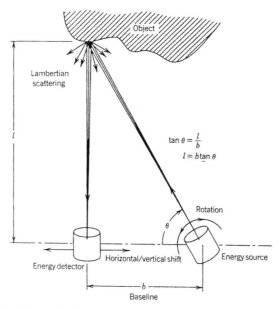

Figure 14. Simple triangulation range sensor. (Copyright © 1983 IEEE)

stripes transparent alternated with two opaque, a fourth with four transparent stripes alternated with four opaque and so on with the last image with the all stripes in one half of the value transparent and all stripes in the other half opaque. Each position established with the first image is associated with a binary code picked up at that position through the other images, thus identifying it uniquely. Figure 16 shows some ranging results of this sensor. Typically, between 500 to 2000 points on the scene one recovered without ranging to background material. Acquisition time is 0.5 second and processing time with sub-pixel accuracy approximately 10 seconds on an IBM 386 (perhaps 4 seconds on a IBM 486).

The Autonomous Systems Laboratory of Canada's National Research Council has developed the most accurate 3D laser scanner to date [64]. Using a white light laser source (He Cd), a high accuracy galvanometer mirror controller, a double sided scanning mirror and a slanted linear CCD array [see Figure 17] it is possible to capture high density (typically 800 x 512) range data with accuracy down to 50 um. Red, green and blue intensity data can also be recovered. Typically, some 20 seconds are involved capturing data from simple objects on a rotating table. The so called 'synchronised scanner' system permits the linear CCD array to be fully utilised to measure disparity without the scan deflection component being added, due to the return beam being reflected off the back side of the scanning mirror. Point at a time rather than a light sheet scanning is used, presumbly to simplify both the positions and intensity determinations. The synchronised scanning system could have been used for a light sheet mode using a 2D camera, and, in fact, a variation of this idea is incorporated into a scanning rangefinder designed at Oxford University's

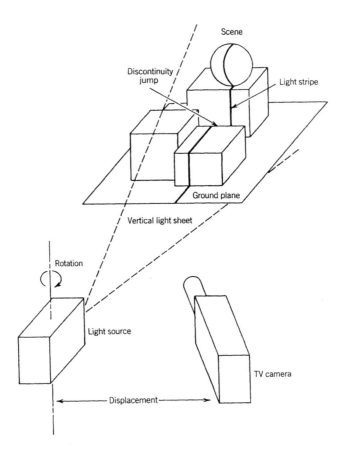

Figure 15. Striped light range sensing. (Copyright © 1983 IEEE)

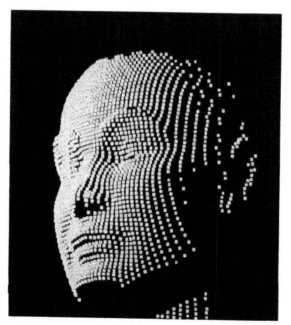

Figure 16. Active coded stereopsis range sensor example result.

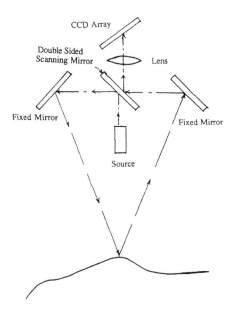

Figure 17. NRC synchronised range scanner.

Engineering Science Department [65].

In the Oxford stripe light rangefinder system [65], both the outward sheet of light and the return image are via the same scanned mirror. Each stripe captured by a 2D CCD camera is therefore "corrected" to allow for the beam scan component so that disparity variations can be detected across the whole width of the camera, thus improving the resolution.

A potentially very fast ranging system has been devised at Carnegie Mellon University [66]. The simple realisation that, if each pixel of a sensing array could record the beam position of a scanned sheet of light at the instant of its image crossing that pixel, 3D fixing data has been captured, revolutionises the way in which rangefinders can be built. If, for example, each pixel captured an analog sample fo the saw tooth signal which represents the sheet beam scanning mirror position at the instant a bright signal flashes past that pixel, subsequent reading of the array of analog values would permit a corresponding array of range values to be determined, since the geometry of the scanning system is known. If such "smart pixel" circuitry were built with sufficiently high density in VLSI, scanning rates up to 20,000 scans per second could be achieved. The miniturisation aspect is also of considerable value.

At Monash University's Intelligent Robotics Research Centre a system inspired by the above notion has been constructed using a standard mirror scanner and 2D CCD camera system. Typically, 128 x 128 range pictures can be captured every two seconds for ranges between .5 to 3.5 metres. This system has been used for mobile robot environmental mapping experiments [67]. A typical scanning result is shown in Figure 18. This apparatus is inexpensive and easy to use.

5. Some Early 3D Processing Ideas

It is not appropriate to delve too deeply into the details of various stages of processing which might follow range sensing, as such is the central subject of other chapters. However, for a sense of completion to round out this chapter, a simple example of 'early processing' of range data may prove sufficient [68].

The Monash University Active Coded Stereopsis rangefinder [63] normally uses a black/white CCD camera; however, if a colour camera is used it is easy to capture both the range and the surface colour data in a registered way since range is extracted from image data. Each (x, y, z) triplet in Cartesian space recovered by the range sensor can be associated with a (r, g, b) colour triplet of a small patch on the surface of the scanned objects. The pixel adjacency propertries of the range data which arise from the image based method of its extraction can be used to triangulate the data into surface patches which can be assumed planar. Cross products of the vectors (in 3D) which correspond to the edges of these triangles at vertices produce surface normal information which can be shown in a needle diagram [Figure 20]. Now each point in Cartesian space which represents the position of a surface element can be represented as a nine component vector. The data set can be represented as

$$ \mathbf{S}_i = \{a, y, z, r, g, b, n_x, n_y, n_z\}, i = 1, 2, \ldots n $$

for n points in space, where (x, y, z) gives position, (r, g, b) colour intensity components, and (n_x, n_y, n_z) the unit normal vector components.

Clustering in this nine dimensional space or any subset of it can produce meaningful segmentations such as spatially distinct objects (x, y, z space) patches of common colour (r, g, b space) or surface normal directionality (n_x, n_y, n_z space) or combinations of these [Figure 19]. This provides a very rich set of "semantic free" partitions upon which to base scene segmentation components including grip-site detection, object recognition, pose, support and adjacency relationships, etc. Convex hulls of spatially distinct components might be used to extract shape and support conditions; various formulae involving surface normal directionalities can be used to determine curvature and help define planar regions and 3D edges, and so on. If "semantic binding" is delayed until the above type of generalised information is nicely structured, the whole scene analysis task may well be made more robust and simpler as well. It is not difficult to invent some relatively simple methods by which domain specific knowledge can guide higher level stages of analysis; it is also the case that domain constraints can dictate the most appropriate properties to extract from the low level data; in the end, an appropriate balance between "top-down" and "bottom-up" approaches must be struck if the whole analysis process is to be both reliable and fast.

Figure 18. Monash University (Intelligent Robotics Research Centre) Scanning Range Sensor example.

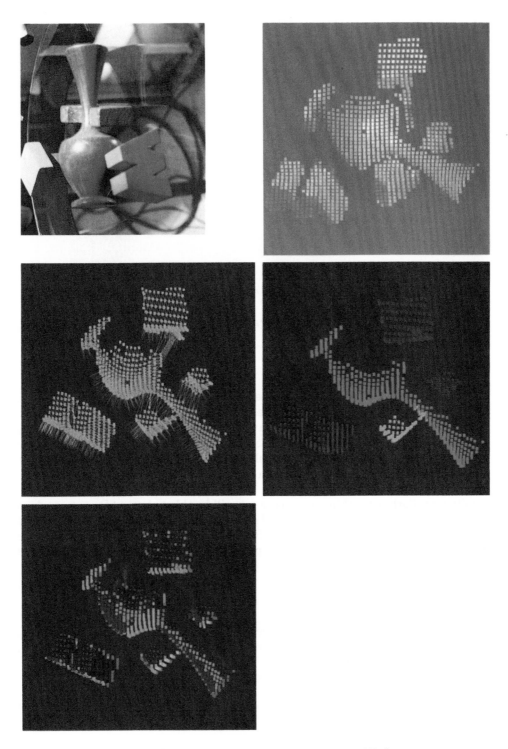

Figure 19. Shape and colour property segmentations using 3D data.

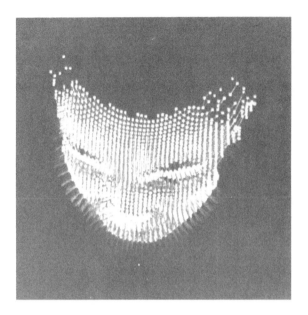

Figure 20. Surface normal needle diagram.

6. Conclusions

Exploration of new methods of range sensing will continue for quite some time yet. However, for most Computer Vision researchers, the availability of a compact, fast, reliable, safe, accurate and inexpensive range sensing device "off the shelf" would be of enormous value since it would permit them to concentrate their efforts on the subsequent stages of analysis to realise integrated working 3D vision systems. These would be of considerable value in support of applications in manufacturing, agriculture, mining, service industries, space exploration, nuclear industries and healthcare. Perhaps within a few years one will be able to buy a 3D camera which supplies registered colour intensity and range data at rates in excess of the current frame rates of standard video cameras. Such an eventuality is well within the scope of the probable and would lead to an explosion of 3D vision systems of tremendous utility in the above application fields.

REFERENCES

1   Jarvis, R.A., A Perspective on Range Finding Techniques for Computer Vision, *IEEE Transactions on Pattern Analysis and Machine Intelligence*. Vol. PAMI-5, No.2., March 1983, pp.122-139.
2   Jarvis, R.A. , Sensors, Distance Measurement, invited entry article for Encyclopaedia of Robotics, John Wiley and Son, published March 1988.
3   Jarvis, R.A., A Semantic-Free Approach to 3D Robot Colour Vision, invited

chapter for book edited by Prof. Takeo Kanade, published by Kluwer Academic Publishers, 1987, pp.565-609.

4 Horn, B.K.P., Focussing, Project MAC, AI Memo. 160, M.I.T., Cambridge, Mass, May 1968.

5 Jarvis, R.A., Focus Optimization Criteria for Computer Image Processing, *Microscope 24*, 163-180 (1976).

6 Tang, K.W. and Jarvis, R.A., Range from Focus - The Direct Method Revisited, *Proc. Australian Pattern Recognition Society Conference on Digital Image Computing* 4-6 December 1991, pp.443-450.

7 Pentland, A.P., A New Sense for Depth of Field, *IEEE Trans. Pattern Anal. Machine Intelligence*, Vol. PAMI-9, No.4, July 1987, pp. 523-531.

8 Jarvis, R.A., Range from Out-of-Focus Blur, *Proc. A.I.'88 - Australian Joint Artificial Intelligence Conference*, Adelaide, 15-18 Nov., 1988, pp.356-372.

9 Marr, D. and Hildreth, E.C., Theory of Edge Detection, *in Proc. R. Soc. London B*, Vol. 207, 1980, pp. 187-217.

10 Shi H., Naghdy F. and Cook C.D., Three-Dimensional Modelling by Parallel Layers of an Object Using Coloured Liquid, *Proc. Conference on Digital Image Computing : Techniques and Applications*, 4-6 Dec., 1991, Melbourne, Australia, pp.97-103.

11 Jarvis R.A., An Autonomous Mobile Robot in a Rangepic World, invited for presentation at *the 2nd International Conference on Automation, Robotics and Computer Vision*, 15-18 Sept., 1992, Singapore.

12 Kolagani N., Fox H.S. and Beidberg D.R. Photometric Stereo Using Point Light Sources, *Proc. IEEE International Conference on Robotics and Automation*, Nice, France, May 12-14, 1992, pp. 1759-1764.

13 Gibson, J.J., The Senses Considered as Perceptual Systems, Houghton-Miffin, Boston, Mass., 1966.

14 Bajcsy, R. and Lieberman, L., Texture Gradient as a Depth Cue, *Comput. Graphics Image Processing 5*, pp.52-67 (1976).

15 Haralick, R.M., Shanmugan, K. and Dinstein, I.H., Textual Features for Image Classification, *IEEE Trans. Syst., Man, Cybern.* SMC-3, 610-621 (Nov. 1973).

16 Weszka, J.S., Dyer, C.R. and Rosenfield, A., A Comparative Study of Texture Measures for Terrain Classification, *IEEE Trans. Syst., Man, Cybern.* SMC-6, 269-285 (Apr. 1976).

17 Moravec H.P., Visual Mapping by a Robot Rover, *Proc. 6th Int. Joint Conf. Artificial Intell.*, 1979, pp.598-620.

18 Yoshida K. and Hirose S., Real-Time Stereo Vision with Multiple Arrayed Camera, *Proc. IEEE International Conference on Robotics and Automation*, Nice, France, May 12-14, 1992, pp.1765-1770.

19 Baker, H.H., Edge Based Stereo Correlation, *Proc. ARPA Image Understanding Workshop*, Univ. Maryland, Apr. 1980.

20 Marr, D and Poggio, T., Computational Approaches to Image Understanding, M.I.T., A.I. Lab., see also *Proc. R. Soc. Longon B*, vol. 204, pp. 301-328, 1979.

21 Marr, D. and Hindreth, E.C., Theory of Edge Detection, in *Proc. R. Soc. London B*, vol. 207, pp. 187-217, 1980.

22 Hildreth, E.C., Edge Detection in Man and Machine, *Robotics Age*, pp. 8-14, Sept./Oct. 1981.

23 Nishihara, H.K. and Larson, N.C., Toward a Real Time Implementation of the Marr-Poggio Stereo Matcher, in *Proc. Image Understanding Workshop*, Lee Bauman, Ed., 1981.

24 Cheung C.C. and Brown W.A., 3D Measurement Using Three Camera Stereopsis, *Symp. on Optical and Opto-electronic Eng.*, Cambridge, Mass., paper 850-17, 1987.

25 Faugeras O.D., How Can Vision Make Mobile Robots Come True, *Proc. International Symposium and Exposition on Robots*, 6-10 Nov, 1988, Sydney, Australia, pp.1430-1461.

26 Jarvis, R.A., Quad-Vision Ranging for Robotic Application, *Proc. 4th Australian Joint Conference on Artificial Intelligence*, November 21-23, 1990, Hyatt Regency, Perth, Western Australia, pp.682-698.

27 Jarvis, R.A., Quad Vision Ranging for Outdoor Mobile (Agricultural) Robots, *Proc. 2nd IARP Workshop on Robots in Agriculture and the Food Industry*, 17-18 June 1991. Genoa, Italy, pp.129-140

28 Levine, M.D., O'Handley, D.A. and Yagi, G.M., Computer Determination of Depth Maps, *Comput. Graphics Image Processing*, vol. 2, pp. 134-150, 1973.

29 Yakimovsky, Y. and Cunningham, R., A System for Extracting Three- Dimensional Measurements from a Stereo Pair of TV Cameras, *Comput. Graphics Image Processing*, vol. 7, pp. 195-210, 1978.

30 Fua P., Combining Stereo and Monocular Information to Compute Dense Depth Maps that Preserve Depth Discontinuities, *Proc. 12th IJCAI Conf.* 24-30 Aug.'91, Sydney, Australia, pp.1292-1298.

31 Inoue H., Tachikawa T. and Inaba M., Robot Vision with a Correlation Chip for Real-Time Tracking, Optical Flow and Depth Map Generation, *Proc. IEEE International Conference on Robotics and Automation*, Nice, France, May 12-14, 1992, pp.1621-1626.

32 Wang, H. and Brady J.M. Corner Detection for 3D Vision Using Array Processors. In *BARNAIMAGE 91*, Barcelona, Sept. 1991. Springer-Verlag.

33 Kitchen, L. and Rosenfeld, A. Gray-Level Corner Detection. *In Pattern Recognition Letters*, volume 1, pages 95-102, 1982.

34 Burt, P. and Julesz, B., A Disparity Gradient Limit for Binocular Fusion. *Science*, 208:615-617, 1980.

35 Pollard, S.B., Mayhew, J.E.W. and Frisby, J.P. PMF: A Stereo Correspondence Algorithm Using a Disparity Gradient Limit. *Perception*, 14:449-470, 1985.

36 Wang H. and Brady M., A Structure-from-Motion Algorithm for Robot Vehicle Guidance, *Proc. IEEE Symposium on Intelligent Vehicles*, Detroit, June 30 - July 2, 1992.

37 Wang H. and Brady M., Parallel Implementation of DROID and Performance Evaluation, Report No. OUEL 1926/92, University of Oxford, Dept. of Engineering Science.

38 Charnley, D. and Blisset, R., Surface Reconstruction from Outdoor Image Sequences. *Image and Vision Computing*, 7(1):10-16, 1989.

39 Burger W. and Bhanu B., On Computing a 'fuzzy' Focus of Expansion for Autonomous Navigation, *Proc. IEEE Computer Society Conf. on Computer Vision and Pattern Recognition*, San Diego, California, 1989, pp.563-568.

40 Frazier J. and Nevatia R., Detecting Moving Objects from a Moving Platform, *Proc. IEEE International Conference on Robotics and Automation*, Nice, France, May 12-14, 1992, pp.1627-1633.

41 Manyika J.M., Treherne I.M. and Durrant-White H.F., A Modular Architecture for Decentralised Sensor Data Fusion, *Proc. 2nd Workshop on Sensor Fusion and Environmental Modelling, International Advanced Robotics Program*, 2-5 Sept. 1991, Oxford University, Session 6B, pp.1-15.

42 Jarvis R.A., Eye-in-Hand Robotic Vision, *Robots in Australia's Future Conference*, Perth, W.A., 13-16 May 1986, pp.150-161.

43 Elfes A., Occupancy Grids: A Probabilistic Framework for Robot Perception and Navigation, Ph.D. Thesis, Electrical and Computer Engineering Dept./Robotics Institute, Carnegie-Mellon University, May 1989.

44 Elfes A., Dynamic Control of Robot Perception Using Multi-Property Inference Grids, *Proc. IEEE International Conference on Robotics and Automation*, Nice, France, May 12-14, 1992, pp.2561-2567.

45 Skolnik M.I., Introduction to Radar Systems, McGraw-Hill, 1980.

46 Durrant-Whyte H.F., Port Automation Technology, Proc. Singapore 92, March 1992.

47 Duda, R.O. and Nitzan, D. Low-Level Processing of Registered Intensity and Range Data, *Proc. 3rd Int. Joint Conf. Artificial Intell.*, 1976.

48 Nitzan, D., Brian, A.E. and Duda, R.O., The Measurement and Use of Registered Reflectance and Range Data in Scene Analysis, *Proc. IEEE 65*, 206- 220 (Feb. 1977).

49 Jarvis, R.A., A Laser Time-of-Flight Range Scanner for Robotic Vision, *IEEE Trans.PAMI*, Vol. PAMI-5, No.5 Sept. 1983, pp.505-512.

50 Lewis, R.A. and Johnston, A.R., A Scanning Laser Rangefinder for a Robotic Vehicle, *Proc. 5th Int. Joint Conf. Artificial Intell.*, 1977, pp. 762-768.

51 Odetics 3D Laser Imaging Radar data sheets, Odetics Inc., 1515 South Manchester Ave., Anaheim, CA92802-2907.

52 Eagle 3004 System data sheets, Digital Optronics Corporation, 5554 Port Royal Road, Suite 202, Springfield, Virginia 22151-2303.

53 Environmental Research Institute of Michigan (ERIM) 3D rangefinder described in: Zuk D., Pont F., Franklin R. and Dell' Eva M., A System for Autonomous Land Navigation, *Proc. Active Systems Workshop*, Naval Postgraduate School, Moterey, California, 1985.

54 Horn, B.K.P., Shape from Shading: A Method for Obtainig the Shape of a Smooth Opaque Object from One View, M.I.T., Project MAC, MAC TR-79, Nov. 1970.

55 Horn B.K.P. and Brooks M.J. (Editors) *Shape from Shading*, The M.I.T. Press, 1989.

56 Ikeuchi, K. and Horn, B.K.P., An Application of the Photometric Stereo Method, in *Proc. 6th Int. Joint Conf. Artificial Intell.*, Tokyo, Japan, 1979, pp.413-415.

57 Jarvis, R.A., Range from Brightness for Robotic Vision, *4th International Conf. on Robot Vision and Sensory Controls*, London, October 1984, proceedings pp.165-172.

58 Jarvis, R.A. A Computer Vision and Robotics Laboratory, *IEEE Computer* 15(6), 8-24 (June 1982).

59 Gower S., Kennedy P. and Holzer A., A New Approach to 3-D Artificial Vision, *Proc. Conf. on Robots in Australia's Future*, 13-16 May, 1986, Perth, Western Australia, pp.163-174.

60 Boyer, K.L. and Kak, A.C., Color-Encoded Structured Light for Rapid Active Ranging. *IEEE Transactions on Pattern Anlysis and Machine Intelligence*, PAMI-9(1):14-28, January 1987.

61 Carrihill, B. and Hummel, R., Experiments with the Intensity Ratio Depth Sensor. *Computer Vision, Graphics and Image Processing*, 32:337-358, 1985.

62 Vuylsteke, P. and Osterlinck, A., 3-D Perception with a Single Binary Coded Illumination Pattern. In SPIE Vol 728, *Optics, Illumination and Image Sensing for Machine Vision*, pages 195-202, Cambridge, MA, USA, 30-31 October, 1986.

63 Alexander, B.F. and Ng, K.C., 3-D Space Measurement by Active Triangulation Using an Array of Coded Light Stripes, *Proc. S.P.I.E. Conf. on Optics, Illumination and Image Sensing for Machine Vision*, II, Cambridge, Mass., 11p., 1987.

64 Littlehales C. and Rioux M., White Light Magic, *Iris Universe (Magazine)*, No.18, 1992, pp.24-27.

65 Reid I., Recognizing Parameterized Objects from Range Data, Report No. OUEL 1918/92, University of Oxford, Dept. of Engineering Science.

66 Kanade T., Gruss A. and Carley L.R., A VLSI Sensor Based Rangefinding System, Preprint of *the 5th International Symposium of Robotics Research*, August 28-31, 1989, Tokyo, Japan, pp.383-390.

67 Badcock, J.M., Dun, J.A., Ajay, K., Kleeman, L. and Jarvis, R.A. An Autonomous Robot Navigation System - Integrating Environmental Mapping, Path Planning, Localisation and Motion Control. Accepted by *Robotica*.

68 Jarvis, R.A., 3D Shape and Surface Colour Sensor Fusion for Robot Vision - Property Segmentation Using Clustering, *Proc. 2nd International Advanced Robotics Program Workshop on Sensor Fusion and Environmental Modelling*, Session 3A, 2-5 Sept. 1991, Oxford, England.

We thank the following publishers for granting us permission to reproduce the figures listed below:

Figures 5, 10, 11, 12, 13, 14, 15: **John Wiley & Sons, Inc.**
From the *Encyclopedia of Robotics* edited by R.C. Dorf. Published 1988. All figures, except figure 13 appeared earlier in Vol. PAMI5, No. 2 of the IEEE PAMI Trans., March 1983. Copyright © 1983 IEEE.
Figure 13: **Kluwer Academic Publishers**
From *Three Dimensional Machine Vision* edited by T. Kanade. Published 1987.

Three-Dimensional Object Recognition Systems
A.K. Jain and P.J. Flynn (Editors)
© 1993 Elsevier Science Publishers B.V. All rights reserved.

Feature Extraction for 3-D Model Building and Object Recognition[1]

Frank P. Ferrie, Shailendra Mathur, and Gilbert Soucy[a]

[a] Computer Vision and Robotics Laboratory
McGill Research Center for Intelligent Machines
3480 University Street
Montréal, Québec H3A 2A7
Canada

Abstract

Progress in surface recovery algorithms and the availability of reliable range sensors has led to increased interest in three-dimensional modelling, recognition, and analysis. This chapter explores one aspect of this problem - the recovery of surface features necessary to carry out such tasks. Many researchers have found it convenient to cast feature representation in terms of differential geometry, largely due to the intuitive interpretation of features computed from common forms, e.g. mean and Gaussian curvatures, and the view invariance afforded by such representations. However, a common problem has been the reliable computation of these features, particularly where directional properties are concerned.

The focus of this chapter is a class of reconstruction algorithms, originally devised by Sander and Zucker, that can successfully recover the local structure of a surface described in terms of differential geometry. We demonstrate the robustness of the so-called *curvature consistency* algorithm when adapted to range images and then present an extension to the original work that also permits accurate localization of surface discontinuities. Because of its formulation in terms of curvature, the resulting algorithm is both view invariant and can successfully recover surface features given initial estimates of either depth or surface orientation. As such it is a useful tool for three-dimensional image analysis.

1. Introduction

Tasks such as object recognition and three-dimensional model building are dependent on either the explicit recovery or the implicit inference of 3-D features encoded in sensor data. For example, features corresponding to surface discontinuities and singularities and extremal points in surface curvature are important cues in determining the parts decomposition of an object [26, 16, 18, 14, 15]. Other surface features, e.g. mean and Gaussian curvatures, singularities in principal

---

[1]This research was supported in part by the Natural Sciences and Engineering Research Council of Canada under Grant OGPIN 011.

direction fields, etc., provide cues from which instances of specific objects can be recognized [7, 2, 10, 5, 6, 35]. Directional attributes such as the principal directions of curvature on a surface, are of particular importance as they provide information that is essential in order to group individual feature points into boundary contours and surface patches [14, 15]. They can also provide a basis from which to determine correspondence between features when considering an analysis based on multiple views [36, 37].

Most efforts to date have been aimed at the recovery of specific features, usually in the context of a given sensor, e.g. range data [2, 10, 20, 6]. The approachs most often used involve local estimation of surface properties either through operator-based methods or surface approximation. These methods are analogous to feature detection in conventional images and suffer from many of the same problems, namely the sensitivity of local operations to noise and quantization error, and limitations in the operators and approximations used. Researchers have addressed these problems in various ways, e.g. adaptive windowing techniques [3], robust statistics [33], constraint minimization strategies [6], incorporation of sensor models [20], again paralleling the techniques and strategies of conventional image analysis. The net result has been success at recovering scalar properties associated with differential forms (e.g. mean and Gaussian curvatures, principal curvature magnitudes [9]), but the recovery of directional properties, i.e. directions of principal curvatures, is difficult without considering more global approaches to feature recovery.

A strategy that embodies this kind of global view is that of visual reconstruction [4]. With respect to 3-D feature extraction, the idea is to first recover a stable representation of the underlying surface and then determine features of interest from this intermediate form. The effect is to shift the computational burden from feature extraction to surface recovery as the latter problem is generally *ill-posed* [31]. Early work in this area was motivated by the stereo interpolation problem [22, 38, 39] and has progressed steadily towards a more general framework for image and surface reconstruction [4]. Ideally, the result of applying an appropriate reconstruction algorithm to a set of data (e.g. points sampled from a surface with a laser rangefinder) would be a piecewise-smooth interpolation of that data which would form the basis for subsequent computation. This notion of smoothness is often related to the kind of interpolation model that underlies a particular reconstruction algorithm (e.g. membrane, thin plate, etc. [4]). However, we argue that the process of reconstruction should not be considered independently of surface representation. For example, a thin plate interpolation model might be appropriate in the context of a polyhedral surface, but can lead to difficulty with surfaces of higher order (e.g. arbitrary fractures on a smoothly curving surface).

Any smoothing procedure should be designed to preserve the features of interest while minimizing the undesired effects of noise. Sander and Zucker [34, 35] were the first to formulate the surface reconstruction problem in terms of differential geometry for computed tomography (C.T.) images. Their approach was cast in a variational framework where the objective was to minimize a functional form related to a minimum variation of curvature. Certain non-linear aspects of this approach bear some resemblance to relaxation labelling processes [27], hence it is sometimes referred to as *variational relaxation*. The success of the resulting curvature consistency algorithm in recovering surface features (directional properties in particular) made it particularly appealing as a reconstruction algorithm for range data [14, 15]. However, because it is formulated in terms of surface curvature, the algorithm is also well-suited to surface reconstruction from estimates of local surface orientation, e.g. shading [12, 13].

This chapter is largely about our efforts to apply the curvature consistency framework to problems of surface reconstruction in range data analysis. Our motivation is the recovery of three-dimensional models of solid shape for which the recovery of surface features forms an important basis [14, 15, 23]. This context also imposes a number of constraints which have resulted in the evolution of the original Sander and Zucker algorithm apart from the mapping from C.T. to range and intensity images. Real objects are at best piecewise-smooth. The pencil sharpener shown in Figure 1a contains both curved and planar surfaces separated by orientation discontinuities. To recover the geometric model shown in Figure 1c, features corresponding to orientation and depth discontinuities and extremal values of curvature that comprise the object's part boundaries (Figure 1b) must be accurately recovered, requiring explicit treatment of discontinuities and planar patches. Another aspect of the solid shape problem is that multiple views of an object are often required to deal with uncertainty [40, 41]. Although not discussed in this chapter, a similar framework can be applied to the temporal domain to recover local motion parameters [36, 37].

The structure of this chapter is as follows. Section 2 provides a short introduction to concepts from differential geometry, the features used for recognition and analysis, and local approximation methods for their estimation. Some of the deficiencies of these methods are demonstrated by various examples of feature estimation on range data acquired with a laser rangefinder. Section 3 describes the curvature consistency algorithm. It closely follows the formulation of [34, 35] for C^2 surfaces but is adapted for surfaces of the form $z = f(x, y)$ and incorporates some changes to the updating rules that improve estimation of surface direction in the vicinity of umbilic points. The resulting algorithm corrects many of the deficiencies of the local methods, but does not preserve the structure of surface discontinuities - a major shortcoming with respect to feature localization. Finally, Section 4 shows how the algorithm can be augmented to overcome the latter problem by incorporating a discontinuity model in the updating procedure. Edges formed by ramps and jump discontinuities are correctly preserved and are largely immune to changes in scale. The final algorithm is robust, efficient, and largely independent of user-specified parameters.

2. Local Representation of a Surface

Analysis begins with either a collection of discrete points $\mathcal{R}_{xy}$ or local surface orientations $\mathcal{O}_{xy}$ sampled through direct measurement or estimated from a variety of Shape-from-X procedures. Since quantities of interest are often defined in terms of continuous mathematics, some form of interpolation is required to characterize

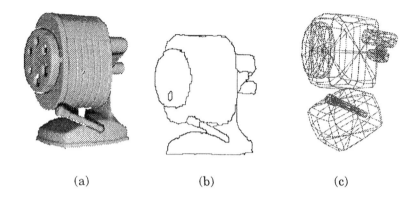

(a)   (b)   (c)

Figure 1. (a) Shaded image of a pencil sharpener rendered from a depth map. (b) Part boundaries resulting from surface analysis. (c) Wireframe rendering of the geometric model resulting from the complete analysis.

the local behaviour of the underlying surface S in the vicinity of each sample point P. Its precise form depends on what must be represented locally.

2.1. Differential Geometry

Differential geometry provides a convenient basis for describing the local behaviour of a surface S in the vicinity of some particular point P [25, 9], and has been widely used in the Computer Vision literature as a tool for describing surfaces [7, 1, 2, 10, 5, 18, 19]. Most approaches assume that surfaces can be adequately modelled as being at least *piecewise-smooth*, i.e. smooth patches separated by discontinuities in orientation and depth, where the amount of smoothness depends on what attributes need to be made explicit at P. To represent orientation and curvature, for example, the function used to represent S in the vicinity of P must be at least of class C^2 (twice differentiable). One commonly used function is a parabolic quadric of the form $w = au^2 + cv^2$, defined on a coordinate frame with origin at P with the w axis parallel to the surface normal N_P at P (Figure 2).

The relation of this local representation to surface curvature is explained with the aid of Figure 2. Two auxiliary planes are shown, T_P, the plane tangent to S at P and π_{N_P}, the plane orthogonal to T_P containing the normal N_P. As π_{N_P} is rotated about N_P, it intersects S in a contour referred to as a normal section; the curvature of the section at P is called the *normal curvature*, κ_{N_P}. Thus for any orientation $\mathbf{v}$ in T_P, there is a corresponding normal curvature $\kappa_{N_P}(\mathbf{v})$. The orientation of the $< u, v, w >$ frame of the local representation is chosen such that the u and v axes align with two special directions on S at P. These are the directions for which $\kappa_{N_P}(\mathbf{v})$ takes on maximum and minimum values, κ_{M_P} and $\kappa_{\mathcal{M}_P}$, and are referred to as the principal directions M_P and $\mathcal{M}_P$ respectively [9]. The scalar quantities κ_{M_P} and $\kappa_{\mathcal{M}_P}$ are similarly referred to as the principal curvatures at P. Following the

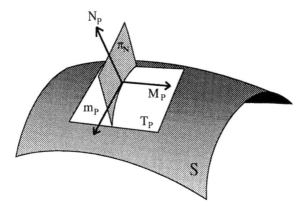

Figure 2. Local surface representation - the augmented Darboux Frame.

convention of [34, 35], we refer to $\mathcal{D}(P) = (P, M_P, \mathcal{M}_P, N_P, \kappa_{M_P}, \kappa_{\mathcal{M}_P})$ collectively as the augmented Darboux frame at P.

The utility of features computed from $\mathcal{D}(P)$ as surface descriptors is well-known to the computer vision community, e.g. [7]. Mean, $\mathcal{H}_P = \frac{\kappa_{M_P} + \kappa_{\mathcal{M}_P}}{2}$, and Gaussian, $\mathcal{K}_P = \kappa_{M_P} \times \kappa_{\mathcal{M}_P}$, curvatures can describe the geometry of S in the vicinity of P in a way that is invariant to parameterization - useful for purposes of recognition for example [2]. Extremal values of κ_{M_P} and $\kappa_{\mathcal{M}_P}$, e.g. negative local minima, are characteristic of part boundaries [26] and can be used to infer partitioning contours [18, 14, 15] or serve as landmarks for correspondence [17, 36, 37]. Gaussian curvature can be used to distinguish the rigid vs. non-rigid motion of an object moving under deformation [21]. The problem is how to reliably estimate $\mathcal{D}(P)$ from measurements or estimates of depth and orientation.

2.2. Local Estimation of the Darboux Frame

The general method employed to estimate $\mathcal{D}(P)$ is to characterize S in the vicinity of P and compute $M_P, \mathcal{M}_P, N_P, \kappa_{M_P}$, and $\kappa_{\mathcal{M}_P}$ from the this local approximation. Methods for doing so are varied, ranging from operator-based methods to functional approximation, analogous to feature analysis in intensity images. The approach described here is typical of local methods and is intended to illustrate some of the problems encountered and motivate the need for a more elaborate procedure. Without any loss in generality, the context will be data acquired with a laser rangefinding system; the same methodology holds for any estimate of surface depth[2].

Surfaces are acquired as a grid of discrete points $z(i, j)$. For each P_i centered in an $n \times n$ window the surface is locally parameterized with a parabolic quadric of the form $h(u, v) = au^2 + buv + cv^2$, using appropriate techniques such as least-squares [28]. Recall that this parameterization assumes a coordinate frame centered at P_i

[2]And surface orientation through a slightly different formulation [13].

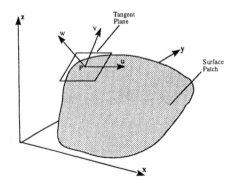

Figure 3. Tangent plane coordinate frame at a point P on S.

aligned with the surface normal N_{P_i}. The additional term in b is added to account for the unknown rotation of this local frame with respect to N_{P_i}. Thus to determine the coefficients of $h(u, v)$, the local neighbourhood of P_i must first be transformed into the coordinates of T_{P_i} as shown in Figure 3. The latter operation involves a planar fit to the $n \times n$ window centered at P_i to determine T_{P_i}, followed by a linear transformation that maps each point in the window to the coordinate frame of P. While this might seem somewhat involved, an alternative parameterization in terms of the global coordinate frame, $< x, y, z >$, can lead to numerical instabilities when estimating quadric coefficients on highly foreshortened surfaces.

A further optimization is often useful in the presence of orientation and depth discontinuities on S. One can attempt to minimize their effect by judiciously sampling $z(i, j)$ to avoid points with large orientation and depth gradients[3]. Figure 4 shows examples of support neighbourhoods corresponding to two pixels lying on the vicinity of a surface discontinuity. In such cases, estimation is somewhat more complex as parameter values must be extrapolated from the interior regions as shown. A tiling algorithm is used to abut support neighbourhoods against putative surface discontinuities.

Once the quadric parameters a, b, and c are estimated for each P_i, the quantities M_{P_i}, $\mathcal{M}_{P_i}$, N_{P_i}, $\kappa_{M_{P_i}}$, and $\kappa_{\mathcal{M}_{P_i}}$ are given by equation 26 in Appendix A. In practice one can obtain reasonable estimates of $\kappa_{M_{P_i}}$, $\kappa_{\mathcal{M}_{P_i}}$, $\mathcal{K}_{P_i}$, and $\mathcal{H}_{P_i}$ from local approximation, but these quantities are very sensitive to the effects of noise and quantization error. A qualitative measure of the stability of these estimates is given by the coherency of the K-H sign map [2], a labelling of a surface S according to the sign pairings of the Gaussian ($\mathcal{K}$) and mean ($\mathcal{H}$) curvatures, i.e., $\mathcal{K}_{P_i} : (-, 0, +) \times \mathcal{H}_{P_i} : (-, 0, +)$. Figure 5a shows the surface map of a toy unicorn from a standardized database of range images available through the National Re-

---

[3]These can be determined locally using appropriate operators, e.g. [8], in conjunction with the procedure used to compute T_{P_i}.

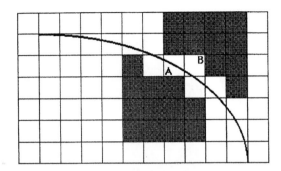

Figure 4. Placement of support neighbourhoods in the vicinity of surface discontinuities. The square grid represents a discrete sampling of the surface and the solid line marks the interpolated position of a discontinuity. Pixels that straddle the contour are placed above or below it according to the division of their area. The pixel marked B has most of its area above the discontinuity and its support neighbourhood is marked by the shaded upper pixels. The lower shaded pixels show the support neighbourhood of A, a pixel that has most of its area below the discontinuity.

search Council of Canada [32]. The image has been averaged down to a 256×256 resolution at 12 bits/pixel. Figure 5b shows the resulting K-H sign map obtained from a local approximation with a 5×5 window using the procedure described above. Even though the laser rangefinder used in acquiring this data has low noise, the result is still sensitive to quantization error and seriously degrades with the addition of noise.

The situation is worse for the directional quantities M_{P_i} and $\mathcal{M}_{P_i}$. Figure 5c shows the maximum principal directions of curvature computed from the same local approximation that produced the result of Figure 5b. The result is an almost random distribution of directions on the surface which is inconsistent with its smooth structure. Better results could have been obtained with additional smoothing, e.g. by approximating S at P_i over a larger window, but this raises the additional questions of how much smoothing to apply, whether to do so uniformly, and the effect on the resulting direction field. The problem of feature recovery must now be considered in the larger context of surface reconstruction[4].

3. Curvature Consistency

Curvature consistency refers to a class of surface reconstruction algorithms originally developed by Sander and Zucker [34, 35] in the context of surface recovery in

---

[4]For a comparison of different operators in the estimation of local curvature, see the article by Flynn and Jain in [19].

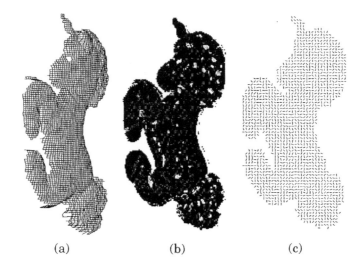

(a)   (b)   (c)

Figure 5. (a) The surface map of a toy unicorn obtained with a laser rangefinder. (b) K-H sign map computed from the surface map by local approximation. (c) Maximum principal direction field computed from the surface map by local approximation.

C.T. imagery. It is formulated in terms of differential geometry and can be viewed as a second stage of processing applied to a set of augmented Darboux frames $\mathcal{D}(P_i)$ estimated by local methods. The idea is to smooth S while preserving the local structure described by $\mathcal{D}(P_i)$. This is posed as a variational problem in which the objective is to minimize a functional form related to a minimum variation of curvature, somewhat analogous to the co-circularity constraint described in [30].

There are three ingredients to the problem formulation:

- A local description for the surface at a point P.

- A *transport model* which describes how this local description changes as it is moved (transported) to an adjacent point[5] Q_α and vice-versa.

- A function which describes how the local description at P is updated so as to be compatible (consistent) with the descriptions of its local neighbours once they have been moved from Q_α to P by the transport mechanism.

The local description of S at P has already been discussed; it is the augmented Darboux frame $\mathcal{D}(P)$. A transport model is required as a means of incorporating the constraint of minimum curvature variation, and an updating function as a means of enforcing it.

---

[5]The following convention is used with respect to subscripts: i, as in P_i, refers to a particular point on a surface S; α, as in Q_α, refers to an element of the local neighbourhood of a point P_i.

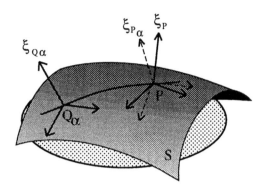

Figure 6. Transporting ξ_{Q_α} from Q_α to P. The transport model in this example is a parabolic quadric (shown occluded by the surface S).

3.1. Transport Model

Minimum variation of curvature between two adjacent points P and Q on S amounts to an assumption of locally constant curvature. To be consistent with the notation of Sander and Zucker[6], call the frame at a point P, ξ_P, and the frame at any of its adjacent neighbours Q_α, ξ_{Q_α} (Figure 6). Transporting the frame ξ_{Q_α} from Q_α to P involves extrapolation along a surface model that enforces the desired constraint of locally constant curvature. The resulting frame at P is called ξ_{P_α} and is an estimate of what the surface at P should look like according to the description at Q_α under the transport constraint.

There are many possibilities for a transport model. The only requirement is that the surface of this model embeds the constant curvature constraint along an arc joining P and Q_α. Figure 7 shows three such possibilities that approximate this constraint to varying degrees. For relatively dense samplings the approximation rendered by a parabolic quadric (Figure 7a) is sufficient[7], but extrapolation over a wider neighbourhood is often better served by the more elaborate models shown in Figures 7b and c.

The computational procedure required to transport each ξ_{Q_α} to P is explained with the aid of Figure 8. For each frame to be transported, P is projected onto the corresponding transport surface of each ξ_{Q_α}, which is a parametric function that can be determined from the parameters of ξ_{Q_α}. Appendix B shows how such a function is determined for a parabolic quadric model given M_{P_i}, $\mathcal{M}_{P_i}$, N_{P_i}, $\kappa_{M_{P_i}}$, and $\kappa_{\mathcal{M}_{P_i}}$ corresponding to a point P_i. Once this function and the projection of P onto it are determined, ξ_{P_α} is completely determined. In the example shown in Figure 8, the

---

[6]To avoid confusion, we usually use ξ_P to refer to the augmented Darboux frame at P in the context of parallel transport and updating, otherwise $\mathcal{D}(P)$ is used.

[7]In fact, one can often get away with a *planar* transport (i.e. a simple shift of the coordinates of Q_α in ξ_{Q_α} in this case).

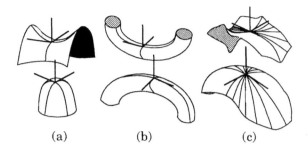

(a)   (b)   (c)

Figure 7. Three different possibilities for a transport model. (a) Parabolic quadrics approximate the constant curvature constraint only in the vicinity of the origin. (b) A toroidal transport model has constant curvature along the equator and along the arcs tangent to it. (c) A still more complex analytical surface with constant curvature in the directions shown in the figure.

frames at points Q_1, Q_4, and Q_5 are transported along their respective quadrics to the points of intersection with each projection of P. There are a number of technical details regarding the precise form of projection, e.g. closest point (perpendicular) vs. projection along the normal to P, N_P, but these are beyond the scope of this chapter. The interested reader is referred to [34, 35] and [29] for details.

3.2. Updating Rules

Given the set ξ_{P_α} determined by the transport model, the task is now to compute a maximum likelihood estimate of ξ_P that minimizes variation subject to the following constraints on M_P, $\mathcal{M}_P$, and N_P,

$$(N_P \cdot N_P) = 1 \quad (M_P \cdot M_P) = 1 \quad (M_P \cdot N_P) = 0. \tag{1}$$

As formulated in [34, 35], the minimization consists of two terms corresponding to (1) the surface normal N_P and principal curvatures κ_M and $\kappa_{\mathcal{M}}$, and (2) the principal direction M_P. Since $\mathcal{M}_P$ is orthogonal to both M_P and N_P, it need not be considered. To simplify the analysis, each is minimized independently. The first term, E_1, follows directly from [34, 35]:

$$E_1 = \sum_{\alpha=1}^{n} \|N_P - N_{P\alpha}\|^2 \quad + \quad (\kappa_M - \kappa_{M_{P\alpha}})^2 + (\kappa_\mathcal{M} - \kappa_{\mathcal{M}_{P\alpha}})^2$$
$$+ \quad \lambda((N_P \cdot N_P) - 1) \tag{2}$$

where $\xi_P = (P, \kappa_{M_P}, \kappa_{\mathcal{M}_P}, M_P, \mathcal{M}_P, N_P)$ and $\xi_{P_\alpha} = (P_\alpha, \kappa_{M_{P\alpha}}, \kappa_{\mathcal{M}_{P\alpha}}, M_{P\alpha}, \mathcal{M}_{P\alpha}, N_{P\alpha})$. Using standard methods, one obtains the following updating functionals for $\mathcal{N}_P$,

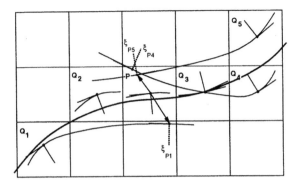

Figure 8. Projection and transport. The frames at points Q_1, Q_4, and Q_5 are transported to P in two steps. First, a transport surface is determined for each frame using its corresponding parameters. Then P is projected onto each surface and the corresponding ξ_{P_α} determined from the corresponding parametric function and projection point.

κ_{M_P}, and $\kappa_{\mathcal{M}_P}$:

$$N_P^{(i+1)} = \frac{\left(\sum_{\alpha=1}^n N_{x_{P_\alpha}}^{(i)}, \sum_{\alpha=1}^n N_{y_{P_\alpha}}^{(i)}, \sum_{\alpha=1}^n N_{z_{P_\alpha}}^{(i)}\right)}{\sqrt{\left(\sum_\alpha N_{x_{P_\alpha}}^{(i)}\right)^2 + \left(\sum_\alpha N_{y_{P_\alpha}}^{(i)}\right)^2 + \left(\sum_\alpha N_{z_{P_\alpha}}^{(i)}\right)^2}} \tag{3}$$

$$\kappa_M^{(i+1)} = \sum_{\alpha=1}^n \frac{\kappa_{M_{P_\alpha}}^{(i)}}{n} \qquad \kappa_{\mathcal{M}}^{(i+1)} = \sum_{\alpha=1}^n \frac{\kappa_{\mathcal{M}_{P_\alpha}}^{(i)}}{n} \tag{4}$$

where the superscript i refers to the current iteration step. Because M_P and $\mathcal{M}_P$ are *directions*, there is a 180° ambiguity in orientation. For this reason the formulation of minimization term E_2 needs to be re-cast from that described in [34, 35]. We avoid the ambiguity by minimizing the difference of directions in the tangent plane at point P as follows. Express M in tangent plane coordinates as

$$M_P = \bar{b}_1 \cos\theta + \bar{b}_2 \sin\theta, \ (0, 2\pi) \ \text{ such that} \quad \begin{matrix} 1) \\ 2) \\ 3) \end{matrix} \quad \begin{matrix} \bar{b}_1, \bar{b}_2 \in T_P \\ \|\bar{b}_1\| = \|\bar{b}_2\| = 1 \\ (\bar{b}_1 \cdot \bar{b}_2) = 0. \end{matrix} \tag{5}$$

$$E_2 = \min_\theta \sum_{\alpha=1}^n \left[1 - (M_P(\theta) \cdot M_{P_\alpha})^2\right]. \tag{6}$$

$\mathcal{M}_P^{(i+1)}$ is found by substituting the value of θ that minimizes (6), back into (6). Again, using standard methods, one obtains the following updating functional for θ:

$$\theta^{(i+1)} = \tan^{-1}\left[\frac{(A_{22}-A_{11})+\sqrt{(A_{11}-A_{22})^2+4A_{12}^2}}{2A_{12}}\right], \tag{7}$$

$$A_{ij} = \sum_{\alpha=1}^n (M_{P_\alpha} \cdot \bar{b}_i)(M_{P_\alpha} \cdot \bar{b}_j).$$

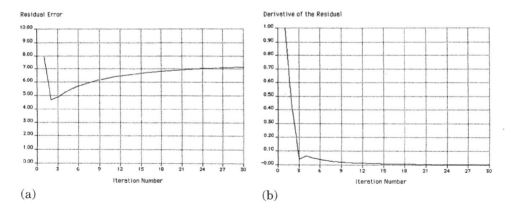

Figure 9. (a) R_S and (b) R'_S as a function of iteration number.

Note that this also determines the solution for $\mathcal{M}_P^{(i+1)}$ for the reason cited earlier.

3.3. The Curvature Consistency Algorithm

The resulting algorithm[8] for refining the set of augmented Darboux frames $\{\mathcal{D}(P)\}$ follows directly from the updating functionals in the previous section. Control over iteration is maintained by tracking the convergence of a composite measure, R_S, which is the sum of local difference measures computed over the surface,

$$R_S^{(i)} = \sum_j R_j(\xi_P^{(i)}, \xi_{P\alpha}^{(i)}) = \sum_j E_{j1}^{(i)} + E_{j2}^{(i)}, \quad P_j \in S. \tag{8}$$

A discussion of convergence properties is beyond the scope of this chapter and is treated elsewhere [29, 34, 35]. However, we have observed empirically over a large number of experiments a consistent behaviour in R_S. It will rapidly reach an absolute minimum (Figure 9a), and then converge asymptotically over a large number of iterations to some final value. The absolute minimum in Figure 9a corresponds to a first stable interpretation of S. In the limit, the final state will eventually correspond to a surface with uniform curvature properties[9]. The problem is to determine at which point to terminate the iteration, a situation somewhat analogous to selection of the λ parameter in regularization algorithms [31]. Both situations represent weightings between the initial interpretation of data and the structure implied by their respective local surface models [29].

Our strategy, like that of Sander and Zucker [34, 35], is to use rate of change of R_S, R'_S, as the criterion for termination. As can be seen from Figure 9b, the boundary between rapid and asymptotic change of R_S is clearly delineated. In practice, the algorithm is allowed to iterate until the difference $|R_S^{(i)} - R_S^{(i-1)}|$ falls

---

[8]Which we refer to as the Curvature Consistency algorithm.
[9]The specific form obtained will be dependent on the extrapolation model used, e.g. a sphere in the case of a parabolic patch.

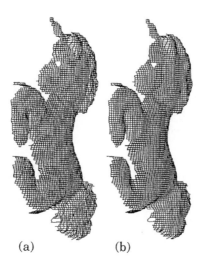

(a)   (b)

Figure 10. (a) The surface map of the toy unicorn shown earlier. (b) The map after 5 iterations of the curvature consistency algorithm.

below a specified threshold. We have found that stable results are generally realized within 5 iterations.

That the algorithm works can be demonstrated by applying it to the example shown earlier in Figure 5. First, it is interesting to compare surfaces before and after reconstruction to get a qualitative appreciation for the amount of smoothing involved. Figure 10a shows the surface map of the toy unicorn shown earlier in Figure 5a, and Figure 10b the reconstructed surface after 5 iterations of the curvature consistency algorithm. As can be seen, the amount of smoothing performed by the algorithm is minimal, yet feature recovery is significantly enhanced according to the results shown next in Figures 11 and 12. Figure 11a shows the initial K-H map shown earlier in Figure 5a and Figure 11b the map corresponding to the reconstructed surface. The characteristics of the resulting map are now consistent with the smooth nature of the surface and largely correct[10]. A more indicative result is obtained by comparing Figures 12a and b which show the maximum principal direction fields computed before and after reconstruction. Whereas the field in Figure 12a appears to be almost random, the structure is correctly recovered in Figure 12b.

4. Adaptive Localization of Discontinuities

The curvature consistency algorithm has been shown to be very effective in recovering the local structure of smooth surfaces obtained from C.T. data [34, 35], range

---

[10]To the degree that we could validate the result by hand.

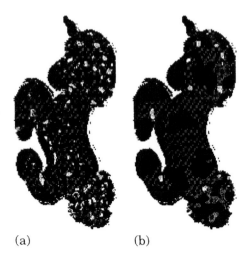

(a)   (b)

Figure 11. (a) KH sign map computed locally (shown earlier). (b) KH sign map after 5 iterations of the curvature consistency algorithm.

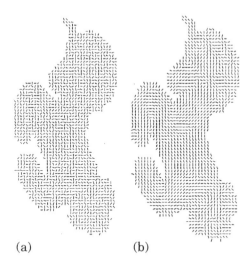

(a)   (b)

Figure 12. (a) Principal direction field computed locally (shown earlier). (b) Direction field after 5 iterations of the curvature consistency algorithm.

images [11, 14, 15], and shape-from-shading [13]. However, because the transport model does not include an explicit representation for discontinuities on S, the algorithm cannot recover the local structure in the vicinity of such a discontinuity[11]. As can be seen from the updating functions associated with ξ_P, there is no mechanism for inhibiting propagation across P and Q_α in the event of a discontinuity. This is demonstrated in Figure 13a which shows a synthetically generated step edge with Gaussian noise added. Figures 13b and 13c show the effects of the curvature consistency algorithm on the edge after 7 and 15 iterations respectively. It seems apparent that a solution to this problem requires a modification of the updating procedure that includes a model of local continuity such that the flow of information can be inhibited when it is not satisfied. Such a model is presented in the this section along with a mechanism for adapting the updating procedure.

We tackle the problem of smoothing over discontinuities by modifying our transport model to incorporate the possibility of a discontinuity. This is accomplished by modifying the updating procedure to weight contributions from the surrounding neighbourhood according to how well these frames support a model of local continuity. This model is based on the following conjecture:

> *Consider the local neighbourhood of a central point P represented by ξ_P. If P is part of a structural discontinuity with respect to any element of ξ_P (where the magnitude of the discontinuity is greater than the additive noise), then there are likely to be other neighbours Q_α whose values ξ_{Q_α} "support" or are consistent with ξ_P. Whereas if P is an isolated noise point, ξ_P will likely be inconsistent with ξ_{Q_α}, hence it will not have support from its neighbours in keeping its present values.*

The idea is to embed a mechanism in the updating procedure that can characterize whether a point is part of a structural discontinuity or a noise point. Points characterized as being on structural discontinuities would only receive support from those local neighbours with compatible values, otherwise the updating would proceed as normal. This characterization involves learning about the neighbourhood support at each iteration of the algorithm. We will next show how this mechanism can be implemented by a recursive filter using an approach similar to that used by Heel in [24].

4.1. Using Recursive Estimation

The general idea behind recursive estimation is:

1. There is a prediction stage, where using some model, the value of a particular quantity and its error variance are predicted as to what they will look like in the next iteration.

2. An updating stage uses the predictions weighted by the predicted error variance, and the new inputs weighted by their error variances to get a new estimate of the quantity.

---

[11]Recall that we are mainly concerned with C^0 and C^1 discontinuities.

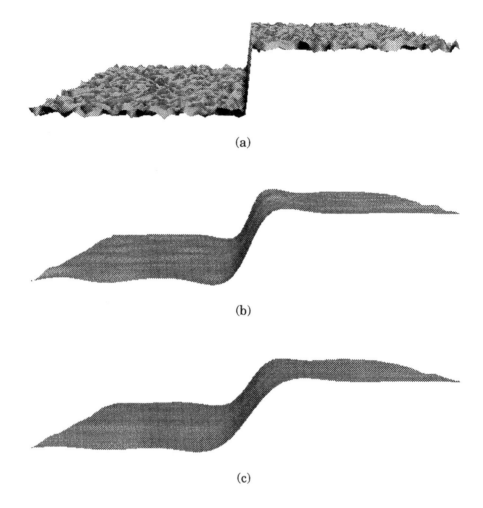

(a)

(b)

(c)

Figure 13. (a) Gaussian noise added to a synthetic step (b) Distortion in the step after 7 iterations of curvature consistency algorithm. (c) and after 15 iterations.

The curvature consistency algorithm can also be thought to have these two stages, where starting with an initial estimate of the surface fit, each neighbouring frame ξ_{Q_α} *predicts* the frame ξ_P corresponding to the central point P of the local neighbourhood for the next iteration. ξ_P is then *updated* using the equations described in Section 3. The only difference between the updating stages in curvature consistency and the recursive filter is in the weighting of these quantities by their error variances. As an illustration of this concept, consider the position vector $P \in \xi_P$ where $P = (x, y, h(x, y))$, and $h(x, y)$ represents S in the form of a range map. The prediction stage can be represented as

$$\hat{P}_-^{(i+1)} = f(\xi_{Q_\alpha}{}^{(i)}), \tag{9}$$

where the quantity $\hat{P}_-^{(i+1)}$ represents the predicted value of P in the next iteration, ξ_{Q_α} represents the set of frames corresponding to each neighbour of P, and the function f represents the application of parallel transport. Similarly, the updating stage in curvature consistency can be represented as

$$\hat{P}_+^{(i+1)} = \sum_{\alpha=1}^{n} \frac{\hat{P}_{\alpha-}^{(i+1)}}{n}, \tag{10}$$

where the quantity $\hat{P}_+^{(i+1)}$ represents the updated value of P.

Equations (9) and (10) would take on the form of a recursive filter if we included new inputs, but more importantly, error variance weighting in (10). However we do not have a new source of inputs. Including error variance weighting effectively means that the role that a particular neighbour Q_α plays in the update of P will depend on how well it predicted P in the previous iteration. Hence, instead of weighting all the neighbour prediction contributions to the update of P equally by $1/n$ as in (10), they should now be weighted by some function of their prediction error $\hat{p}_{\alpha-}^{(i+1)}$. Equation (10) then becomes

$$\hat{P}_+^{(i+1)} = \sum_{\alpha=1}^{n} \frac{\hat{P}_{\alpha-}^{(i+1)}}{f(\hat{p}_{\alpha-}^{(i+1)})}. \tag{11}$$

This also provides the mechanism to learn whether a given point P has sufficient support from its neighbours and thus avoid smoothing over discontinuities. Consider a point P at the center of a $M \times M$ window and positioned at an ideal step edge as shown in Figure 14a. Clearly the neighbouring points to the right will better predict P than the the ones on the left. As a result, the weights of the points on the left will diminish through successive iterations and ultimately have minimal impact on the updating of P, whereas the points on the right will have more influence in the updating of P and hence refine P according to their ξ_{Q_α} parameters. Now consider the case where P is positioned on an isolated noise discontinuity, as shown in Figure 14b. Here all neighbours will have equally poor prediction errors resulting in a uniform distribution of weights among them. As a result the normal curvature consistency updating will be applied with the central noise point being pulled closer to the level of its neighbours after successive iterations. This same strategy can be applied to each of the updating functions in Section 3.

(a)   (b)

Figure 14. A 1-D case of window with M = 5, (a) showing a height distribution in an ideal step discontinuity, (b) showing an ideal model of a noise point.

To conform to the recursive filtering concept, we modify the updating functions to include weighting by a function of the prediction error of each quantity. We (a) consider only the previous and present states of the filter, and (b) have no new source of information on each iteration. For one particular iteration, just the prediction error variance for each neighbour Q_α is calculated, i.e.,

$$\hat{p}_{\alpha_-}^{(i+1)} = \left(\xi_{P_+}{}^{(i)} - \xi_{P_{\alpha_-}}{}^{(i)} \right)^2 \tag{12}$$

The weight for each neighbour, w_α is computed by first taking $w_\alpha = f(\hat{p}_\alpha^{(i+1)})$ and then normalizing all the weights in the local neighbourhood such that they sum to 1.

4.2. Form of the Weighting Function

The weighting function used in standard recursive filtering is of the form $w_\alpha = \frac{1}{\hat{p}_\alpha^{(i+1)}}$. As a consequence of (12) it has the form shown in Figure 15. The convergence properties of the algorithm using this weighting function are not desirable, i.e., as the prediction error decreases, the weighting tends to go to infinity. Also, if by chance on a smooth surface one of the neighbours predicts the midpoint quite closely, it will get nearly all the weight in the next iteration, depriving the whole neighbourhood from playing a role in the updating. This phenomenon allows the surface to "freeze" around this particular neighbour. The idea then is to remove the convergence to infinity near zero-error values and put a finite extent on the weights. In addition, there should be some control over how much smoothing is applied.

One way of doing this is to model the weighting function as a Gaussian (Figure 16). The weights for the elements of $\mathcal{D}(P)$ to be used in the next iteration are then calculated as

$$w_\alpha = W_\alpha / \sum_{k=1}^{n} W_k \tag{13}$$

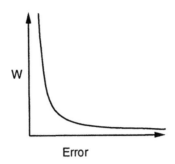

Figure 15. Distribution of weights according to the function $w_\alpha = 1/\hat{p}_\alpha$.

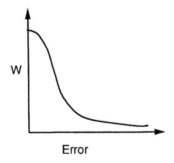

Figure 16. Distribution of weights according to a Gaussian error function.

$$W_\alpha = (2\pi\sigma^2/\gamma)^{-1/2} exp(-\frac{\xi_{Q_\alpha}^2}{2\sigma^2/\gamma}) \tag{14}$$

$$\sigma^2 = \sum_{\alpha=1}^{n} \frac{(\xi_{Q_\alpha} - \xi_Q)^2}{n} \tag{15}$$

$$\xi_Q = \sum_{\alpha=1}^{n} \frac{\xi_{Q_\alpha}}{n} \tag{16}$$

A small variance, σ, of the Gaussian will usually give better results than a larger one as it better approximates $1/x^2$ for most of the extent, hence it can be reduced by dividing σ^2 by a scalar parameter γ. The σ^2/γ will basically control the convergence properties of the algorithm. The larger the σ^2/γ, the more the algorithm will behave like the original curvature consistency algorithm.

A comparison between the original and modified algorithms using a noisy simulated step edge is shown in Figure 17 using $\gamma = 8$ and a 5×5 window. As can be seen by comparing the figures, both algorithms produce results which are locally consistent (i.e. the structure of the frame fields). However, the modified algorithm has the desirable property of also correctly localizing the step edge. When applied to a 'V' shaped (wedge) discontinuity (Figure 18a), the original algorithm, distorts the edge by making it tend towards a parabolic shape instead of the correct 'V' shape. This results in the odd deformation that we see in Figure 18b. The modified algorithm however better preserves the 'V' edge as can be seen by comparing Figures 18b and 18c[12].

Figure 19 shows the algorithms applied to a real image acquired with a laser rangefinder (Figure 19a). After 8 iterations, the original algorithm blurs away most of the edges and completely loses some of the finer curvature discontinuities (Figure 19b). The modified algorithm, on the other hand, preserves most of the edges and structure (Figure 19c).

4.3. The Modified Curvature Consistency Algorithm

One problem in using the original curvature consistency algorithm is to establish when to stop iterating. As described earlier in Section 3, one approach is to track the global error of fit between iterations [35, 29, 14]; when it falls below a particular threshold, the process is stopped. The same approach can be taken for the modified algorithm as well, but the difference is that it actually converges much faster.

In the modified curvature consistency algorithm, almost all the neighbours in the updating procedure initially play a large role in the refinement, avoiding local minimas. Once a refined, locally smooth and consistent fit is obtained, any further iterations tend to make the local surface regions converge in isolation from the neighbouring regions in the global structure. The result is that the surface converges to the appropriate local structures in the data, but neighbouring structures are sometimes broken by artifactual discontinuities. However we find that in most applications this convergence property is more desirable than that of the original algorithm.

During the updating process, sometimes the prediction error of one of the neighbours becomes almost zero, i.e, the updated midpoint corresponds exactly to a neighbouring estimate. This may happen even when neighbours are inconsistent with one another. But the consequence in the following iteration is that this neighbour is given maximum weight, precluding the contributions from the rest of the local neighbourhood. This behaviour is clearly undesirable, so we prevent its occurrence by including an "iso-surface" check on the neighbour responsible and the mid-point of the window.

An iso-surface check determines whether two points are on the same surface or not. The idea behind using the iso-surface check is as follows. Ideally, a neighbour should have prediction error zero for the mid-point of the window not only when P is correctly predicted by Q, but also when Q is predicted by P. This is the ideal case to which the curvature consistency algorithm should converge and is fairly easy to implement. If the iso-surface check fails, then Q is given zero weight just for

---

[12]The observant reader will notice slight artifactual discontinuities introduced by the modified algorithm in Figures 17c and fig.18c. These are discussed in a following section along with solution to this problem.

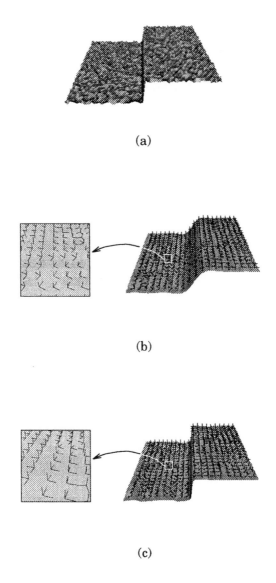

(a)

(b)

(c)

Figure 17. (a) Gaussian noise added to a synthetic step edge. (b) Reconstructed surface (with Darboux frames superimposed) obtained after 15 iterations of the original algorithm. (c) Reconstructed surface (with Darboux frames superimposed) obtained after 15 iterations of the modified algorithm.

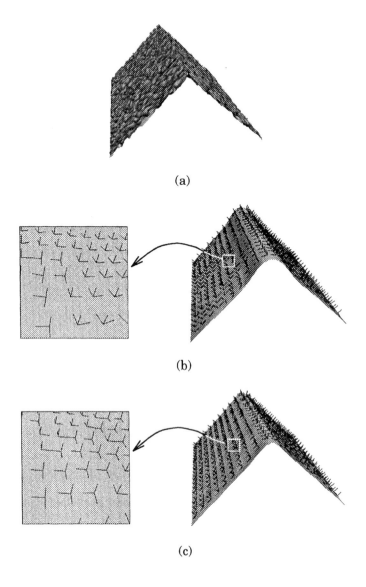

(a)

(b)

(c)

Figure 18. (a) Gaussian noise added to a synthetic wedge (roof edge). (b) Reconstructed surface (with Darboux frames superimposed) obtained after 15 iterations of the original algorithm. (c) Reconstructed surface (with Darboux frames superimposed) obtained after 15 iterations of the modified algorithm.

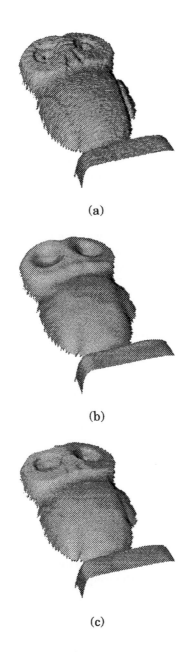

(a)

(b)

(c)

Figure 19. (a) Range image of an owl sculpture acquired with the NRCC/McGill laser rangefinder. (b) Reconstructed surface after 8 iterations of the original algorithm. (c) Reconstructed surface after 8 iterations of the modified algorithm.

the next iteration to allow contributions from P's other neighbours. This strategy appears to work quite well.

The idea of discontinuities in data is intuitively connected to the concept of scale. At one scale some surface structure may be considered important while at a larger scale, to be undesirable noise and smoothed over. In several applications such as volumetric fitting to range data, segmentation (parts decomposition) is performed on the reconstructed surface. One approach is to look for surface features marked by extremal values of curvature (e.g. negative local minima, concave discontinuities, etc.) [26, 14, 15]. Parts are then segmented along boundaries comprised of these features. It is thus important that the reconstruction algorithm be largely scale invariant with respect to correctly localizing discontinuities.

The different localization properties of the original and modified algorithms are shown for two different sampling windows in Figures 20 and 21. Notice that the original algorithm progressively blurs away the location of the step discontinuity with increasing scale (Figures 20b and 21b), whereas the modified algorithm does a much better job at preserving the edge position (Figures 20c and 21c).

5. Discussion and Conclusions

In this chapter we have examined different methods of estimating local surface properties cast in terms of differential geometry. These ranged from local approximation to full surface reconstruction using a class of algorithms based on the concept of minimizing the local variation of curvature, i.e. curvature consistency. The motivation for this reconstruction approach is that it induces a smoothing that minimizes the effects of noise and quantization error while preserving the local structure of a surface as reflected by the augmented Darboux frame at each sample point P. Because the basic representation is locally defined, the algorithm is largely view invariant (with the exception of the embedding coordinate system). We have taken advantage of this fact in reconstructing surfaces across multiple views [36, 37].

This chapter also addressed a major shortcoming of the original algorithm, the lack of any provision to deal with smoothing across surface discontinuities. The modifications introduced in Section 4 partially resolve this problem; surface discontinuities are preserved, but no attempt is made to identify them as such. Current work is underway in our laboratory to resolve this remaining problem by examining what can be reliably inferred from the weights associated with each point upon termination of the algorithm. Finally, the modifications also improve the convergence properties of the algorithm. The decoupling introduced by the weights limits the progressive smoothing of the original algorithm and, as shown in the last example presented, reduces sensitivity to scale.

We would conclude that reliable estimation of features based on differential geometry is indeed feasible; the question is at what cost. The computations associated with different variations of the curvature consistency algorithm are expensive. For example, complete reconstruction of a 256×256 range image with 5 iterations of the algorithm requires approximately one hour of cpu time on a Silicon Graph-

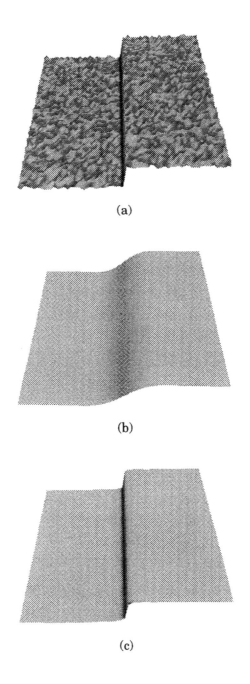

(a)

(b)

(c)

Figure 20. (a) Original noisy synthetic step, (b) after 15 iterations of the original algorithm using 7 X 7 window, (c) after 15 iterations of the modified algorithm using 7 X 7 window.

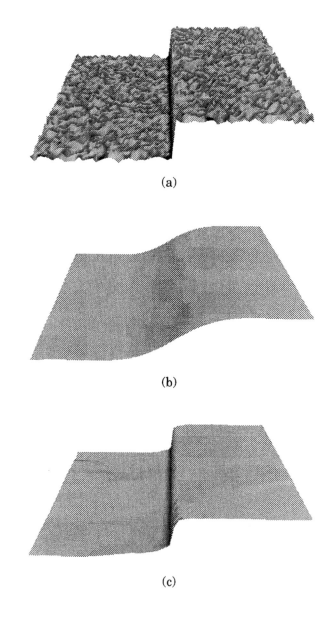

(a)

(b)

(c)

Figure 21. (a) Original noisy synthetic step, (b) after 15 iterations of the original algorithm using 9 X 9 window, (c) after 15 iterations of the modified algorithm using 9 X 9 window.

ics 4D/35 workstation. At least two different strategies are possible for reducing this burden: the use of multi-grid methods as has been shown in [38, 39], and a re-formulation of how parallel transport is implemented. The computations associated with the latter process amount to the majority of operations performed (mostly in transformations between different coordinate frames). It turns out that different mathematical formulations are available which can reduce the computational burden to a significant degree.

Referring back to Figure 1, it has long been our contention that much of the difficulty in computing different representations for shape is the difficulty in obtaining reliable surface descriptions at the early stages of computation. Hopefully further progress in surface reconstruction and analysis will contribute to the work of others such as those cited in other chapters of this book.

A. $\mathcal{D}(P)$ in terms of quadric parameters

Consider a local approximation of S of the form $\bar{x}(u, v) = (u, v, h(u, v))$. The following derivatives are obtained:

$$
\begin{aligned}
&\bar{x}_u = (1, 0, h_u) \quad \bar{x}_v = (0, 1, h_v) \\
&\bar{x}_{uu} = (0, 0, h_{uu}) \quad \bar{x}_{uv} = (0, 0, h_{uv}) \quad \bar{x}_{vv} = (0, 0, h_{vv}).
\end{aligned}
\tag{17}
$$

The normal to the surface $\bar{x}(u, v)$ is then given by

$$
N = \frac{\bar{x}_u \times \bar{x}_v}{|\bar{x}_u \times \bar{x}_v|} = \frac{(-h_u, -h_v, 1)}{\sqrt{1 + h_u^2 + h_v^2}}.
\tag{18}
$$

Now let

$$
\begin{aligned}
e &= - < N_u, \bar{x}_u > = < N, \bar{x}_{uu} > = \frac{h_{uu}}{\sqrt{1 + h_u^2 + h_v^2}} \\
f &= - < N_v, \bar{x}_u > = - < N_u, \bar{x}_v > = \frac{h_{uv}}{\sqrt{1 + h_u^2 + h_v^2}} \\
g &= - < N_v, \bar{x}_v > = < N, \bar{x}_{vv} > = \frac{h_{vv}}{\sqrt{1 + h_u^2 + h_v^2}},
\end{aligned}
\tag{19}
$$

and

$$
\begin{aligned}
E &= < \bar{x}_u, \bar{x}_u > = 1 + h_u^2 \\
F &= < \bar{x}_u, \bar{x}_v > = < \bar{x}_v, \bar{x}_u > = h_u h_v = h_v h_u \\
G &= < \bar{x}_v, \bar{x}_v > = 1 + h_v^2.
\end{aligned}
\tag{20}
$$

where $< \bar{a}, \bar{b} >$ denotes the inner product of $\bar{a}$ and $\bar{b}$. It can subsequently be shown [9] that

$$
\begin{aligned}
dN \begin{pmatrix} du \\ dv \end{pmatrix} &= \begin{pmatrix} \frac{fF - eG}{EG - F^2} & \frac{gF - fG}{EG - F^2} \\ \frac{eF - fE}{EG - F^2} & \frac{fF - gE}{EG - F^2} \end{pmatrix} \begin{pmatrix} du \\ dv \end{pmatrix} \\
&= \begin{pmatrix} a_{11} & a_{12} \\ a_{21} & a_{22} \end{pmatrix} \mathbf{v} \\
&= A \mathbf{v},
\end{aligned}
\tag{21}
$$

where the eigenvalues of A are the principal curvatures κ_M and $\kappa_{\mathcal{M}}$, and the eigenvectors of A are the principal directions M and $\mathcal{M}$, i.e.,

$$-\kappa_M = \tfrac{1}{2}\left(a_{11} + a_{22} - \sqrt{(a_{11} - a_{22})^2 + 4a_{12}a_{21}}\right),$$
$$-\kappa_{\mathcal{M}} = \tfrac{1}{2}\left(a_{11} + a_{22} + \sqrt{(a_{11} - a_{22})^2 + 4a_{12}a_{21}}\right), \tag{22}$$

where $\kappa_1 \geq \kappa_2$, and provided that $(a_{11} - a_{22})^2 + 4a_{12}a_{21} \geq 0$.

$$M = \begin{cases} \left(a_{12}, -\tfrac{1}{2}(a_{11} - a_{22} + \sqrt{(a_{11} - a_{22})^2 + 4a_{12}a_{21}})\right) & a_{11} \geq a_{22}, \\ \left(\tfrac{1}{2}(a_{11} - a_{22} - \sqrt{(a_{11} - a_{22})^2 + 4a_{12}a_{21}}), a_{21}\right) & a_{11} < a_{22}, \end{cases}$$

$$\mathcal{M} = \begin{cases} \left(\tfrac{1}{2}(a_{11} - a_{22} + \sqrt{(a_{11} - a_{22})^2 + 4a_{12}a_{21}}), a_{21}\right) & a_{11} \geq a_{22}, \\ \left(-a_{12}, -\tfrac{1}{2}(a_{11} - a_{22} - \sqrt{(a_{11} - a_{22})^2 + 4a_{12}a_{21}})\right) & a_{11} < a_{22}. \end{cases} \tag{23}$$

Now consider the particular case of determining $\mathcal{D}(P)$ for a local parameterization, (u, v, w), centered at P such that the w component is aligned with the surface normal N_P, where $\bar{x}(u, v) = (u, v, au^2 + buv + cv^2)$. The first and second fundamental form coefficients are respectively

$$\begin{matrix} E = 1 & F = 0 & G = 1, \\ e = 2a & f = v & g = 2c. \end{matrix} \tag{24}$$

In this case A has a particularly simple form, i.e.,

$$A = \begin{pmatrix} -2a & -b \\ -b & -2c \end{pmatrix}, \tag{25}$$

from which the elements of $\mathcal{D}(P) = (P, \mathcal{M}_P, M_P, N_P, \kappa_{M_P}, \kappa_{\mathcal{M}_P})$ are as follows (tangent plane coordinates).

$$\mathcal{M}_P = \begin{cases} \left(a - c + \sqrt{(a-c)^2 + b^2}, b\right) & a \geq c \\ \left(b, -\left(a - c + \sqrt{(a-c)^2 + b^2}\right)\right) & a < c \end{cases}$$

$$M_P = \begin{cases} \left(-b, a - c + \sqrt{(a-c)^2 + b^2}\right) & a \geq c \\ \left(a - c - \sqrt{(a-c)^2 + b^2}, b\right) & a < c \end{cases} \tag{26}$$

$$N_P = (0, 0, 1)$$

$$\kappa_{M_P} = a + c + \sqrt{(a-c)^2 + b^2}$$

$$\kappa_{\mathcal{M}_P} = a + c - \sqrt{(a-c)^2 + b^2}$$

B. Determining ξ_{P_α} from ξ_{Q_α}

The following is adapted from the original work of Peter Sander in [34].

An estimate of ξ_P, ξ_{P_α}, can be obtained from ξ_{Q_α} by extrapolating along a parabolic quadric S_{Q_α} determined from ξ_{Q_α}. This, in effect, can be thought of as a description of how ξ_P "looks" according to neighbouring frame ξ_{Q_α} [34]. The minimization algorithm works by updating $\xi_P^{(i+1)}$ with the least mean square error fit to these estimates subject to constraints relating to the local model of the surface. In this section we outline how ξ_{P_α} is obtained from ξ_{Q_α} ([34], Appendix B). To be consistent with Sander's notation, S_{Q_α} is referred to as s_α.

For a particular frame ξ_{Q_α},

$$s_\alpha = \left(u, v, \frac{1}{2} \left(\kappa_{M\alpha} u^2 + \kappa_{M\alpha} v^2 \right) \right), \tag{27}$$

for which the local tangent plane basis vectors are

$$\begin{aligned}
P(u,v) &= \tfrac{\partial s}{\partial u}(u,v) = s_u(u,v) = (1, 0, \kappa_{M\alpha} u), \\
Q(u,v) &= \tfrac{\partial s}{\partial v}(u,v) = s_v(u,v) = (0, 1, \kappa_{M\alpha} v),
\end{aligned} \tag{28}$$

in the parameterization $(M_\alpha, \mathcal{M}_\alpha, N_\alpha)$. In this basis the first and second fundamental form coefficients are

$$\begin{aligned}
E(u,v) &\quad =< s_u, s_u >= \kappa_{M\alpha}^2 u^2 + 1, \\
F(u,v) &\quad =< s_u, s_v >= \kappa_{M\alpha} \kappa_{M\alpha} uv, \\
E(u,v) &\quad =< s_v, s_v >= \kappa_{M\alpha}^2 v^2 + 1, \\
e(u,v) &=< N_g, s_{uu} >= \frac{\kappa_{M\alpha}}{\sqrt{\kappa_{M\alpha}^2 u^2 + \kappa_{M\alpha}^2 v^2 + 1}}, \\
f(u,v) &\quad =< N_g, s_{uv} >= 0, \\
e(u,v) &=< N_g, s_{vv} >= \frac{\kappa_{M\alpha}}{\sqrt{\kappa_{M\alpha}^2 u^2 + \kappa_{M\alpha}^2 v^2 + 1}},
\end{aligned} \tag{29}$$

and

$$N_g(u,v) = \frac{1}{\sqrt{\kappa_{M\alpha}^2 u^2 + \kappa_{M\alpha}^2 v^2 + 1}} (-\kappa_{M\alpha} u, -\kappa_{M\alpha} v, 1) \tag{30}$$

is the surface normal at P. The differential of the surface normal at (u,v) is given by

$$dN \begin{pmatrix} du \\ dv \end{pmatrix} = \begin{pmatrix} a_{11} & a_{12} \\ a_{21} & a_{22} \end{pmatrix} \begin{pmatrix} du \\ dv \end{pmatrix}, \tag{31}$$

where the $a_{ij}(u,v)$ are

$$\begin{aligned}
a_{11} &= \tfrac{fF - eG}{EG - F^2} = -\frac{\kappa_{M\alpha} \kappa_{M\alpha}^2 v^2 + \kappa_{M\alpha}}{(\kappa_{M\alpha}^2 u^2 + \kappa_{M\alpha}^2 v^2 + 1)^{\frac{3}{2}}}, \\
a_{12} &= \tfrac{gF - fG}{EG - F^2} = -\frac{\kappa_{M\alpha} \kappa_{M\alpha}^2 uv}{(\kappa_{M\alpha}^2 u^2 + \kappa_{M\alpha}^2 v^2 + 1)^{\frac{3}{2}}}, \\
a_{21} &= \tfrac{eF - fE}{EG - F^2} = -\frac{\kappa_{M\alpha}^2 \kappa_{M\alpha} uv}{(\kappa_{M\alpha}^2 u^2 + \kappa_{M\alpha}^2 v^2 + 1)^{\frac{3}{2}}}, \\
a_{22} &= \tfrac{fF - gE}{EG - F^2} = -\frac{\kappa_{M\alpha} \kappa_{M\alpha} u^2 + \kappa_{M\alpha}}{(\kappa_{M\alpha}^2 u^2 + \kappa_{M\alpha}^2 v^2 + 1)^{\frac{3}{2}}}.
\end{aligned} \tag{32}$$

For any point $(u, v) \in s_\alpha$, the principal curvatures are the eigenvalues of dN,

$$
\begin{aligned}
-\kappa_1 = \lambda_1 &= \tfrac{1}{2}\left(a_{11} + a_{22} - \sqrt{(a_{11} - a_{22})^2 + 4a_{12}a_{21}}\right), \\
-\kappa_2 = \lambda_1 &= \tfrac{1}{2}\left(a_{11} + a_{22} + \sqrt{(a_{11} - a_{22})^2 + 4a_{12}a_{21}}\right),
\end{aligned}
\tag{33}
$$

where $\kappa_1 \geq \kappa_2$, and provided that $(a_{11} - a_{22})^2 + 4a_{12}a_{21} \geq 0$. The principal directions can be found from the eigenvectors of dN,

$$
\bar{x}_1 =
\begin{cases}
\left(a_{12}, -\tfrac{1}{2}(a_{11} - a_{22} + \sqrt{(a_{11} - a_{22})^2 + 4a_{12}a_{21}})\right) & a_{11} \geq a_{22}, \\
\left(\tfrac{1}{2}(a_{11} - a_{22} - \sqrt{(a_{11} - a_{22})^2 + 4a_{12}a_{21}}), a_{21}\right) & a_{11} < a_{22},
\end{cases}
$$

$$
\bar{x}_2 =
\begin{cases}
\left(\tfrac{1}{2}(a_{11} - a_{22} + \sqrt{(a_{11} - a_{22})^2 + 4a_{12}a_{21}}), a_{21}\right) & a_{11} \geq a_{22}, \\
\left(-a_{12}, -\tfrac{1}{2}(a_{11} - a_{22} - \sqrt{(a_{11} - a_{22})^2 + 4a_{12}a_{21}})\right) & a_{11} < a_{22}.
\end{cases}
\tag{34}
$$

Note that the resulting frame determined from the above equations is in the basis (P, Q, N_g) and thus cannot be compared directly to ξ_P. Thus, the parameters of the transformation T_w which maps (P, Q, N_g) to $(M, \mathcal{M}, N)$ must be determined. This involves three rotations, two which rotate N_g into alignment with N such that the tangent planes are parallel, and one which rotates (P, Q) into alignment with $(M, \mathcal{M})$. Since ξ_P is known in $(M, \mathcal{M}, N)$ at $s_\alpha|_{(0,0)}$, the three rotations are relatively straightforward to determine. Thus, ξ_{P_α} for a particular ξ_{Q_α} can be found by moving along s_α in the direction of ξ_P in the manner just outlined.

REFERENCES

1 H. Asada and M. Brady. The curvature primal sketch. *IEEE Trans. PAMI*, 8:2–14, 1986.

2 P. Besl and R. Jain. Segmentation through symbolic surface description. In *Proceedings IEEE Conf. Computer Vision and Pattern Recognition*, pages 77–85, Miami Beach, Florida, June 1986.

3 P. J. Besl, J. B. Birch, and L. T. Watson. Robust window operators. In *Second International Conference on Computer Vision*, Tampa, Florida, Dec. 1988. IEEE Computer Society.

4 A. Blake and A. Zisserman. *Visual Reconstruction*. MIT Press, Cambridge, Massachusetts, 1987.

5 R. Bolles. On optimally combining pieces of information, with application to estimating 3-d complex-object position from range data. *IEEE Tranactions on Pattern Analysis and Machine Intelligence*, PAMI-8(5):619–638, 1986.

6 P. Boulanger. Label relaxation technique applied to the topographic primal sketch. In *Proceedings, Vision Interface 1988*, pages 158–162, Edmonton, Canada, June 1988.

7 M. Brady, J. Ponce, A. Yuille, and H. Asada. Describing surfaces. *Computer Vision, Graphics, and Image Processing*, 32:1–28, 1985.

8 J. Canny. A computational approach to edge detection. *IEEE Transactions on Pattern Analysis and Machine Intelligence*, 8(6):679–698, 1986.

9   M. do Carmo. *Differential Geometry of Curves and Surfaces*. Prentice-Hall, Inc., Englewood Cliffs,New Jersey, 1976.

10 T. Fan, G. Medioni, and R. Nevatia. Description of surfaces from range data using curvature properties. In *Proceedings IEEE Conf. Computer Vision and Pattern Recognition*, pages 86–91, Miami Beach, Florida, June 1986.

11 F. Ferrie and J. Lagarde. On computing stable surface descriptions from range images. In *Proceedings 5th International Conference on Image Analysis*, Positano, Italy, September 20-22 1989.

12 F. Ferrie and J. Lagarde. Robust estimation of shape from shading. In *Proceedings 1989 Topical Meeting on Image Und. and Machine Vision*, pages 24–27, Cape Cod, Massachusetts, June 1989.

13 F. Ferrie and J. Lagarde. Curvature consistency improves local shading analysis. *Image Understanding*, 55(1):95–105, Jan. 1992.

14 F. Ferrie, J. Lagarde, and P. Whaite. Darboux frames, snakes, and super-quadrics: Geometry from the bottom-up. In *Proceedings IEEE Workshop on Interpretation of 3D Scenes*, pages 170–176, Austin, Texas, Nov. 27-29 1989. IEEE Trans. PAMI - accepted for publication.

15 F. Ferrie, J. Lagarde, and P. Whaite. Recovery of volumetric object descriptions from laser rangefinder images. In *Proceedings First European Conference on Computer Vision*, Antibbes, France, April 1990.

16 F. Ferrie and M. Levine. Piecing Together the 3-D Shape of Moving Objects: An Overview. In *Proceedings IEEE Conf. on Computer Vision and Pattern Recognition*, pages 574–584, San Francisco, CA., June 1985.

17 F. Ferrie and M. Levine. Integrating information from multiple views. In *Proceedings of the IEE Computer Society Workshop on Computer Vision*, pages 117–122, Miami Beach, Florida, Dec. 1987. Computer Society of the IEEE, IEEE Computer Society Press.

18 F. Ferrie and M. Levine. Deriving Coarse 3D Models of Objects. In *IEEE Comp. Soc. Conf. on Computer Vision and Pattern Recognition*, pages 345–353, University of Michigan, Ann Arbor, Michigan, June 1988.

19 P. Flynn and A. Jain. On reliable curvature estimation. In *Proceedings IEEE Conf. Computer Vision and Pattern Recognition*, pages 110–116, San Diego, California, June 4-8 1989.

20 G. Godin and M. Levine. Structured edge map of curved objects in a range image. In *Proceedings IEEE Comp. Soc. Conf. on Computer Vision and Pattern Recognition*, San Diego, California, June 4-8 1989.

21 D. Goldgof, H. Lee, and T. Huang. Motion analysis of nonrigid surfaces. In *Proceedings of the IEEE Computer Society Conference on Computer Vision and Pattern Recognition*, pages 374–380, Ann Arbor, Michigan, June 1988. Computer Society of the IEEE, IEEE Computer Society Press.

22 W. Grimson. *From Images to Surfaces*. MIT Press, Cambridge, MA., 1981.

23 A. Gupta and R. Bajcsy. Integrated approach for surface and volume segmentation of range images using biquadrics and superquadrics. In *SPIE Applications of Artificial Intelligence X*, pages 210–227, Orlando, Florida, April 20-24 1992.

24 J. Heel. Temporally integrated surface reconstruction. In *Proc. Third Inter-*

national Conference on Computer Vision, pages 292–295, Osaka, Japan, Dec. 1990. Computer Society of the IEEE, IEEE Computer Society Press.

25 D. Hilbert and S. Cohn-Vossen. *Geometry and the Imagination*. Chelsea, New York, 1952.

26 D. Hoffman and W. Richards. Parts of recognition. *Cognition*, 18:65–96, 1984.

27 R. Hummel and S. Zucker. On the foundation of relaxation labeling processes. *IEEE Trans. PAMI*, 6:267–287, 1983.

28 R. Johnson and D. Wichern. *Applied Multivariate Statistical Analysis*. Prentice-Hall, Englewood Cliffs, New Jersey, 1982.

29 J. Lagarde. Constraints and their satisfaction in the recovery of local surface structure. Master's thesis, Dept. of E.E., McGill Univ., 1989.

30 P. Parent and S. Zucker. Curvature consistency and curve detection. *J. Opt. Soc. Amer., Ser. A*, 2(13), 1985.

31 T. Poggio, V. Torre, and C. Koch. Computational vision and regularization theory. *Nature*, 317(26):314–319, Sept. 1985.

32 M. Rioux and L. Cournoyer. The NRCC three-dimensional image data files. National Research Council of Canada, CNRC No. 29077, June 1988.

33 G. Roth and M. Levine. Segmentation of geometric signals using robust fitting. In *Proceedings 10th International Conference on Pattern Recognition*, pages 826–831, Atlantic City, New Jersey, jun 1990.

34 P. Sander. *Inferring Surface Trace and Differential Structure from 3-D Images*. PhD thesis, Dept. Elect. Eng., McGill University, Montréal, Québec,Canada, 1988.

35 P. Sander and S. Zucker. Inferring differential structure from 3-d images: Smooth cross sections of fiber bundles. *IEEE Trans. PAMI*, 12(9):833–854, 1990.

36 G. Soucy and F. Ferrie. Multi-view surface reconstruction using curvature and motion consistency. In *PROC. Second International Workshop on Sensor Fusion and Environmental Modelling*, Oxford, U.K., September 2-6 1991.

37 G. Soucy and F. Ferrie. Motion and surface recovery using curvature and motion consistency. In *PROC. Second European Conference on Computer Vision*, Santa Margheri-ta Ligure, Italy, May 18-23 1992.

38 D. Terzopoulos. *Multiresolution Image Processing and Analysis*, chapter Multi-Level Representation of Visual Surfaces: Variational Principles and Finite Element Representations. Springer-Verlag, New York, 1983.

39 D. Terzopoulos. *Advances in Computational Vision*, chapter Multiresolution algorithms in computational vision. Ablex, New-Jersey, 1984.

40 P. Whaite and F. Ferrie. From uncertainty to visual exploration. *IEEE Transactions on Pattern Analysis and Machine Intelligence*, 13(10):1038–1050, Oct. 1991.

41 P. Whaite and F. Ferrie. Uncertain views. In *PROC. IEEE Computer Society Conference on Computer Vision and Pattern Recognition*, Champaign, Illinois, June 15-18 1992.

Three-Dimensional Object Recognition Systems
A.K. Jain and P.J. Flynn (Editors)
© 1993 Elsevier Science Publishers B.V. All rights reserved.

Three-Dimensional Surface Reconstruction: Theory and Implementation

Robert L. Stevenson[a] and Edward J. Delp[b]

[a] Laboratory for Image and Signal Analysis
Department of Electrical Engineering
University of Notre Dame
Notre Dame, Indiana 46556 USA

[b] Computer Vision and Image Processing Laboratory
School of Electrical Engineering
Purdue University
West Lafayette, Indiana 47907 USA

Abstract

Clues about the three-dimensional shape of an object can be obtained from many types of sensory inputs. Three-dimensional information can be obtained from images of the object by examining the interaction of light with the object's surface, or through examining changes in the image data between frames in a time-sequence of images of a moving object. Surface shape can be measured more directly through the use of ranging sensors which provide exact measurements of the location of points on object surfaces. In order to provide a single consistent idea of the three-dimensional object shape to higher level recognition stages, the various sensory inputs must be combined and gaps in the data must be estimated to provide a complete and accurate representation of three-dimensional shape.

Reconstruction can be cast as an ill-posed inverse problem which must be stabilized using *a priori* information about the surface type and the data collection procedures. In three-dimensional object recognition systems, the prior information is usually the knowledge that the surfaces are piecewise smooth and that the surface shape should not depend on the representation chosen for the surface. This chapter uses this prior information to form a well-posed algorithm for estimating surface shape from various sensory inputs.

1. Introduction

1.1. The Problem

In early computer vision processing, a common task is to extract symbolic descriptive information about objects in a scene from multi-dimensional sensory data. The exact form of the symbolic description (or model) will depend on the ultimate vision application. For example, in tracking problems, the goal is to identify moving objects and to estimate their locations and trajectories. In this simple case, the model may consist of just a sequence of locations which describe the objects' motions. In navigation problems, the goal may be to identify the current location of a mobile vehicle by matching a three-dimensional model of the terrain with a database of sensed terrain information. The generation of the symbolic description from the sensor data is generally not an easy task. Often model parameters cannot be measured directly from the sensor data. For example, in the problem of matching images of three-dimensional objects to elements of a database of three-dimensional models, the sensor data is a two-dimensional intensity projection of the three-dimensional surfaces. The measurement of three-dimensional properties from this two-dimensional data is not straightforward. To overcome this problem it is often desirable to form an *intermediate representation* of the scene. The purpose of this representation is to bridge the gap between the sensor data and the symbolic description. The representation should be such that the symbolic description can be formed from the representation, and the representation should be extractable from the sensor data.

In computer vision applications, we are often dealing with geometric objects in three-dimensional space; therefore, a common intermediate representation is a collection of surfaces in three-dimensional space. A fundamental problem in deriving this intermediate representation is extracting information from the sensor data. Over the years, many algorithms have been developed to extract geometric information about the objects in a scene from various two-dimensional image sensors. A limitation of this type of processing is that only a noisy partial representation can be obtained from the data. That is, only an incomplete and noisy representation can be formed directly from the sensor data. Therefore another processing step is required to fill in the gaps in the representation and to remove noise artifacts in the representation.

In the simplest sense, this subsequent processing can be thought of as a surface interpolation or approximation procedure. The first level of processing generates sparse geometric information describing a surface from which a more complete representation is to be formed. Since the surfaces represent physical properties, the characteristics of the reconstruction process differ from straightforward surface reconstruction in two important areas. First, since the collected data is represented in an arbitrarily chosen coordinate system, the reconstruction process should be invariant to the choice of the coordinate system (except for the transformation relating the coordinate systems). This is important because a change in this intermediate representation may cause a change in the symbolic description that is derived which in affect changes the knowledge that has been gathered about the objects in a scene. This obviously should not happen if all that has occurred is a change in an arbitrarily chosen coordinate system. Secondly, in many reconstruction applications the surface that is being represented may be discontinuous. For example, for a surface that represents a box, there are discontinuities in the surface at the corners of the box.

Another concern when reconstructing surfaces is the computational complexity of the reconstruction algorithms. Computer vision tasks are often performed in dynamic environments. The sensors are often continually providing new data for processing so that decisions about the environment can be constantly updated. Therefore the processing time of an intermediate step needs to be as small as possible. This puts a limitation on the type of processing and algorithms that can realistically be examined for this application.

1.2. A brief history

The problem of interpolating or approximating surfaces has received much attention in the mathematical and statistical literature. Until recently, most research has concentrated on algorithms to fit continuous and smoothly varying curves or surfaces to a set of given data points[9, 25, 33]. Additional constraints on the reconstruction algorithm characteristics (such as invariance of the choice of the co-ordinate system and accommodation of surfaces with discontinuities) has become more important as the approaches have been utilized in more complex applications.

In the early 1980's, Grimson[15] recognized the need to incorporate discontinuity information into the surface reconstruction process, and Terzopoulos was the first to suggest a possible modification to Grimson's algorithm to include discontinuity information[45]. Shiau[34] and Wahba[48] also proposed several methods for reconstructing curves and surfaces with discontinuities. These methods were based on either *a priori* knowledge of the location of discontinuities, or a preprocessing step which essentially performed an edge detection to localize the discontinuities. This approach to the problem is limited because generally there is no *a priori* knowledge of the discontinuity locations, and the detection of discontinuities from sparse noisy data will produce noisy and unreliable estimates of the discontinuity locations.

More recently, several algorithms which include the incorporation of discontinuity information as an integral part of the reconstruction process have been proposed. Weiss[49], Lee[26, 27], and Aloimonos *et al.*[1] have all proposed methods for reconstructing curves with discontinuities. These algorithms work well; however, it is not clear how they can be extended to the problem of reconstructing surfaces with discontinuities. For the problem of reconstructing surfaces, several algorithms based on statistical concepts have been proposed[3, 6, 29]. These approaches are generally very computationally expensive. An algorithm proposed by Sinha and Schunck[36] does not suffer from the limitations mentioned for the previously discussed algorithms; however, since their algorithm is based on characteristics which vary with the choice of coordinate system, the reconstruction will not be invariant to coordinate transforms. In some applications this may not be a concern, but this will be a limitation for many applications in computer vision.

The property of invariance to the choice of coordinate system has been recognized as an important issue in computer vision. The computation of invariant curve reconstructions was reviewed by Malcolm[28]. However, as discussed by the author, all of these algorithms are unstable, and depending on the input data set, the algorithms may diverge. In order to stabilize the reconstructions, an approximately invariant algorithm must be used. One such approximation has been proposed by Blake and Zisserman[3].

1.3. Scope and organization

In this chapter, the emphasis will be on techniques and algorithms which formulate the problem as an ill-posed inverse problem which must be stabilized. *Ad hoc* algorithms are ignored in favor of methods which rely on proven mathematical concepts. Whenever possible, an attempt is made to keep concepts general so that the algorithms can be easily adapted to similar problems in other fields. Section 2 begins by discussing the notation and conventions that will be used throughout the presentation. The nature of the abstract problem is then explored, and regularizing techniques that will be used to make the problem well-posed and stable are presented. Section 3 examines an approximately-invariant algorithm which is guaranteed to be stable. Section 4 examines how *a priori* information concerning discontinuities can be incorporated into the problem, and how the algorithms can be modified so that discontinuities are automatically detected and incorporated. Computational issues are addressed in Section 5 and an example surface reconstruction is shown. Results are summarized and possible future research is proposed in Section 6.

2. Abstract Problem Solution

As a result of their inherent structure, low-level image analysis problems are generally inverse problems, and like most physical inverse problems, the mathematical formulation of the problem statement is ill-posed[31]. The general inverse problem can can be stated by the following: find the parameter field r from the observed finite collection of data $S = \{c_i\}_{i=1}^{M}$. For the problem to be well-posed in the sense of Hadamard[19], the solution must exist, be unique, and depend continuously on the data. In image analysis problems, the observed data S is generally sparse and/or noisy and will not uniquely determine a solution r; hence, the problem is ill-posed[31].

To obtain a unique and stable solution from the data, supplementary information must be used so that the problem becomes well-posed [47]. The basic principle common to all methods is to use *a priori* knowledge of the properties of the inverse problem to resolve conflicts in the estimates, and to restrict the space of possible solutions so that the data uniquely determine a stable estimate. Two techniques which are often used to form a well-posed problem from many ill-posed inverse problems are the methods of Tikhonov regularization[47] and stochastic regularization[29]. This chapter will use the approach of Tikhonov to formulate a well-posed inverse problem. It should be noted that an equivalent computational algorithm can be devised through the stochastic regularization approach.

To make the problem well-posed, a continuous operator (known as a *regularizing operator*) which approximates the inverse operator, is defined. To construct a regularizing operator, $R(\cdot, \cdot)$, Tikhonov and Arsenin[47] introduce a stablizing functional, $[\cdot]$. This stablizing functional provides a measure of the consistency of a particular solution based on the *a priori* assumptions. The stablizing functional is used to define the functional

$$M^\lambda[\mathbf{r}^*, \mathcal{S}] = [\mathbf{r}^*] + \sum_{c_i \in \mathcal{S}} \lambda_i \|\mathcal{A}_i \mathbf{r}^* - c_i\|^2, \tag{1}$$

where $\| \cdot \|$ is a norm, and $\mathcal{A}_i$ denotes the process of acquiring the ith data point and is assumed to be linear. This norm measures the distance between a possible solution and the observed data. This term will be large for solutions that are not near the observed data. Let λ denote the collection of $\{\lambda_i\}_{i=1}^M$. Then, for certain values of λ, the minimization of this functional is a regularizing operator,

$$\mathbf{r}_\lambda = R(\mathcal{S}, \lambda) = \arg\min_{\mathbf{r}^*} \ M^\lambda[\mathbf{r}^*, \mathcal{S}]. \tag{2}$$

This regularized solution, $\mathbf{r}_\lambda$, will be used as the solution estimate, $\hat{\mathbf{r}}$. Note that the regularizing operator for a particular problem is in general not unique; there may exist many operators which stabilize the ill-posed problem. The choice of the particular operator and the value of the regularization parameter λ is based on supplementary information pertaining to the problem. The hardest task is to find a stabilizer which not only yields a unique and stable solution to the inverse problem, but also accurately measures the consistency of the estimate with respect to the true solution.

If $[\cdot]$ is chosen to be quadratic, then it can be shown that the solution space is convex and a unique solution exists[47]. Most applications will therefore define $[\cdot]$ to be some norm or seminorm on the solution space. When this is not the case then the functional may be nonconvex. This makes finding the optimal solution more difficult and ill-posed since there may exist many suboptimal local minimum.

For univariate regularization of scalar functions, Tikhonov proposed a general stabilizer based on the mth-order weighted Sobolev norm. Let $f(x)$ be some scalar univariate function, then a general stablizing functional can be written as

$$\|f\|_m^2 = \sum_{p=0}^m \int_\Re \left(w_p(x) \frac{d^p f(x)}{dx^p} \right)^2 dx, \tag{3}$$

where the $w_p(x)$'s are nonnegative and possibly discontinuous weighting functions[47] and m determines the degree of smoothing. Using such a stabilizer makes the problem well-posed by restricting the space of admissible solutions to the Sobolev space of smooth functions. For multivariate vector-valued functions, Tikhonov's suggestion can be generalized for the n-dimensional case to

$$\|\mathbf{r}\|_m^2 = \sum_{|\mathbf{p}| \leq m} \int_{\Re^n} \left\| \mathbf{w_p}(\mathbf{x}) \left(\frac{\partial^{p_1}}{\partial x_1^{p_1}} \right) \left(\frac{\partial^{p_2}}{\partial x_2^{p_2}} \right) \cdots \left(\frac{\partial^{p_n}}{\partial x_n^{p_n}} \right) \mathbf{r}(\mathbf{x}) \right\|^2 dx, \tag{4}$$

where $\mathbf{p} = <p_1, p_2, ..., p_n>$, $|\mathbf{p}| = p_1 + p_2 + \cdots + p_n$ and $\mathbf{x} = <x_1, x_2, ..., x_n>$. This multivariate weighted Sobolev norm is the basis of many of the stabilizing norms used in low-level image analysis[31]. Stabilizing functionals of this form measure function smoothness and will lead to algorithms which smooth discontinuities. This functional is also quadratic, thus applications which utilize such stabilizers result in convex optimization problems. In this chapter we will examine a more general stablizing functional with the form

$$[\mathbf{r}] = \sum_{|\mathbf{p}| \leq m} \int_{\Re^n} \rho \left(\mathbf{w}_{\mathbf{p}}(\mathbf{x}) \left(\frac{\partial^{p_1}}{\partial x_1^{p_1}} \right) \left(\frac{\partial^{p_2}}{\partial x_2^{p_2}} \right) \cdots \left(\frac{\partial^{p_n}}{\partial x_n^{p_n}} \right) \mathbf{r}(\mathbf{x}) \right) \, d\mathbf{x}, \tag{5}$$

where $\rho(\cdot)$ is some univariate function. We will show that $\rho(\cdot)$ can be chosen so that the resulting regularizing functional has the two desirable properties of being convex and allowing discontinuities. It is easy to show that the stabilizer in equation (5) is convex if and only if the scalar function $\rho(\cdot)$ is convex.

3. Invariant Surface Estimates

In this section, an invariant surface reconstruction problem statement is developed based on regularizing the ill-posed problem with a Tikhonov regularizer.

3.1. Conditions for an invariant solution

Consider a collection of constraints, S_1, and any surface $\mathbf{r}_1^*$. Let the rotated and translated constraints be denoted by S_2 and the rotated and translated surface $\mathbf{r}_2^*$. Invariance of the functional M^λ implies that

$$M^\lambda[\mathbf{r}_1^*, S_1] = M^\lambda[\mathbf{r}_2^*, S_2]. \tag{6}$$

Therefore,

$$M^\lambda[\hat{\mathbf{r}}_1, S_1] = \inf_{\mathbf{r}_1^*} M^\lambda[\mathbf{r}_1^*, S_1] = \inf_{\mathbf{r}_2^*} M^\lambda[\mathbf{r}_2^*, S_2], = M^\lambda[\hat{\mathbf{r}}_2, S_2]. \tag{7}$$

Since M^λ is convex, $\hat{\mathbf{r}}_1$ and $\hat{\mathbf{r}}_2$ are unique and, by (6), $\hat{\mathbf{r}}_2$ is the rotated and translated version of $\hat{\mathbf{r}}_1$.

Therefore the surface reconstruction algorithm will be invariant to rotations and translations. The invariance of M^λ can be achieved by finding an invariant metric on the constraint space and an invariant stabilizer. Therefore the stabilizer, $[\cdot]$, and the metric, $\rho(\cdot, \cdot)$, must be based on invariant surface characteristics so that the functional $M^\lambda[\cdot, \cdot]$ is invariant.

3.2. Invariant metrics on the constraint space

A natural and invariant measure for location constraints is the Euclidean distance from the surface to the constraint location. Therefore, for this type of constraint, a possible metric is

$$\rho_{C_i}(\mathbf{r}(\mathbf{u}_i), \mathbf{c}_i) = \left((x(\mathbf{u}_i) - c_{x,i})^2 + (y(\mathbf{u}_i) - c_{y,i})^2 + (z(\mathbf{u}_i) - c_{z,i})^2 \right)^{1/2}, \tag{8}$$

where the vector components of $\mathbf{c}_i$ are denoted by $c_{x,i}, c_{y,i}$ and $c_{z,i}$. This metric is invariant to rotations and translations since it is the measure of the distance between two points which is an invariant quantity. For orientation constraints, the angle between the normal and the constraint is invariant to rotations and shifts; therefore it can be used as a basis for the metric defined on C_i.

3.3. Invariant characteristics of surfaces

A surface in three-dimensions is uniquely determined by local invariant quantities known as the first and second fundamental forms[2, 17]. These forms can be combined to produce surface invariants which have a physical meaning. Two of the most popular of these invariant are the Gaussian and mean curvatures. The Gaussian and mean curvature can be expressed directly in terms of partial derivatives of the parameterization as

$$K(u,v) = \frac{[\mathbf{r}_{uu}\mathbf{r}_u\mathbf{r}_v][\mathbf{r}_{vv}\mathbf{r}_u\mathbf{r}_v] - [\mathbf{r}_{uv}\mathbf{r}_u\mathbf{r}_v]^2}{\parallel \mathbf{r}_u \times \mathbf{r}_v \parallel^4} \tag{9}$$

$$H(u,v) = \frac{(\mathbf{r}_v, \mathbf{r}_v)[\mathbf{r}_{uu}\mathbf{r}_u\mathbf{r}_v] + (\mathbf{r}_{uu}\mathbf{r}_u)[\mathbf{r}_{vv}\mathbf{r}_u\mathbf{r}_v] - 2(\mathbf{r}_u, \mathbf{r}_v)[\mathbf{r}_{uv}\mathbf{r}_u\mathbf{r}_v]}{2 \parallel \mathbf{r}_u \times \mathbf{r}_v \parallel^3}, \tag{10}$$

where $\mathbf{r}_u$ represents the derivative of the surface with respect to the parameter u, $(\mathbf{x}, \mathbf{y})$ represents the dot product, and $[\mathbf{xyz}]$ represents the triple product. The explicit forms of these equations are

$$K(x,y) = \frac{z_{xx}z_{yy} - z_{xy}^2}{(1 + z_x^2 + z_y^2)^2}, \tag{11}$$

and

$$H(x,y) = \frac{(1 + z_x^2)z_{yy} + (1 + z_y^2)z_{xx} - 2z_x z_y z_{xy}}{2(1 + z_x^2 + z_y^2)^{3/2}}. \tag{12}$$

Another pair of important surface invariants are based on the normal curvature. The normal curvature is the curvature of a line on the surface in a plane that contains the normal at that point, and a tangent vector in a particular direction. The principal curvatures correspond to the minimum and maximum normal curvatures at a point. The direction of these principal curvatures are know as the principal directions. The principal curvatures can be computed from the coefficients in the first and second fundamental forms by

$$\kappa_1 = \frac{(\mathbf{r}_{uu}, \mathbf{n})\cos^2(\theta_1) + 2(\mathbf{r}_{uv}, \mathbf{n})\cos(\theta_1)\sin(\theta_1) + (\mathbf{r}_{vv}, \mathbf{n})\sin^2(\theta_1)}{(\mathbf{r}_u, \mathbf{r}_u)\cos^2(\theta_1) + 2(\mathbf{r}_u, \mathbf{r}_v)\cos(\theta_1)\sin(\theta_1) + (\mathbf{r}_v, \mathbf{r}_v)\sin^2(\theta_1)} \tag{13}$$

$$\kappa_2 = \frac{(\mathbf{r}_{uu}, \mathbf{n})\cos^2(\theta_2) + 2(\mathbf{r}_{uv}, \mathbf{n})\cos(\theta_2)\sin(\theta_2) + (\mathbf{r}_{vv}, \mathbf{n})\sin^2(\theta_2)}{(\mathbf{r}_u, \mathbf{r}_u)\cos^2(\theta_2) + 2(\mathbf{r}_u, \mathbf{r}_v)\cos(\theta_2)\sin(\theta_2) + (\mathbf{r}_v, \mathbf{r}_v)\sin^2(\theta_2)} \tag{14}$$

where θ_1 and θ_2 are the angles between the principal directions and the tangent to the u-parameter curve (i.e. $\mathbf{r}_u$), and n is the surface normal. The principal curvatures can also be computed from the Gaussian and mean curvature and thus are also second order invariants of the surface. The Gaussian and mean curvatures can be computed from the principal curvatures by

$$K(u,v) = \kappa_1 \kappa_2 \tag{15}$$

and

$$H(u, v) = \frac{(\kappa_1 + \kappa_2)}{2}.$$ (16)

If the normal curvature at a point is constant for all directions then all directions are considered principal directions. If $|\kappa_n| = 0$ this is a planar point, if $|\kappa_n| > 0$ this point is known as an umbilical point of the surface. As an example, all the points on the surface of a sphere are umbilical points.

The local characteristics of a surface at a point P on the surface are completely specified by the surface normal, the principal directions and principal curvatures. Following the convention of [32], we refer to the collection of information $(P, n, d_1, d_2, \kappa_1, \kappa_2)$ as the augmented Darboux frame at the point P, $\mathcal{D}(P)$, where d_1 and d_2 are the principal directions.

3.4. An invariant stablizing functional

A *prior* knowledge about the desired type of reconstruction is used to determine an appropriate stabilizer based on invariant characteristics. It is desirable to require that reconstructed surfaces are smooth. Recall that the stabilizer, , is used in the definition of M^λ. To define an invariant stabilizer, an invariant characteristic of the surface is integrated over the surface to form a measure of the surface consistency. The measure is defined so that the more consistent the surface is with our idea of a reconstructed surface, the smaller its value. In this section, an invariant stabilizer based on second-order invariants will be defined and examined.

For smoothly varying surfaces, the surface can be modeled as an ideal thin flexible plate of elastic material[45, 46]. The stabilizer is a measure of the strain energy of the deformed plate. The total energy stored in a deformed thin plate is given in terms of the Gaussian and mean curvatures by [8]

$$[\mathbf{r}] = \int_U (2H^2 - K)dA,$$ (17)

or in terms of the principal curvatures as

$$[\mathbf{r}] = \int_U (\kappa_1^2 + \kappa_2^2)dA.$$ (18)

The reconstructed surface r obtained by the minimization of M^λ with this stabilizer is referred to as the thin-plate spline approximate. For smooth invariant surface reconstruction (18) is an adequate stabilizer, and is used in the remainder of this chapter.

3.5. An invariant problem statement

The problem of reconstructing a surface given a set of constraints $\mathcal{S}$ is now posed as a functional minimization problem. The estimate of the surface is formed by minimizing the functional

$$M^\lambda[\mathbf{r}, \mathcal{S}] = \int \int_U \left[\kappa_1(u, v)^2 + \kappa_2(u, v)^2 \right] \| \mathbf{r}_u \times \mathbf{r}_v \| \, du \, dv + \lambda \sum_{i=1}^{M} \rho_{C_i}(\mathbf{r}(\mathbf{u}_i), \mathbf{c}_i)^2,$$ (19)

where κ_1 and κ_2 are given by (13) and (14), respectively, and ρ_{C_i} is the distance measure from subsection 3.2 and depends on the type of constraint. The explicit form of the functional can be written as

$$M^\lambda[z, S] = \int \int_U \left\{ \frac{[(1+z_y^2)z_{xx} - 2z_x z_y z_{xy} + (1+z_x^2)z_{yy}]^2}{2(1+z_x^2+z_y^2)^{5/2}} - \frac{z_{xx}z_{yy} - z_{xy}^2}{(1+z_x^2+z_y^2)^{3/2}} \right\} dx\, dy$$

$$+ \lambda \sum_{i=1}^{M} \rho_{C_i}(z(x_i, y_i), c_i)^2. \tag{20}$$

The constant λ determines a tradeoff between the smoothness of the surface and closeness of fit of the surface to the data. To show that this problem is well-posed it must be shown that the minimization of this functional exists and is unique.

Unfortunately, even if this problem is well posed, the minimization of this functional will not be an easy task. A functional $p(\mathbf{x})$ on a linear functional space, X, is convex if

1. $p(\mathbf{x}) \geq 0$ for all $\mathbf{x} \in X$;

2. for all $\mathbf{x}, \mathbf{y} \in X, p(\mathbf{x} + \mathbf{y}) \leq p(\mathbf{x}) + p(\mathbf{y})$ (subadditive);

3. for $\alpha \in \Re$, $\alpha \geq 0$, and $\mathbf{x} \in X, p(\alpha\mathbf{x}) = \alpha p(\mathbf{x})$ (positive homogeneous).

The stabilizer in (18) can be shown to be nonconvex by considering property (iii) since

$$[\alpha r] = \frac{1}{\alpha}[r] \tag{21}$$

for any constant α.

Recently, several techniques have been developed to solve nonconvex optimization problems with the above form. Stochastic optimization algorithms, such as simulated annealing[24] can be applied to find the invariant solution. Unfortunately, such algorithms are computationally very expensive[4], and therefore it would be advantageous if a convex stabilizer could be found which closely approximated the nonconvex problem.

3.6. Common convex approximations
This section describes two commonly used approximations to the nonconvex problem which are well-posed. Both of these algorithms use the explicit representation of a surface and are therefore limited to surfaces in three-dimensional space.

3.6.1. Quadratic Variation
The most common approximation made is to use the assumption

$$z_x \approx 0 \qquad z_y \approx 0. \tag{22}$$

Then the Gaussian and mean curvatures can be approximated by

$$H(x, y) \approx \frac{z_{xx} + z_{yy}}{2} \tag{23}$$

$$K(x, y) \approx z_{xx}z_{yy} - z_{xy}^2, \tag{24}$$

and the stabilizer in (18) becomes

$$_1[z] = \frac{1}{2} \int \int_U z_{xx}^2 + 2z_{xy}^2 + z_{yy}^2 \, dx dy. \qquad (25)$$

This stabilizer is not invariant to rigid three-dimensional transforms, and the assumption that z_x and z_y are small is often invalid. This model is usually referred to as the planar-plate model. Several methods have been developed to find the surface which minimizes the functional (18) under the planar-plate model assumptions. Grimson[15] uses the finite difference method to obtain a discrete problem which is then minimized by using the conjugate gradient method. Terzopoulos[45] applies the finite-element method to transform the continuous variational principal into a discrete problem, which is then solved using an efficient multigrid algorithm. The finite element method also has the advantage (over the finite difference method) of being able to be defined on nonrectangular grids. A third method uses the fact that the functional space is a reproducing kernel space. Duchon[10] and Meinguet[30] provide methods for obtaining the continuous function which minimizes the functional using this fact. This method obtains the solution to the continuous problem, and is used extensively by Wahba[48] who posed several surface reconstruction problems in meteorology as functional minimizations of the quadratic variation (25). This approach, however, becomes very computationally demanding when the number of constraints increases.

3.6.2. An approximation to the explicit invariant stabilizer

If the first-order derivatives (z_x and z_y) of z are approximated then a closer approximation to the invariant stabilizer (20) can be made. Define the new stabilizer as

$$_2[z] = \int \int_U 2[A(x,y)z_{xx} - B(x,y)z_{xy} + C(x,y)z_{yy}]^2 - D(x,y)(z_{xx}z_{yy} - z_{xy}^2) dx dy. \qquad (26)$$

If the approximations to z_x and z_y are denoted by $\hat{z}_x$ and $\hat{z}_y$, then the constant functions A, B, C and D are given by

$$A(x,y) = \left[\frac{1 + \hat{z}_y^2}{(1 + \hat{z}_x^2 + \hat{z}_y^2)^{5/4}}\right]^2 , B(x,y) = \left[\frac{2\hat{z}_x\hat{z}_y}{(1 + \hat{z}_x^2 + \hat{z}_y^2)^{5/4}}\right]^2 ,$$

$$C(x,y) = \left[\frac{1 + \hat{z}_x^2}{(1 + \hat{z}_x^2 + \hat{z}_y^2)^{5/4}}\right] , D(x,y) = \frac{1}{(1 + \hat{z}_x^2 + \hat{z}_y^2)^{3/2}}. \qquad (27)$$

Using this stabilizer (26), it can be shown that the problem is well-posed and that the functional is convex. Any of the methods used to minimize the functional with the quadratic variation stabilizer (25) can be modified to form a discrete functional minimization problem with the added weight terms.

To form an estimate of the first-order derivatives, Blake and Zisserman[3] suggest first fitting an invariant weak membrane to the constraints. Since the constraints are obtained only at discrete points, this forms a piecewise planar approximation to the surface. However, the functional minimization problem that arises from fitting the weak membrane to the constraints is nonconvex. Convex planar approximations can be used, such as the one described in the next section.

The surface estimated using $_2$ as the stabilizer is much more robust to variations in the viewpoint than the surface estimated using the quadratic variation, $_1$. However, the inaccuracy in estimating z_x and z_y when either (or both) are large causes undesirable variation in the surface estimate.

3.6.3. An approximately invariant estimate

The problem with the approximations used in the previous section is that unrealistic assumptions are necessary to obtain a convex functional. If the parametric form of the functional is used, a more realistic approximation to (20) can be found. In this section, we describe a two-stage algorithm for constructing a surface estimate which is more robust to variations of the coordinate system. In the first stage, a piecewise planar invariant approximation of the surface is found by minimizing an invariant functional. This piecewise linear surface is used to form an approximate parameterization of the surface. With this approximate parameterization of the surface, it is possible to obtain a valid approximation to the invariant stabilizer, (20), which is well-posed and convex.

The proposed algorithm will be examined in two stages. The first stage forms estimates of several surface characteristics by forming a piecewise planar estimate of the surface. These estimates are used to approximate the nonconvex invariant functional minimization problem with a well-posed convex functional minimization problem which is approximately invariant. The second stage of the algorithm minimizes this convex functional to form the surface estimate $\hat{r}$.

Stage 1: Estimating surface characteristics

In order to estimate important surface characteristics, a piecewise planar estimate of the surface is first computed. This is computed by first constructing a Thiessen triangulation of the constraint points and then smoothing the constraints using a weighted least squares smoothing algorithm. The smoothed constraints and the triangulation information is then used to define a piecewise planar surface approximation. A uniform mesh is placed on this surface, and surface characteristics are estimated at the nodes of the mesh.

A Thiessen triangulation, T, of the surface is found using an algorithm adapted from the work of by A. K. Cline, et al.[7]. This algorithm is guaranteed to terminate with a valid Thiessen triangulation of the surface after a finite number of steps.

The set of constraints S and the triangulation T define a continuous piecewise planar surface estimate (i.e. the surface within a triangle is given by the plane which passes through the three constraint nodes). This surface estimate is first smoothed before the surface characteristics are estimated. The smoothing is performed by using a weighted least squares smoother. An arc in the triangulation can be described by the indices of the two end nodes. Let (e_1, e_2) denote the indices of an arc. The collection of arcs in the triangulation T is denoted by $\mathcal{E}$. Let S^* be a collection of nodes $\{c_i^* \in C_i, i = 1, \ldots, M\}$ which with T describe an arbitrary piecewise planar surface. A smoothed piecewise planar surface is computed by minimizing the functional

$$M^{\lambda_p}[\mathcal{S}^*, \mathcal{S}] = \sum_{(e_1,e_2)\in\mathcal{E}} w_{e_1,e_2} \parallel \mathbf{c}_{e_1}^* - \mathbf{c}_{e_2}^* \parallel^2 + \lambda_p \sum_{i=1}^{M} \parallel \mathbf{c}_i^* - \mathbf{c}_i \parallel^2 \tag{28}$$

with respect to the collection $\mathcal{S}^*$. The constant weight term is given by

$$w_{e_1,e_2} = \frac{1}{\parallel \mathbf{c}_{e_1} - \mathbf{c}_{e_2} \parallel}, \tag{29}$$

and is used so that nodes which are spatially far apart (e.g. across jump discontinuities) have little smoothing effect on each other even if they are adjacent in the triangulation. The function (28) is convex since each term is quadratic. The function (28) is also invariant since each term is based on an invariant quantity (the distance between two nodes). Let the collection of S^* that minimizes (28) be denoted by $\hat{\mathcal{S}} = \{\hat{\mathbf{c}}_i \in C_i, \; i = 1,\ldots,M\}$.

The function (28) is based on the physical model of finding the minimum energy configuration of a set of springs. The first term represents the energy of a collection of springs, one along each arc in the triangulation. The weight term, w_{e_1,e_2}, is related to the spring coefficient; a larger weight corresponds to a stronger spring. The second term in (28) represents the energy of a collection of M springs, each one connecting a constraint location with its corresponding location on the surface. The term λ_p controls the ratio of the spring constants between the two sets of springs.

On this continuous piecewise planar surface estimate a uniform grid (uniform in the XY plane) is placed. At the grid nodes, surface characteristics for the smooth continuous surface estimate are estimated. For points which are outside the convex hull of the constraint points (i.e. not interior to any triangle) the plane of the nearest triangle is extended to form a surface estimate. This forms a uniform mesh of nodes in the XY plane on which to calculate the smooth surface estimate and initialized the surface with the piecewise planar fit.

In order to make the desired approximations to (20), we need estimates of the normal, n, and the principal directions, d_1 and d_2, at each node on the mesh. Note that in order to make the approximation to the explicit invariant stabilizer, (26), we only need to estimate the surface normal at each mesh node. While local least square estimation techniques are usually sufficient to ensure stable estimates of first-order characteristics (e.g. n), they are rarely adequate for the estimation of second-order properties (e.g. d_1, d_2). Our approach, which follows[32], is to first form local estimates of the Darboux frame $\mathcal{D}$ and then to iteratively refine the estimates to obtain a stable reconstruction of $\mathcal{D}[32, 11]$. When the difference between the principal curvatures is small, the estimate of the principal directions becomes unreliable. Therefore, whenever $|\kappa_1 - \kappa_2| < 0.2h$, where h is the distance between samples in the XY plane, it is assumed that the estimate is not accurate enough to be used.

Stage 2: A smooth invariant surface estimate

Using the estimates of $\mathcal{D}(P)$ formed in Stage 1, estimates of r_u, r_v, θ_1, θ_2 and n can be used to compute approximations to the principal curvatures κ_1 and κ_2 by

$$\kappa_1 = \omega_1(r_{uu}, n) + \omega_2(r_{uv}, n) + \omega_3(r_{vv}, n), \tag{30}$$

and

$$\kappa_2 = \omega_4(\mathbf{r}_{uu}, \mathbf{n}) + \omega_5(\mathbf{r}_{uv}, \mathbf{n}) + \omega_6(\mathbf{r}_{vv}, \mathbf{n}), \tag{31}$$

where the ω's are constant functions given by

$$\omega_1 = \frac{\cos^2 \hat{\theta}_1}{(\hat{\mathbf{r}}_u, \hat{\mathbf{r}}_u) \cos^2 \hat{\theta}_1 + 2(\hat{\mathbf{r}}_u, \hat{\mathbf{r}}_v) \cos \hat{\theta}_1 \sin \hat{\theta}_1 + (\hat{\mathbf{r}}_v, \hat{\mathbf{r}}_v) \sin^2 \hat{\theta}_1}, \tag{32}$$

$$\omega_2 = \frac{2 \cos \hat{\theta}_1 \sin \hat{\theta}_1}{(\hat{\mathbf{r}}_u, \hat{\mathbf{r}}_u) \cos^2 \hat{\theta}_1 + 2(\hat{\mathbf{r}}_u, \hat{\mathbf{r}}_v) \cos \hat{\theta}_1 \sin \hat{\theta}_1 + (\hat{\mathbf{r}}_v, \hat{\mathbf{r}}_v) \sin^2 \hat{\theta}_1}, \tag{33}$$

$$\omega_3 = \frac{\sin^2 \hat{\theta}_1}{(\hat{\mathbf{r}}_u, \hat{\mathbf{r}}_u) \cos^2 \hat{\theta}_1 + 2(\hat{\mathbf{r}}_u, \hat{\mathbf{r}}_v) \cos \hat{\theta}_1 \sin \hat{\theta}_1 + (\hat{\mathbf{r}}_v, \hat{\mathbf{r}}_v) \sin^2 \hat{\theta}_1}, \tag{34}$$

$$\omega_4 = \frac{\cos^2 \hat{\theta}_2}{(\hat{\mathbf{r}}_u, \hat{\mathbf{r}}_u) \cos^2 \hat{\theta}_2 + 2(\hat{\mathbf{r}}_u, \hat{\mathbf{r}}_v) \cos \hat{\theta}_2 \sin \hat{\theta}_2 + (\hat{\mathbf{r}}_v, \hat{\mathbf{r}}_v) \sin^2 \hat{\theta}_2}, \tag{35}$$

$$\omega_5 = \frac{2 \cos \hat{\theta}_2 \sin \hat{\theta}_2}{(\hat{\mathbf{r}}, \hat{\mathbf{r}}_u) \cos^2 \hat{\theta}_2 + 2(\hat{\mathbf{r}}_u, \hat{\mathbf{r}}_v) \cos \hat{\theta}_2 \sin \hat{\theta}_2 + (\hat{\mathbf{r}}_v, \hat{\mathbf{r}}_v) \sin^2 \hat{\theta}_2}, \tag{36}$$

$$\omega_6 = \frac{\sin^2 \hat{\theta}_2}{(\hat{\mathbf{r}}_u, \hat{\mathbf{r}}_u) \cos^2 \hat{\theta}_2 + 2(\hat{\mathbf{r}}_u, \hat{\mathbf{r}}_v) \cos \hat{\theta}_2 \sin \hat{\theta}_2 + (\hat{\mathbf{r}}_v, \hat{\mathbf{r}}_v) \sin^2 \hat{\theta}_2}, \tag{37}$$

and a stabilizer can be defined by

$$\int \int_U [\omega_1(\mathbf{r}_{uu}, \mathbf{n}) + \omega_2(\mathbf{r}_{uv}, \mathbf{n}) + \omega_3(\mathbf{r}_{vv}, \mathbf{n})]^2 + [\omega_4(\mathbf{r}_{uu}, \mathbf{n}) + \omega_5(\mathbf{r}_{uv}, \mathbf{n}) + \omega_6(\mathbf{r}_{vv}, \mathbf{n})]^2 du \ dv. \tag{38}$$

This functional is convex and leads to efficient computational solutions. However, when computing the solution using standard techniques a numerical instability can occur. When using iterative methods for computing the solution, two points on the surface can come close together. This can occur because the parameterization curves, u and v move during each update of an iterative solver and one of the curves may cross back on itself. This results in errors in the following iterations, possibly causing the surface parameterization to have nonsingularities (i.e. overlap) which leads to further instability.

To overcome this numerical instability, after each iteration the surface is resampled onto a uniform grid, which keeps the surface points from coming close to each other. The iteration and resampling can be combined into one step. At iteration k an update vector $\delta\mathbf{r}_{(k)}(i,j)$ is computed for each point $\mathbf{r}_{(k)}(i,j)$ in the representation. Instead of setting $\mathbf{r}^{(k+1)} = \mathbf{r}^{(k)} + \delta\mathbf{r}^{(k)}$, use $\delta\mathbf{r}^{(k)}$ to define a planar approximation to the new surface. The planar surface patch contains the point $\mathbf{r}^{(k)} + \delta\mathbf{r}^{(k)}$ and is orthogonal to the update vector $\delta\mathbf{r}^{(k)}$. Using this approximation, the change along the Z-axis is computed by

$$\delta z'(k)(i,j) = \frac{\| \delta\mathbf{r}^{(k)}(i,j) \|^2}{\delta z^{(k)}(i,j)} \tag{39}$$

where $\delta z^{(k)}$ is the Z-axis component of $\delta\mathbf{r}^{(k)}$. Thus instead of updating $\mathbf{r}^{(k)}$ with $\delta\mathbf{r}^{(k)}$, the update is computed by

$$\mathbf{r}^{(k+1)}(i,j) = \mathbf{r}^{(k)} + <0, 0, \delta z'^{(k)}(i,j)>. \tag{40}$$

Since the second-order derivatives of x and y are zero on a uniform grid, and since the update vector $\delta r^{(k)}$ is always in the normal direction, the stabilizer can be reduced to the much simplified form

$$3 = \int\int_U [\omega_1 z_{xx} + \omega_2 z_{xy} + \omega_3 z_{yy}]^2 + [\omega_4 z_{xx} + \omega_5 z_{xy} + \omega_6 z_{yy}]^2 du dv. \tag{41}$$

Since both terms are quadric the functional is convex.

There are many methods by which the minimization of this functional can be performed. All of the methods used by[15, 3, 45, 46, 10, 30] can be adapted to solve this minimization problem. For simplicity, we choose to discretize the problem using finite difference methods and we choose to perform the minimization using the conjugate gradient method. See [15] for details of this method.

4. Discontinuous Surface Estimates

In order to incorporate the *a priori* knowledge that discontinuities exist into the reconstruction process we either examine modifications of the stablizing function in Tikhonov regularization or the prior distribution function in stochastic regularization. In this section we examine several previously proposed modifications to these functions and the resulting characteristics of the reconstruction algorithms.

4.1. Controlled-continuity stabilizers

If the location and type of discontinuity is known, this information can easily be incorporated into the stablizing functional through the use of the weight terms, $w_{p(x)}$, in the Sobolev norm (4). For an mth order discontinuity ($m = 0$ is a jump discontinuity, $m = 1$ is a discontinuity in the first derivative, etc.) that occurs at the location (x'), set

$$w_p(x') = 0, \quad \text{for } |p| \geq m$$

and for all other locations set the weight term to one. This controls the order of the continuity at discontinuities[45]. When used in this fashion, the weight term, $w_p(x)$, is referred to as a line process since it indicates where lines (i.e. edges) exist in the parameter field. Since the knowledge of the location and type of discontinuity is rarely given in most applications, the weight term cannot usually be set prior to computing the estimate. Therefore, the weighting function must also be estimated. Approaches based on estimating the weighting function and the reconstructed parameter field separately have been proposed[16, 26, 36, 37] as well as approaches based on estimating both the weight function and the parameter field together[45, 3].

The techniques proposed in [16, 26] make hard decisions about the value of the weight function, i.e. w_p is set either to 1 or 0. This preprocessing step is essentially performing discontinuity detection on sometimes sparse and noisy data. Since this type of preprocessing requires a decision (i.e. a threshold), these techniques will be prone to unstable reconstructions. This occurs since small perturbations of the data when the decision parameter is near the threshold can result in a different decision being made, and consequently a drastically different parameter field being estimated. A slightly more robust technique was proposed by Sinha and Schunck[36, 37]. Their method allowed the weighting function to take on any value greater than 0. They perform an initial fit without any discontinuities and then set the weight term to be inversely proportional to the first derivative of the initial estimate. This has the effect of creating a region where the weight term is small near discontinuities and large in regions where the parameter field is smooth. This type of soft decision making will result in a more stable estimate than making a hard decision. The disadvantage is that near discontinuities, where the weight term is small, the noise removal properties of the spline fitting will be defeated. That is, the estimating procedure will produce noisy estimates of the parameter field near the discontinuities. To overcome this problem they introduce another parameter which can be used to make the decision harder, i.e. to make the region where the weight term is near zero smaller. Of course, making a harder decision will result in a more unstable estimate.

To form estimates for $w_p(x)$ in conjugation with $r(x)$ we cannot simply minimize (1) with respect to $w_p(x)$ and $r(x)$ when using the Sobolev norm (4). Doing this would result in the trivial solution $w_p(x) = 0$ and $r(x)$ being only uniquely determined at points where there are constraints.

4.2. Estimating the line process

In order to estimate $w_p(x)$ simultaneously with $r(x)$, we need to augment our prior information with some knowledge about the form of $w_p(x)$. This can be done by adding a term to $[\cdot]$ which measures the consistency of a given $w_p(x)$[45]. That is we minimize a functional of the form

$$M^\lambda[r^*, w_p^*, \mathcal{S}] = w[r^*, w_p^*] + \sum_{c_i \in \mathcal{S}} \lambda_i \|\mathcal{A}_i r^* - c_i\|^2, \tag{42}$$

where $w[\cdot, \cdot]$ has the form

$$w[r^*, w_p^*] = \mathcal{E}(r^*, w_p^*) + \mathcal{D}(w_p^*). \tag{43}$$

The functional $\mathcal{E}(\cdot, \cdot)$ will have a form such as in equation (4) and $\mathcal{D}(\cdot)$ will measure the consistency of the line process. Unfortunately even for simple $\mathcal{D}(\cdot)$ the functional minimization problem becomes nonconvex. This is a problem not only because of the increased computation required for a solution but also because the mathematical problem is no longer well-posed.

Blake and Zisserman proposed a $\mathcal{D}(\cdot)$ which counted the number of discontinuities for a particular $w_p(x)$ and used a stabilizer of the form of (4) for $\mathcal{E}(\cdot, \cdot)$. They showed that this was equivalent to a using standard Tikhonov regularizing functional with a stabilizer of the form of (5) with $w_p(x) = 1 \ \forall \ x$ and $\rho(\cdot)$ of the form

$$\rho_1(x) = \begin{cases} x^2, & x \leq T, \\ T^2, & x > T. \end{cases} \tag{44}$$

This choice for $\rho(\cdot)$ is nonconvex and thus results in a nonconvex and ill-posed functional minimization problem.

Similar schemes for estimating the line process were proposed by Geman and Geman [14] and by Marroquin[29]. Their approach also results in a nonconvex and ill-posed functional minimization problems. Several novel techniques have been proposed to overcome some of the computational complexity associated with such an approach[13, 20]. While the accuracy of signal estimators based on nonconvex optimization may be adequate, the stability of such an approach will be poor. Small noise in the data can dramatically change the result. Even for the same input data and the same functional to minimize, different optimization algorithms can compute very different results.

4.3. Convex and stable approaches

As was discussed in the last section, in order for the regularization problem to be well-posed the stablizing functional must be convex. The question then becomes, can a convex stablizing functional be devised such that discontinuities in the parameter field are accurately reconstructed. In this section we will examine using a stabilizer functional of the form of (5) with some convex $\rho(\cdot)$. The weight terms will be set to some constant value throughout the entire field, that is $w_p(x) = w_p$. Since the class of convex functionals is very large, we will begin by first discussing other desirable characteristics (besides convexity) which are desirable for the functional $\rho(\cdot)$.

Besides being convex the functional should be symmetric since the sign of a particular term should not change the importance of its magnitude. The most important characteristic of course is that discontinuities should be allowed to form. In the previous section the weight term, w_p, was used to allow discontinuities, the weight was changed to deemphasize regions where the derivative is large (i.e. regions of discontinuities). Regions of discontinuities can also be deemphasized by modifying $\rho(\cdot)$ so that it is less than the squared term, i.e. $\rho(x) < x^2$ for large values of x. Finally, we would like a parameter, T, which controls the degree of smoothness of the reconstruction. We will denote the parameterization of $\rho(\cdot)$ by $\rho^T(\cdot)$. As T varies from some T_1 to T_2 the estimated parameter field varies from a smooth reconstruction to a reconstruction that allows more discontinuities. Mathematically, this is equivalent to the condition that $\rho^T(\cdot)$ decreases monotonically as T varies from T_1 to T_2 for all x. To better understand this condition, recall that the stabilizer $[\cdot]$ measures the consistency of a particular function with our *a priori* information. If we want a parameter which controls the degree to which discontinuities are allowed (or conversely the degree of smoothness imposed), then for any particular function $r_1(x)$ if the parameter T is varied to allow more discontinuities then the consistency measure should decrease. If the $\rho^T(\cdot)$ functional does not decrease monotonically as T is varied then it is easy to devise a function for which the consistency measure will increase as the degree of allowed discontinuities is increased.

A class of convex stabilizers which has attracted some attention can be characterized by the functional

$$\rho^T(x) = \begin{cases} |x|^p, & x \leq T, \\ \left(|x| + \left(\frac{p}{q}T^{p-1}\right)^{\frac{1}{q-1}} - T\right)^q + T^p - \left(\frac{p}{q}T^{p-1}\right)^{\frac{q}{q-1}}, & x > T, \end{cases} \tag{45}$$

where $p, q \geq 1$ and generally $q \leq p$. This choice is based on varying the degree of smoothness imposed at different scales. Below some threshold T one weighting function is used while above that threshold a weighting function that increases less rapidly (i.e. less smoothness imposed) is used. By choosing $p > q$, we smooth small scale noise while retaining large scale discontinuities in the parameter field. The constants in equation (45) are chosen so that the weight function is convex and continuous and so that the influence function is continuous.

4.3.1. A scale-invariant stabilizer

Bouman and Sauer[5] and Harris[20] examine the case where $1.0 \leq p \leq 2.0$ and $T = \infty$, that is the special case of (45) when

$$\rho_2^p(x) = |x|^p. \tag{46}$$

When $p = 2.0$ the regularizing functional will be the standard Tikhonov regularizer with a quadratic stabilizer and thus will estimate smooth parameter fields. For $p = 1.0$ the corresponding estimator is the sample median and will allow discontinuities. To control the degree that discontinuities are allowed, they use the parameter p. This choice for $\rho^p(\cdot)$ satisfies the first three of our desirable characteristics but not the fourth. Therefore this choice for a stablizing functional will not allow consistent adjustment of the allowable degree of discontinuity. When the data is sparse, the estimate formed with this stabilizer also appears to be very sensitive to the selection of the discontinuity parameter. This occurs because it is not possible with the single parameter to provide sufficient smoothness necessary for interpolation while retaining discontinuities in the estimate. In the application in which Bouman and Sauer are interested, the data is dense and they chose the discontinuity parameter to be between 1.0 and 1.2. In this case satisfactory estimates can be computed. Thus, while in general this may not be a good stabilizer, in some applications it may work satisfactorily.

4.3.2. Huber minimax functional

The second special case that we will examine was motivated by work done by Huber in robust statistics[22]. He was interested in computing smooth estimates when outliers existed in the data set. He proposed to weight the outliers in the data with a functional of the form

$$\rho_3^T(x) = \begin{cases} x^2, & x \leq T, \\ T^2 + 2T|x - T|, & x > T, \end{cases} \tag{47}$$

where T varies from $+\infty$ to 0. This is the special case of (45) for when $p = 2.0$ and $q = 1.0$. This function quadratically below the threshold and linearly above the threshold. Thus, small scale noise is smoothed with a least squares smoother while large scale discontinuities are weighted only linearly. For $T = \infty$ we get a smooth estimate. As T approaches zero more discontinuous regions are allowed. This functional as been applied by Stevenson and Delp to the problems of curve[38–40] and surface[41, 44] reconstruction and by Shulman and Herve[35] for the problem of estimating discontinuous optical flow fields.

4.3.3. General case

There is of course an infinite number of variations on the choice of p and q that can be examined. The correct choice will depend on the particular application and the *a priori* information that is known. If a good quality reconstruction is available or can be modeled, then p and q can be chosen statistically by fitting the statistical model to the available reconstruction. In most cases p should be chosen to be near 2.0 to provide least squares type smoothing of the noise and q chosen to be around 1.0 to reconstruct the discontinuities as sharply as possible with a convex stabilizer. The threshold T will depend on the scale at which discontinuities are present. Since there is a smooth transition in the weight function at the threshold the quality of the estimate is not very sensitive to the selection of T and the smaller the quantity $|p - q|$ is the less sensitive the estimate will be to this parameter.

5. COMPUTATIONAL ISSUES

The previous section presented a mathematically well-posed technique for estimating parameter fields with discontinuities. The technique results in a convex but nonquadratic functional minimization problem. This section examines several computational issues of the resulting mathematical problem statement. The mathematical problem will first be discretized, then digital computational techniques will be examined.

5.1. Discrete problem statement

The finite difference method is utilized to discretize the continuous mathematical problem. The function to be estimated, $r(x)$, is discretized on a regular fine-grid mesh. That is $r(x)$ is sampled uniformly along each of its variables. Assume that N samples are taken along each of the variables. This sampling process results in a finite set of nodal variables which will be represented by $r(x_i)$ where x_i represents the indexed vector quantity $< x_{1,i_1}, x_{2,i_2}, ..., x_{n,i_n} >$ and each of the indices, i_j vary in the range of 1 to N. The discrete approximation to a first order derivative is

$$\frac{\partial r(x_i)}{\partial x_j} = r(x_{1,i_1}, ..., x_{j,i_j+1}, ...x_{n,i_n}) - r(x_{1,i_1}, ..., x_{j,i_j}, ...x_{n,i_n}) \tag{48}$$

Second order terms are given by the discrete approximations

$$\frac{\partial^2 r(x_i)}{\partial x_j^2} = r(x_{1,i_1}, ..., x_{j,i_j+1}, ...x_{n,i_n}) - 2r(x_{1,i_1}, ..., x_{j,i_j}, ...x_{n,i_n})$$
$$+ r(x_{1,i_1}, ..., x_{j,i_j-1}, ...x_{n,i_n}) \tag{49}$$

and

$$\frac{\partial^2 r(x_i)}{\partial x_j \partial x_k} = r(x_{1,i_1}, ..., x_{j,i_j+1}, ...x_{k,i_k+1}, ...x_{n,i_n}) - r(x_{1,i_1}, ..., x_{j,i_j}, ...x_{k,i_k+1}, ...x_{n,i_n})$$
$$- r(x_{1,i_1}, ..., x_{j,i_j+1}, ...x_{k,i_k}, ...x_{n,i_n}) + r(x_{1,i_1}, ..., x_{j,i_j}, ...x_{k,i_k}, ...x_{n,i_n}). \tag{50}$$

Similarly, higher order terms can also be written. If the finite number of nodal variables are collected and written as a vector r_d then the discrete differential operators can be written as a matrix multiplication, that is

$$\left(\frac{\partial_1^p}{\partial x_1^{p_1}}\right)\left(\frac{\partial_2^p}{\partial x_1^{p_2}}\right)\cdots\left(\frac{\partial_n^p}{\partial x_1^{p_n}}\right)r(x_i) = A_{i,p}r_d \tag{51}$$

where the matrix $A_{i,p}$ depends on the location in the mesh and the order of the discontinuity. Most of the terms in $A_{i,p}$ are zero, the only nonzero terms being given by the difference equation which approximate the derivatives, e.g. equations (48,49,50). Using these approximations the generalized stabilizer function (5) has the discrete form

$$_d[r_d] = \sum_{|p|<m}\sum_{i_1=1}^{N}\cdots\sum_{i_n=1}^{N}\rho(w_p A_{i,p}r_d). \tag{52}$$

Since the acquisition process $A_i r(x)$, is assumed to be linear, the process can be also written as a matrix multiplication, which we will write as $A_i r_d$. The resulting discrete regularization functional can be written as

$$M_d^\lambda(r_d, S) = \sum_{|p|<m}\sum_{i_1=1}^{N}\cdots\sum_{i_n=1}^{N}\rho(w_p A_{i,p}r_d) + \sum_{c_i \in S}\lambda_i\|A_i r_d - c_i\|^2, \tag{53}$$

where $\rho(\cdot)$, w_p, $A_{i,p}$, $\|\cdot\|$, λ, and A are based on our *a priori* information about the application and S is the collection of data. To form the discrete parameter field estimate, this functional is minimized with respect to the nodal variables r_d.

5.2. Digital descent methods

The most prevalent class of algorithms for digital convex functional minimization are based on iterative techniques, where the update at each iteration monotonically decreases the functional value. Let r_d^k denote the function value at the kth iteration. At each iteration the function is updated by

$$r_d^{k+1} = r_d^k + \alpha^k p^k, \tag{54}$$

where the vector p^k is the direction of the update and the scalar α^k determines the size of the step taken in that direction. Since our function is convex, any of the descent based methods will converge to the optimal solution given any starting vector r_d^0. However overall computational time will depend on the initial guess r_d^0, and the scheme for choosing the update vector p^k and step size α^k. For a particular application, computation time can be dramatically reduced if there exists some quick technique for forming a rough estimate to be used as the initial guess[23, 43]. For example, in the curve reconstruction problem an initial estimate can be formed quickly by fitting a piecewise linear estimate to the data. Similarly, for the surface reconstruction problem a piecewise planar surface can provide an initial guess. Forming this initial estimate is especially helpful when the data is sparse. In applications where the data is sparse and when it is not possible to quickly form an initial estimate, multigrid techniques can be used to improve the computation time[18].

While there are many methods for choosing the update vector $\mathbf{p}^k$, the conceptually simplest minimization methods choose $\mathbf{p}^k$ from among the coordinate vectors $\mathbf{e}^1, ..., \mathbf{e}^{N^n}$. This results in univariate relaxation where at each iteration only one component of $\mathbf{r}_d^k$ changes. Another intuitive choice for the update vector is the direction of steepest descent, that is the direction for which the functional will decrease the fastest. The direction of steepest descent is the direction for which

$$-\frac{\nabla M_d^\lambda[\mathbf{r}_d^k, \mathcal{S}]^T \mathbf{p}}{\|\mathbf{p}\|} \tag{55}$$

takes on its minimum value as a function of $\mathbf{p}$. If the l_2-norm is used for $\|\cdot\|$ then the direction of steepest descent will be the negative of the gradient vector, i.e.

$$\mathbf{p}^k = -\nabla M_d^\lambda[\mathbf{r}^k, \mathcal{S}]. \tag{56}$$

The choices for $\mathbf{p}^k$ discussed thus far are not guaranteed to converge to a solution in a fixed number of iterations. For linear systems of equations (quadratic optimization problems) the conjugate gradient method has been shown to converge in a bounded number of steps[21]. This minimization method computes a conjugate basis for the linear system which are used for the update vectors. With this choice for the update vectors, the iterative process can be shown to converge in one cycle through the basis set. For nonquadratic functional optimization problems, generalized conjugate gradient algorithms have been proposed[12].

Once an update vector is chosen, the next step is to compute the size of the step, α^k, which will be taken in that direction. The maximal decrease for a given $\mathbf{p}^k$ occurs when α^k is chosen so as to minimize the functional along that direction,

$$\alpha^k = \arg\min_{\alpha \in \mathcal{R}} M_d^\lambda[\mathbf{r}_d^k + \alpha \mathbf{p}^k, \mathcal{S}]. \tag{57}$$

This results in a univariate nonquadratic minimization problem. In many applications, including the examples presented here, this one-dimensional minimization problem can be approximated by minimization of the osculating parabola.

$$\alpha^k = \arg\min_{\alpha \in \mathcal{R}}\{M_d^\lambda[\mathbf{r}_d^k, \mathcal{S}] + \alpha \nabla M_d^\lambda[\mathbf{r}_d^k, \mathcal{S}]^T \mathbf{p}^k + \frac{1}{2}\alpha^2 \mathbf{p}^{k^T} \nabla^2 M_d^\lambda[\mathbf{r}_d^k, \mathcal{S}]\mathbf{p}^k\}. \tag{58}$$

In this case α^k is well-defined and given by

$$\alpha^k = \frac{\nabla M_d^\lambda[\mathbf{r}_d^k, \mathcal{S}]^T \mathbf{p}^k}{\mathbf{p}^{k^T} \nabla^2 M_d^\lambda[\mathbf{r}_d^k, \mathcal{S}]\mathbf{p}^k}. \tag{59}$$

One of the advantages of these methods is in the inherent parallelism associated with these algorithms. This can be exploited by a mesh of computational nodes which can greatly reduce the total computation times[42].

As an example of the type of surfaces that are reconstructed in this type of work, a typical collection of sparse three-dimensional data is shown in Figure 1. This range data was produced by a Technical Arts 100X scanner (White scanner) at Michigan State University's Pattern Recognition and Image Processing Lab. The reconstruction of the surface using the techniques described in this chapter is shown in Figure 2.

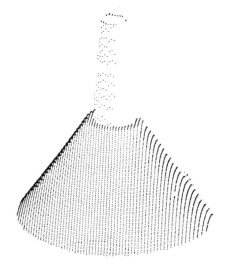

Figure 1. Original constraints, S

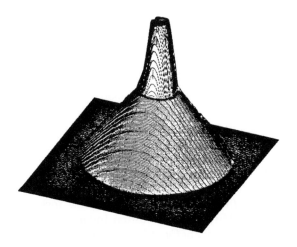

Figure 2. Reconstructed surface, $\hat{f}$

6. Conclusions

This chapter has presented a mathematically well-posed method for estimating invariant parameter fields with discontinuities. The proposed method is based on regularization theory where the consistency measure is nonquadratic, but convex. The convexity of the functional is important from both a mathematical and computational viewpoint. Since the surfaces used in computer vision correspond to physical surfaces in the real world it is important that algorithms which are used to estimate these surfaces from raw data can account for realistic surfaces. Finally, some computational issues were discussed and an example reconstruction was shown.

REFERENCES

1   J. Aloimonos and D. Shulman "Learning early-vision computations," *Journal of the Optical Society of America A,* Vol. 6, No. 6, June 1989, pp. 908-919.

2   P. J. Besl and R. C. Jain, "Invariant Surface Characteristics for 3D Object Recognition in Range Data", *Computer Vision, Graphics and Image Processing,* Vol. 33, No. 1, pp. 33-80, 1986.

3   A. Blake and A. Zisserman, *Visual reconstruction,* Cambridge, Massachusetts: MIT Press, 1987.

4   A. Blake, "Comparison of the Efficiency of Deterministic and Stochastic Algorithms for Visual Reconstruction", *IEEE Transactions on Pattern Analysis and Machine Intelligence,* Vol. 11, No. 1, pp. 2-12, January 1989.

5   C. Bouman and K. Sauer, "A generalized Gaussian image model for edge-preserving MAP estimation," submitted to the *IEEE Transactions on Medical Imaging.*

6   C. Chu and A. Bovik, "Visible surface reconstruction via local minimax approximation," *Pattern Recognition,* Vol. 21, No. 4, 1988, pp. 303-312.

7   A. K. Cline and R. L. Renka, "A Storage-Efficient Method for Construction of a Thiessen Triangulation", *Rocky Mountain Journal of Mathematics,* Vol. 14, No. 1, pp. 119-139, Winter 1984.

8   R. Courant and D. Hilbert, *Methods of Mathematical Physics,* New York: Interscience Publishers, Inc. Vol. I, 1953.

9   J. C. Davis, *Interpolation and Approximation,* New York, NY: Blaisdell, 1963.

10 J. Duchon, "Splines Minimizing Rotation-Invariant-Semi-Norms in Sobolev Spaces", *Constructive Theory of Functions of Several Variables,* pp. 85-100, 1976.

11 F. P. Ferrie, J. Lagarde and P. Whaite, "Darboux Frames, Snakes and Superquadrics: Geometry from the Bottom-Up", *Proceedings of the IEEE Workshop on Interpretation of 3D Scenes,* pp. 170-176, Austin, TX, November 19-27, 1989.

12 R. Fletcher, and C. Reeves, "Function minimization by conjugate gradients," *Computer Journal,* vol. 7, pp. 149-154, 1964.

13 D. Geiger and F. Girosi, "Parallel and deterministic algorithms for MRFs: surface reconstruction and integration," *IEEE Transactions on Pattern Analysis*

and *Machine Intelligence,* vol. 13, 401-412, 1991.

14 S. Geman and D. Geman, "Stochastic relaxation, Gibbs distributions, and the Bayesian restoration of images," *IEEE Transactions on Pattern Analysis and Machine Intelligence,* vol. PAMI-6, no. 6, pp. 721-741, November 1984.

15 W. E. L. Grimson, *From images to surfaces: a computational study of the human early visual system,* Cambridge, MA: MIT Press, 1981.

16 W. E. L. Grimson and T. Pavlidis, "Discontinuity detection for visual surface reconstruction," *Computer Vision, Graphics and Image Processing,* vol. 30, pp. 316-330, 1985.

17 H. W. Guggenheimer, *Differential Geometry,* New York: Dover Publications, Inc. 1977.

18 W. Hackbusch, *Multi-Grid Methods and Applications,* Berlin, Germany: Springer-Verlag, 1985.

19 J. Hadamard, *Lectures on the Cauchy Problem in Linear Partial Differential Equations,* New Haven, Connecticut: Yale University Press, 1923.

20 J. G. Harris, "Analog Models for Early Vision," Ph.D. Thesis, California Institute of Technology, May 1991.

21 M. R. Hestenes and E. Stiefel, "Methods of conjugate gradients for solving linear systems," *Journal of Research of the National Bureau of Standards,* vol. 49, no. 6, pp. 409-427, December 1952.

22 P. J. Huber, *Robust Statistics,* New York: John Wiley & Sons, 1981.

23 J. Jou and A. C. Bovik, "Improved initial approximation and intensity-guided discontinuity detection in visible-surface reconstruction," *Computer Vision, Graphics and Image Processing,* vol. 4, no. 3, pp. 292-325, September 1989.

24 S. Kirkpatrick, C. D. Gelatt and M. P. Vecchi, "Optimization by Simulated Annealing", *Science,* Vol. 220, No. 4598, pp. 671-680, May 13, 1983.

25 P. Lancaster and K. Salkauskas *Curve and Surface Fitting,* London: Academic Press, 1986.

26 D. Lee and T. Pavlidis, "One-dimensional regularization with discontinuities," *IEEE Transactions on Pattern Analysis and Machine Intelligence,* Vol. 10, No. 6, November 1988, pp. 822-829.

27 D. Lee, "Some computational aspects of low-level computer vision," *Proceedings of the IEEE,* Vol. 76, No. 8, August 1988, pp. 890-898.

28 M. A. Malcolm, "On the computation of nonlinear spline functions," *SIAM Journal of Numerical Analysis,* Vol. 14, 1977, pp. 254-282.

29 J. L. Marroquin, S. Mitter, T. Poggio, "Probabilistic solution of ill-posed problems in computational vision," *Journal of the American Statistical Association,* Vol. 82, No. 397, March 1987, pp. 76-89.

30 J. Meinguet, "Multivariate Interpolation at Arbitrary Points Made Simple", *Journal of Applied Mathematics and Physics,* Vol. 30, pp. 292-304, 1979.

31 T. Poggio, V. Torre and C. Kock, "Computational vision and regularization theory," *Nature,* vol. 317, no. 26, pp. 314-319, September 1985.

32 P. T. Sander and S. W. Zucker, "Inferring surface trace and differential structure from 3-D Images," *IEEE Transactions on Pattern Analysis and Machine Intelligence,* vol. 12, no. 9, September 1990, pp. 833-854.

33 L. L. Schumaker, "Fitting surfaces to scattered data," in *Approximation Theory II*, G. G. Lorentz, C. K. Chui, and L. L. Schumaker, eds., New York: Academic Press, 1976, pp. 203-268.

34 J. H. Shiau, *Smoothing Spline Estimation of Functions with Discontinuities*, PhD Thesis, University of Wisconsin-Madison, 1985.

35 D. Shulman and J. Y. Hervé, "Regularization of discontinuous flow fields," *Proceedings of the IEEE Workshop on Visual Motion*, Irvine, CA, March 20-22, 1989, pp. 81-86.

36 S. S. Sinha and B. G. Schunck, "Discontinuity preserving surface reconstruction," *Proceedings of IEEE Conference on Computer Vision and Pattern Recognition*, San Diego, California, June 4-8, 1989, pp. 229-234.

37 S. S. Sinha and B. G. Schunck, "Surface approximation using weighted splines," *Proceedings of the IEEE Conference on Computer Vision and Pattern Recognition*, Lahaina, Maui, Hawaii, June 3-6, 1991.

38 R. L. Stevenson and E. J. Delp, "Invariant recovery of curves in m-dimensional space from sparse data," *Journal of the Optical Society of America A*, vol. 7, no. 3, pp. 480-490, March 1990.

39 R. L. Stevenson, "Invariant reconstruction of curves and surfaces with applications in computer vision," Ph.D. Thesis, Purdue University, 1990.

40 R. L. Stevenson and E. J. Delp, "Fitting curves with discontinuities," *IEEE International Workshop on Robust Computer Vision*, Seattle, WA, Oct 1-3, 1990, pp. 127-136.

41 R. L. Stevenson and E. J. Delp, "Viewpoint invariant recovery of visual surfaces from sparse data," *Proceedings of the IEEE Third International Conference on Computer Vision*, Osaka, Japan, December 4-7, 1990, pp. 309-312.

42 R. L. Stevenson, G. B. Adams, L. H. Jamieson and E. J. Delp, "Three-dimensional surface reconstruction on the AT&T Pixel Machine," *Proceedings of the 24th Asilomar Conference on Signals, Systems, and Computers*, November 5-7, 1990.

43 R. L. Stevenson, and E. J. Delp, "Viewpoint invariant recovery of visual surfaces from sparse data," *IEEE Transactions on Pattern Analysis and Machine Intelligence*, Vol. 14, No. 9, September 1992, pp. 897-909.

44 R. L. Stevenson and E. J. Delp, "Surface reconstruction with discontinuities," *Proceedings of the SPIE Conference on Computer Vision and Graphics II*, Boston, Massachusetts, November, 1991.

45 D. Terzopoulos, "Regularization of inverse visual problems involving discontinuities," *IEEE Transactions on Pattern Analysis and Machine Intelligence*, Vol. PAMI-8, No. 4, July 1986, pp. 413-424.

46 D. Terzopoulos, "The computation of visible-surface representations", *IEEE Transactions on Pattern Analysis and Machine Intelligence*, Vol. PAMI-10, No. 4, pp. 417-438, July 1988.

47 A. N. Tikhonov and V. Y. Arsenin, *Solutions of Ill-Posed Problems*, Washington, D. C.: V. H. Winston & Sons, 1977.

48 G. Wahba, *Spline Models for Observational Data,* Philadelphia, PA: SIAM Press, 1990.

49 I. Weiss, "Shape reconstruction on a varying mesh," *Proceedings of the DARPA Image Understanding Workshop,* February 1987, pp. 749-765.

Three-Dimensional Object Recognition Systems
A.K. Jain and P.J. Flynn (Editors)
© 1993 Elsevier Science Publishers B.V. All rights reserved.

CAD-Based Object Recognition in Range Images Using Pre-compiled Strategy Trees

Farshid Arman[a] and J. K. Aggarwal[b]

[a]Siemens Corporate Research
755 College Road East
Princeton, New Jersey 08540 USA

[b]Computer and Vision Research Center
Department of Electrical and Computer Engineering
University of Texas at Austin
Austin, Texas 78712 USA

Abstract

This chapter considers the problem of recognizing an object in a given scene using a three-dimensional model of the object. The scene may contain several objects, each arbitrarily positioned and oriented. A laser range scanner is used to collect three-dimensional data points from the scene. The data collected is segmented into surface patches, and the segments are used to calculate various 3-D surface properties needed to identify the desired object. The CAD models are designed using the commercially available CADKEY and accessed via the industry standard IGES file format. The models are analyzed off-line to derive various geometric features, their relationships, and their attributes. A strategy for identifying each model is then automatically generated and stored. The strategy is applied at run-time to complete the task of object recognition. The goal of the generated strategy is to select the model's geometric features in the sequence which may best be used to identify and locate the model in the scene. The generated strategy is guided by several factors, such as the computational cost of estimating the features in the collected data, the reliability of the estimation, the detectability, and the relationships of the features. We present examples of the generated strategies and their use in recognition of parts in scenes containing multiple objects.

1. Introduction

The chapter addresses the problem of CAD-based object recognition. The objective is to use an object's CAD model to locate it in a scene containing several overlapping objects. The solution to this problem involves several steps (see Fig. 1), including steps to match the representations derived from the collected data to that of the given geometric model. Currently, most model-based vision systems use various pruning and searching strategies to reduce the total number of possible matches; however, most of these techniques depend on the expertise of the designer, the class of objects considered, and various search methods. This chapter introduces a novel scheme to systematically derive a matching strategy from a geometric model. The given models are 3–D CAD descriptions, and the 3–D data is collected using a laser range scanner.

The desired object in the scene is identified by matching the model surfaces inferred from the given CAD model to the segmented surface patches in the scene. Using the CAD model a recognition strategy is compiled which dictates which model surface to locate first, followed by its neighbors, and the second surface and its neighbors, and so on. If the desired model feature is not found in the scene due to partial occlusion or viewpoint, the strategy may dictate a next-best feature. Additionally, some model features may not be detectable because of their geometry, or the effects of low level processing techniques — such as smoothing — applied to the collected data. In such cases, the recognition strategy is used to automatically identify difficult-to-detect model features and to disregard those features at matching, thus enhancing the performance of the system.

To compile the strategy, the features of the CAD model are used to construct a tree, referred to as the *recognition tree* [3, 5]. The features form the leaves of the tree, and similar features are grouped to form the parent nodes. The features of the model topologically related are connected in the tree as well. Each connection in the the tree is assigned a weight representing such factors as the visibility, detectability, the frequency of occurrence, and the topology of the features. Once the recognition tree for a particular CAD model has been compiled off-line, the corresponding recognition strategy is derived on-line by using the leaves of the tree which represent the features of the CAD model and the weights assigned to each tree link.

The object recognition system presented here differs from previous ones in several respects (for a complete review of other object recognition systems see [4]). First, unlike most previous object recognition systems, the CAD system used to design the models is not an experimental one, but, a commercial 3-D CAD system, *CADKEY*, used widely in the industry ([8]). Second, IGES, Initial Graphics Exchange Specification, has been used as an interface to the CAD system to infer the geometric information necessary for object recognition. IGES is an industry standard developed by the National Institute of Standards and Technology to allow for the *transfer* (or interchange) of CAD models between various commercial CAD systems. Using *CADKEY* and IGES decreases the dependency of the vision system on any particular CAD modeler and increases its applicability. Third, the model description is used to systematically derive a matching strategy from a geometric model. Such a methodology enables the vision system to recognize a larger set of objects, without requiring any constraints or conditions even when a new model is added to the database. By using the recognition strategy, all possible matching combinations of sensory features and model features need not be considered. More importantly, by "pre-compiling" a recognition strategy, the vision system may perform more efficiently at run-time since less time is spent on model analysis in task execution. Fourth, using the recognition tree, the system is able to predict which CAD features will be not be detectable in scene. During recognition, the system will not expect to find those features, and so will not spend time during the search process. Finally, the matching strategy does not rely significantly on moment-based and boundary-based features, and unlike many previous approaches to object recognition, our system does not impose an unconditional one-to-one matching of sensory features and model features. Thus, the oversegmentation of surface patches and the partial occlusion of objects are easily tolerated. The presented system has been implemented and successfully tested using 17 models in numerous scenes.

The chapter is organized as follows: Section 2 provides an overview of the system presented here. Section 3 describes the recognition tree derived from each CAD model. Section 4 illustrates the details of the recognition strategy compilation, and section 5 describes the segmentation and other on-line procedures of the system. Section 6 presents the application of the recognition strategies to several scenes, and Section 7 presents the concluding remarks.

2. System Overview

The vision system presented here is divided into two components – on-line and off-line. Each component is further divided into several modules: data collection, low level image processing, data description, CAD model, model analysis, and CAD description (see Fig. 1).

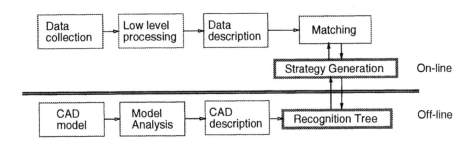

Figure 1. The modules of the proposed model-based object recognition system.

In the first module of the on-line component of the system, 3–D data is collected using a laser range scanner from a scene containing several overlapping objects. The collected data is then segmented, partitioning the pixels into several disconnected sets, each representing a homogeneous surface patch [10]. Each segmented surface patch is then classified by surface type as planar, concave, or convex. Surface attributes such as surface normal, surface area, the axis symmetry, and neighborhood relationships are then derived to form the description of the input data.

In the first module of the off-line component of the system, a commercial CAD system, *CADKEY*, is used to model numerous planar and non-planar (simple quadric) objects. Some of the object are off-the-shelf items and some have been constructed for experimental purposes (Fig. 2). The CAD models are accessed using IGES. The IGES description of the CAD model is analyzed to derive the features of the model and their relationships. Several surface attributes are then calculated for each CAD surface forming the CAD model description. The CAD surfaces are then used to organize the *recognition tree* (see Fig. 3). To identify a model in the given scene and to derive a recognition strategy, the corresponding recognition tree of the desired model is traversed. The goal of the generated strategy is to select the model's geometric features in the *sequence* which may best be used to identify and locate the model in the scene.

3. Recognition Tree

The organization of the recognition tree is as follows: Beginning from the leaves representing the features of the model, similar features are grouped to form a parent node in the tree, referred to as the **Feature Type** node. Similar feature types are grouped to form the next higher level, the **Feature Class** node. The feature class nodes suggest a representation scheme to best describe the feature types. For example, all planar regions in a CAD model may be grouped under the **planar** feature type, which is in turn grouped under the **surface** feature class. The models used in this research have surface information only; therefore, the **Surface** branch of the recognition tree has been implemented (Fig. 3). More abstract properties of the CAD model may be represented at the higher levels of the tree.

The recognition tree is used to rank the model features. Many factors must be considered in ranking of the model features. First and foremost is the choice of feature representation. Each representation has advantages and disadvantages, such as the computational cost of deriving features using the representation from the input data, the accuracy and uniqueness of the description, etc. The second factor to consider in ranking the features is the repetition of a particular feature [9]. Features which appear frequently are much less susceptible to occlusion and hence more "valuable."[1] The third factor is whether a feature is visible, or detectable, using the chosen representation, sensor type, and the applied low level processes, such as noise removal techniques. The last factor to consider is the relationships of the feature with its neighbors. Features with high ranking neighbors should be ranked higher than similar features with low ranking neighbors. This is necessary when the feature in focus is occluded or otherwise undetectable; in such cases the neighbors of the feature may be needed to localize the object. These factors are built into the organization of the recognition tree by assigning a weight to each

---

[1]Unique features are more important once such a feature has been located in a scene. Then, the feature may be used to isolate a particular model in a database of models; however, the problem at hand is to locate a *given* model in a scene. In such cases, a unique feature may easily be occluded; thus, it is less "valuable."

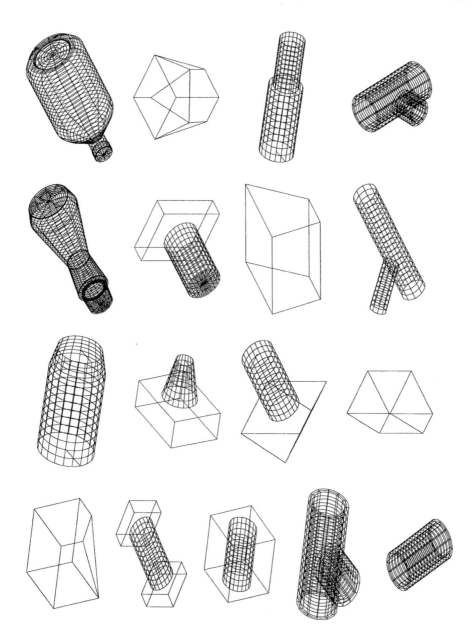

Figure 2. The database of CADKEY models.

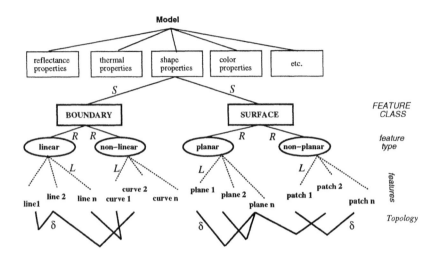

Figure 3. Recognition Tree where only the shape properties have been delineated explicitly. Symbols represent the weights assigned to each link.

branch of the tree.

The first level of the tree addresses the most fundamental considerations, the possible choices in representing shape properties depicted in the recognition tree as the *Feature Class* nodes. The possible choices considered are surface-based and boundary-based schemes. Of the factors listed above, accuracy and the cost of deriving the feature descriptions are used to assign weights to the first level of the recognition tree. The second level of the recognition tree further considers the choice of representation schemes. Shapes described using boundary-based representation may be sub-divided into linear and non-linear segments, and surface-based representations may similarly be subdivided into planar and non-planar patches. The factors used to assign weights to this level of the tree are the number of features and the feature repetition. For example, if there are more non-planar features in a model, then the non-planar features should be given a higher ranking than the planar features since a search for planar regions may not be rewarding. Furthermore, feature repetition is also a factor in deciding between planar and non-planar features. A planar feature repeated many times is valuable as a localization tool, since there is a lower probability of the feature's being occluded. In this example, then, the planar representation should be given a higher priority, and these features have a higher ranking. The last level of the tree focuses on feature detectability and topology to rank the CAD model's features. A detectability function is defined for each feature type. Simple features, such as a line, use the line length. More complex features, such as the surface patch, combine many properties, e.g. surface

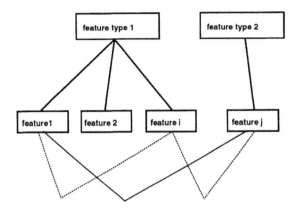

Figure 4. Virtual links in the recognition tree. The detectability of feature i is zero; therefore, its neighbors, feature 1 and feature j, are connected bypassing feature i. The dotted connections reflect the true topology, the solid connections represent the virtual links.

area, the number of cavities in the surface. Features of the CAD model that are physical neighbors are connected in the recognition tree to reflect the topology, and each connection is assigned a weight representing the detectability of the neighbor. If the neighbor is has zero detectability, then the features are connected, bypassing the common neighbor (see Fig. 3); such links are referred to as virtual links. Virtual links effectively simplify the CAD model by ignoring relationships that are not detectable either due to the sensor or to the low level processes such as segmentation. By *a priori* realizing such relationships in the CAD model, the vision system may perform more efficiently during the matching stage since no search time is spent on undetectable features and relationships.

The recognition tree example in figure 5 is a simple object with only two cylinders connected at an angle. The analysis of the CAD model has yielded two non-planar surfaces. The numbers on each edge of the tree are the calculated weights. The numbers in the boxes (leaf nodes) are the surface numbers, and the numbers below each box are the calculated attributes: surface classification, surface area, visibility ν, and detectability δ (shown as d) [1]. The topology of the two surfaces is preserved in the recognition tree by connecting the two leaf nodes by solid lines. The second example (fig. 6) introduces a more complex object. In the shampoo bottle used for this example, two conical surfaces form the main body (surface numbers 3 and 5 in the tree). The two surfaces are connected by a fillet which is detected as surface number 4. The visibility of this surface is zero [1]; therefore, the weight assigned to this surface is zero, and the surface numbers 3 and 5 are connected using a virtual link. During recognition, the system will not expect to find surface 4, and

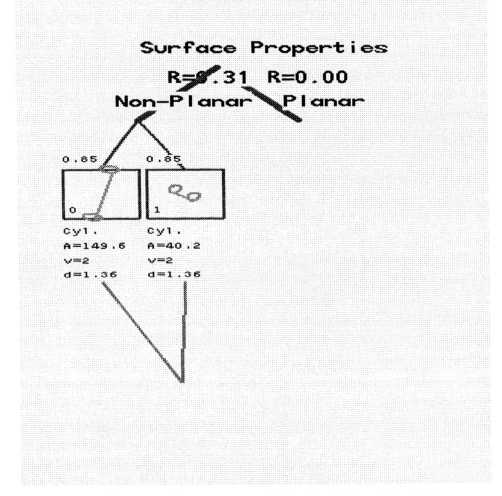

Figure 5. The recognition tree generated for the CAD model on the top.

will expect to locate surfaces 3 and 5 as neighbors during the search process.

4. Recognition Strategy Generation

The strategy is a series of filters that are invoked using the weights assigned to each link of the tree and by using the leaves of the tree. A filter is a set of conditions, which depend upon the model feature, used to compare the model features and the features from the scene. Specifically, there are two classes of filters. Unary filters examine unary constraints such as matching surface classifications (convex, concave, planar) and surface area. Since objects may be partially occluded and since data points are from a single view, the surface area of a surface patch from the scene is always expected to be less than the corresponding surface area of the model. For planar surface types, the number of edges of the face and the angles between the edges of the face are also used as filter conditions [2]. Binary filters check for constraints between two neighboring surfaces (binary constraints). The angle between the representative vectors of the two surfaces in the model is compared to the corresponding two surfaces in the scene. In addition, if the surfaces are non-planar, the ratios of the approximated curvatures are used as a constraint. The curvature values are not used directly, since the approximated values are very sensitive to noise and are often inaccurate [7]. Rather, our experiments have shown that the ratios of curvatures of two surfaces (applied to cylindrical surface types only) are an effective constraint when used in addition to other constraints. Following segmentation (section 5), the connected surface patches are grouped into several sets. When a filter is issued, each member of each set is compared to the filter conditions. If at least one member passes the constraints (two members for binary filters), then the set is allowed to continue. Otherwise, the set is deactivated.

Given the recognition tree of a given model, the recognition strategy is derived by the following algorithm:

```
1:    for all features          choose features according to weights
2:        FIND best feature      use unary filters
3:        for all neighbors      choose neighbors according to weights
4:            FIND best neighbor use unary filters
5:            examine relationship use binary filters
        next neighbor
    next feature
```

In Step 1, features are chosen in the order of the weight L (weight assigned to each leaf); in case of ties, surface areas are used. In Step 2, unary filters are used, and the sets which satisfy the filter conditions are passed on to the next step. In Step 3, the best neighbor of the first feature is chosen based upon L, and a second unary filter is issued (Step 4). The sets from the previous steps are tested against the constraints of this filter as before. Next, in Step 5, binary constraints are issued to check for the relations of the model feature and its chosen neighbor against the corresponding surface patches that passed the earlier two filters. Again, only the sets that satisfy the filter conditions are allowed to continue. This process continues

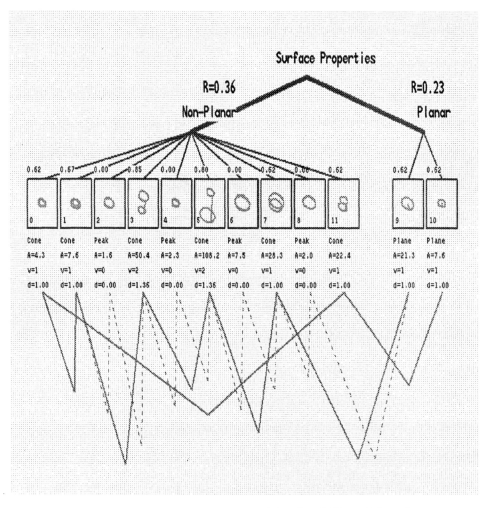

Figure 6. The recognition tree generated for the CAD model on the top.

for all other neighbors in order of the assigned L. At any stage, if none of the sets pass a particular filter (i.e., the model feature is not found in the scene), then the system breaks from the loop; in other words, the next best neighbor or the next best feature is chosen, and the search continues. As the system progresses through the succession of recognition filters, the number of candidate sets diminishes rapidly. If more than one set remains after all filters have been exhausted, then the set which has passed more binary filters is considered the successful match. If no sets have passed at least one binary filter, then no matches exist.

5. On-Line Processing

The range images are obtained from a Technical Arts laser range scanner [11] equipped with a translation stage where the objects are placed and scanned. The data points are segmented into a set of homogeneous surface patches [10] and several attributes are calculated.

The segmentation of the input data is achieved in two steps. The first is the low level segmentation module, where the local properties are extracted from the given input data and clustered into homogeneous regions. This module gives an initial over-segmented output. These over-segmented regions are merged in the second module using the final surface representations.

The low level segmentation module uses local information in a pyramidal clustering scheme. The clustering is performed on seven properties for each point in the range image. The seven properties are the surface normal vector and six intensity images generated by assuming a light source from the x-direction, the y-direction, etc.[2] In pyramidal clustering, the data is initially divided arbitrarily forming several clusters, and a representative value is derived for each cluster using the corresponding cluster members. Then the members of each cluster are moved from cluster to cluster iteratively to minimize the differences between the members and those of their cluster representative values. The pyramidal algorithm is an iterative procedure that clusters pixels with similar properties into groups in a hierarchical manner. Each of the seven images generated by the pre-processing stage is used by pyramidal algorithms *independently*, to result in seven initial segmentations of the input data. The seven segmentation outputs are then "added," resulting in a maximally partitioned image.

The segmentation resulting from the first stage represents a grouping of the local properties. The second stage of the procedure merges these regions based upon the surface representation and a homogeneity criterion described below. The representation used by the high level segmentation module is bivariate polynomials of up to the fifth degree:

$$z(\vec{a}, m, x, y) = \sum_{i+j \leq m} a_{ij} x^i y^j,$$

---

[2]Since the collected data is three dimensional, a 2–D intensity image may be generated by assuming a light source in an arbitrary point in 3–D space. Then, the *cos* of the angle between the vector representing the light source point and the surface normal vector may be used to generate a simulated intensity image.

where $\vec{a}$ is the parameter vector, $m \in \{0, 1, 2, 3, 4\}$ is the order of the surface, and (x, y, z) are the coordinates of the points. The merging stage is performed incrementally, starting with $m = 0$, planar representation of surfaces. All neighboring surface patches are considered for merging as follows: Two adjacent surface patches are approximated using bivariate polynomial of mth degree and then merged if parameters of one of the patches, when used to extrapolate over the neighboring patch, results only in a small error. When $m = 0$, this has the effect that all neighboring surface patches that are planar, with no discontinuities at their common boundary are merged into one patch. Then, m is increased by one, and the process of merging neighboring patches is repeated. This is continued until $m = 4$.

Once the segmentation is achieved, the desired description for each surface patch is then derived. The specific attributes for each surface patch are the surface area, surface type (convex, concave, planar), axis of symmetry (if non-planar), surface normal (if planar), approximation of the radius of curvature (if non-planar), and a list of neighboring patches [6, 1].

6. Experimental Results

This section presents several applications of the above strategies on several scenes (Figures 7 through 12). The segmentation procedure was applied using the same thresholds for all the examples presented; the average time was approximately 15 minutes on an IBM RS/6000-530. The automatic tree construction from the IGES files takes, on the average, approximately 15 seconds. The application of the generated strategy to the segmented image is in order of 2 seconds when the desired model actually exists in the scene, and 3 seconds when the actual model does not exist in the scene. No thresholds were adjusted from example to example. More than forty experiments have been performed using the database of CAD models (Fig. 2) in numerous scenes [1, 5]. The experiments involve objects that are partially occluded by other objects in the scene, objects that are on the top of the pile of objects, and cases where the surfaces were oversegmented. In addition, in instances when the desired CAD model did not exist in the scene, the strategy successfully exited, noting the absence of the desired model from the scene.

7. Conclusion

In this chapter we have addressed the problem of is CAD-based object recognition. Given a CAD model, the corresponding object in the scene is located. In general, the scenes contain several overlapping objects in arbitrary orientation and arrangement. A laser scanner is used to collect 3–D data from the scene. Once segmented into a set of surface patches, various surface attributes are calculated. These attributes are then used to match to the given CAD model in the matching stage. In an earlier, off-line process, the model's surfaces have been automatically derived using the IGES description of the CAD model, and various surface attributes and the relationships among the surfaces have been inferred. Using the derived description, a recognition strategy has been compiled once and stored. At

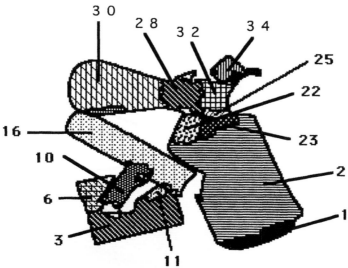

Figure 7. Scene 1: The input image on the top, the segmented output on the bottom.

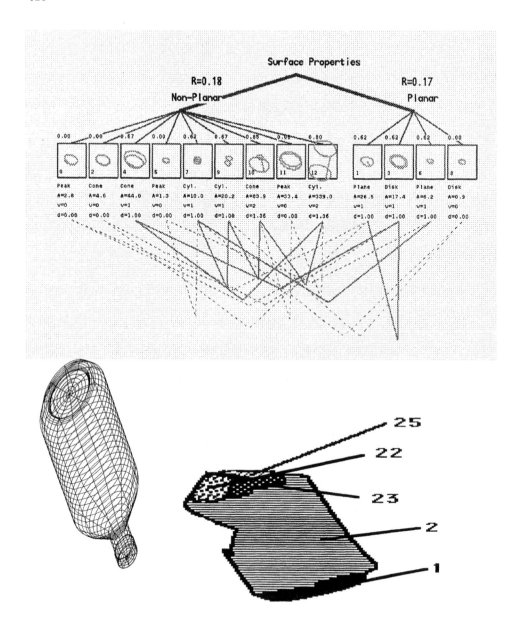

Figure 8. Application of the recognition tree (at the top) to scene 1, the model is at the bottom left, and the identified surfaces are displayed at the bottom right.

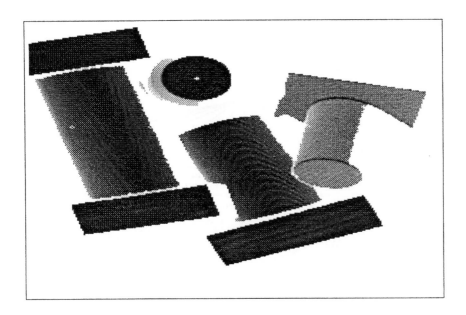

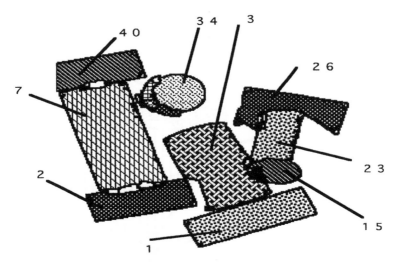

Figure 9. Scene 2: The input image on the top, the segmented output and calculated attributes on the bottom.

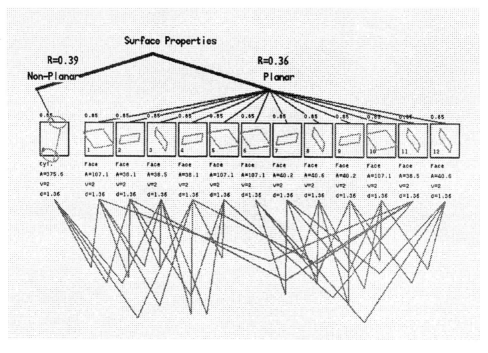

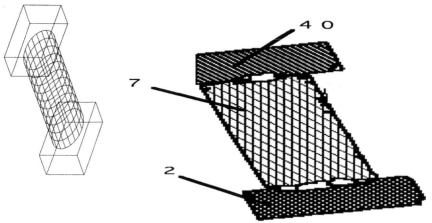

Figure 10. Application of the recognition tree (at the top) to scene 2, the model is at the bottom left, and the identified surfaces are displayed at the bottom right.

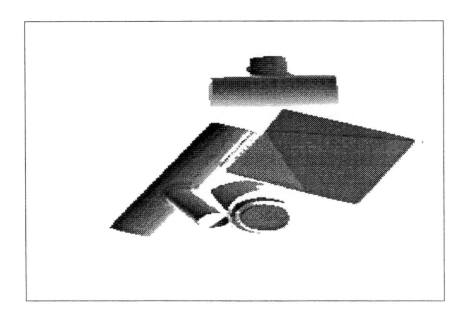

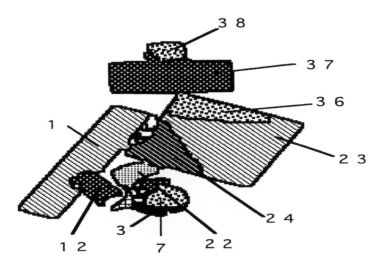

Figure 11. Scene 3: The input image on the top, the segmented output on the bottom.

132

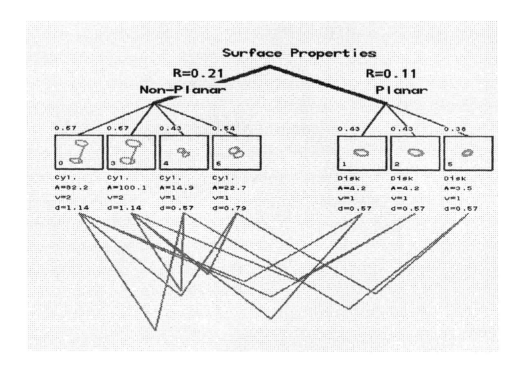

Figure 12. Application of the recognition tree (at the top) to scene 3, the model is at the bottom left, and the identified surfaces are displayed at the bottom right.

recognition time, the strategy is then used to guide the search. The desired object in the scene is identified by matching the surface patches derived from the given CAD model to the segmented surface patches in the scene. The strategy dictates which model feature to locate first, followed by its neighbors, and the second feature and its neighbors, and so on. The generated strategy is guided by several factors, such as feature detectability and the relationships of the features. Furthermore, in cases where the model features may not be detectable due to their geometry or to the effects of low level processing techniques, the recognition strategy is used to disregard the feature at the matching stage, enhancing the performance of the system.

The object recognition system presented here significantly differs from previous systems in several respects. First, unlike many previous object recognition systems, the CAD system used to design the models is not an experimental one; rather, a commercial 3-D CAD system has been used to design the models. Second, IGES has been used as an interface to the CAD system to infer the geometric information necessary for object recognition. Using *CADKEY* and IGES decreases the dependency of the vision system on any particular CAD modeler and increases its applicability. Third, the model description, derived automatically from IGES, is used to systematically derive a matching strategy from a geometric model. By using the recognition strategy, not all the possible matching combinations of sensory features and model features are necessarily considered, increasing the efficiency of the system. And, by "pre-compiling" a recognition strategy, the vision system may perform more efficiently at run-time since less time is spent on model analysis in task execution. Finally, the matching strategy is not significantly dependent on moment-based and boundary-based features, and unlike many previous approaches to object recognition, our system does not impose an unconditional one-to-one matching of sensory features and model features. Thus, the recognition system is not sensitive to the partial occlusion of objects, and the oversegmentation of surface patches is easily tolerated. The effectiveness of the system is demonstrated on numerous examples.

Acknowledgments

We would like to thank CADKEY Inc. for providing the CAD package. This research is supported in part by Army Research Office under contract number DAAL-03-91-G-0050.

REFERENCES

1   F. Arman. *CAD-Based Object Recognition In 3–D Range Images Using Pre-Compiled Recognition Strategy Programs*. PhD thesis, University of Texas at Austin, Department of Electrical and Computer Engineering, 1992.

2   F. Arman and J. K. Aggarwal. Object recognition in dense range images using a CAD system as a model base. In *IEEE conference on Robotics and Automation*, pages 1858–1863, Cincinnati, OH, May 13-18 1990.

134

3 F. Arman and J. K. Aggarwal. Automatic generation of recognition strategies using CAD models. In *IEEE Workshop on Directions in Automated CAD-Based Vision*, pages 124–133, Maui, Hawaii, Jun 2-3 1991.

4 F. Arman and J. K. Aggarwal. Model-based object recognition in dense range images - a review. In *ACM Computing Surveys (in press)*, 1992.

5 F. Arman and J. K. Aggrawal. CAD-based vision: Object recognition in cluttered range images using recognition strategies. *Computer Vision, Graphics, and Image Processing: Image Understanding*, (in press).

6 P. J. Besl and R. C. Jain. Invariant surface characteristics for 3–D object recognition in range images. *Computer Vision, Graphics, and Image Processing*, 33(1):33–80, 1986.

7 P. J. Flynn and A. K. Jain. On reliable curvature estimation. In *proceedings of Computer Vision and Pattern Recognition*, pages 110–116, San Diego, CA, Jun 4-8 1989.

8 P. J. Flynn and A. K. Jain. Bonsai: 3–D object recognition using constrained search. *IEEE transactions on Pattern Analysis and Machine Intelligence*, 13(10):1066–1075, Oct 1991.

9 C. Hansen and T. C. Henderson. Cagd-based computer vision. *IEEE transactions on Pattern Analysis and Machine Intelligence*, 11(11):1181–1193, Nov 1989.

10 B. Sabata, F. Arman, and J. K. Aggarwal. Segmentation of 3–D range images using pyramidal data structures. In *proceedings of International Conference in Computer Vision*, pages 662–665, Osaka, Japan, December 4-7 1990.

11 Technical Arts Corporation. *Technical Arts 100X Users Manual and Application Programming Guide*.

Three-Dimensional Object Recognition Systems
A.K. Jain and P.J. Flynn (Editors)
© 1993 Elsevier Science Publishers B.V. All rights reserved.

Active 3D Object Models

Ruud M. Bolle, Andrea Califano, Rick Kjeldsen, and Rakesh Mohan[a]

[a] Exploratory Computer Vision Group
I.B.M. Thomas J. Watson Research Center
P.O. Box 704
Yorktown Heights, New York 10598 USA

Abstract

We present a new approach for pruning the amount of search needed to match image features to object models. The technique relies on *active networks* which capture various visibility and geometric constraints between features of a model to prune these features from search space during matching. The networks, which can be efficiently implemented in Boolean logic, integrate harmoniously with our previous work in feature recognition and object matching. We propose a method of clustering model features (*vsets*) and four types of constraints which assist in building the networks.

We show, both analytically and empirically, the dramatic reduction in search provided by activation nets.

1. Introduction

Within the field of computer vision, one of the most challenging tasks is recognition (and localization) of three-dimensional objects given sensory data as input. Obviously, such a task cannot be achieved without some sort of internal representation of the 3D entities that can be expected in the scene. To that end, many schemes for 3D object modeling have been proposed; most of these schemes represent objects in terms of their geometric features (see Besl [6], for an overview). Object recognition then entails finding a correspondence between the geometric knowledge embedded in the object representations and some (geometric) description derived from the input data. In general, this matching problem is solved using a search process; i.e., interpretations of the data are found by pairing data features with model features such that a rigid body (or similarity) transformation maps the model features in a consistent way onto the data features (or vice versa). For the most general case where spurious features in the data are allowed, the expected search to match s data features to m model features, is exponential in the problem size. As given in [23], the amount of search is bounded from below and above by

$$O(m2^c + ms) < \text{Problem size} < O(m^2 2^s + ms) \qquad (1)$$

where c is the number of scene features that correctly belong to an object.

Hence, as Grimson [23] notes, "one of the hard parts of the recognition problem is separating out a correct subset of the data from the spurious data." Here, a correct subset of the data is a group of features that belongs to *one* of the objects in the scene. That is, partitioning (segmenting) the data features into sets, such that each set belongs to one of the objects in the scene, is a formidable and potentially time consuming part of the visual recognition task.

Data-driven clues can be used to determine such a partition of data features, e.g., sharp edges in the data. But clearly, an accurate and complete partition of scene features cannot be found without an internal representation of the 3D objects, i.e., model-driven segmentation. Moreover, partitioning the scene features involves matching these features to the features of the models – there seems to be no way to avoid this. Consequently, efficient matching schemes, i.e., schemes that avoid searching the total, exponential size, space of possible solutions, have been a topic of study since the early days of computer vision [4].

Two types of approaches to limiting the search space can be distinguished – geometric and spatial constraints on the scene-to-model feature pairings [23, 26]. As geometric constraints, frequently, unary and binary conditions on feature pairings are applied [23]. Unary constraints involve *one* scene-model pairing, e.g., feature type and properties of the features. For example, in [18] it is shown that simply introducing multiple types of features significantly decreases the amount of search that has to be done. Binary constraints (e.g., the angle between two lines) on the other hand, use relations between features to avoid incorrect pairings of data and model features (see, e.g., [22]).

Spatial constraints on the scene-model pairings is, for example, used by Lowe [30, 31]. A set of spatial heuristics is defined, called *viewpoint consistency constraints*, e.g., parallel features in the image belong to the same object. Groupings of the scene features are based on these heuristics.

In this chapter, we propose *new* model-driven constraints on scene-to-model feature matches –*visibility* constraints, *resolution* constraints, *position* constraints and *configuration* constraints. These constraints are captured in object models using *active networks* which prune the number of model features that scene features can be matched to. The activation a_i of a node in an active network is a Boolean variable indicating that object feature i is part of the search space, hence, can be matched to a scene feature. Nodes in the networks that have been matched to a scene feature, can enable or disable activation of other model feature nodes. Hence, the activation a_i of feature node i is a Boolean function $a_i = f_i(n_1, n_2, \ldots)$, where the n_i are nodes (model features) that are matched to scene features. (This is like our recognition network paradigm, described briefly in Section 2 and in detail in [8]. The networks behave like a distributed *winner take all* networks [20] and settle into one of several stable states that satisfy all the constraints.)

In the active object models, we distinguish various types of constraints:

1. **Visibility constraints**, encoding viewpoint information in the object models – e.g., if feature α of an object is visible, feature β probably is visible (barring obscuration) and feature γ is invisible.

2. **Resolution constraints**, representing resolution information about 3D objects; if feature α of an object is *not* visible then feature α' of finer resolution is probably also not visible.

3. **Location constraints**, indicating that a certain location **p** within an object hypothesis can contain one of m features $\alpha_1, \ldots, \alpha_m$.

4. **Configuration constraints**, which indicate that a certain feature α can be in one of m configurations $C_1, \ldots, C_m$ in an object model hypothesis. (This, e.g., allows for models with nonrigid body constraints.)

A typical example of an object where the latter two constraints are useful is a die. When one of the square planar faces of a die is detected in a scene, this face can contain one of six features – the patterns that correspond to the numbers 1 ... 6; also if a feature has been located on a face, it cannot be on any other visible (or invisible) face (both location constraints). On the other hand, if one of these six patterns (features) has been found in the image, then this pattern can be located on one of the six faces of a die (configuration constraint).

To formalize active models, we introduce the concept of *object viewsets* (referred to as *vsets*). A vset is the generalization of the object face in the polyhedral world, a surface patch and its bounding edges. That is, a vset is a minimal set V of object features, such that, if a specific feature α in V is visible from a viewpoint then any other feature β in V is visible with high probability. The active network of an object is compiled by partitioning the model features into a collection of vsets of multiple resolution levels. That is, the object surface is described as vsets, where vsets can be neighboring one another and occluding one another. In the network, vsets and features within vsets are connected through links that represent the constraints.

Active object models significantly help reduce the size of the search space by fusing information to form consistent interpretations in terms of viewpoint, resolution, and other information of object features.

The active object models are used within a vision system, briefly introduced in Section 2. The models contain both geometric information (Section 3) and constraint information (Section 4). In Section 5, we describe how these models are used in our indexing scheme. Finally, Sections 6 and 7 present an analysis of the dynamic behavior of the networks and experimental results, respectively.

2. A framework for recognition

Figure 1 represents an overview of our approach; the reader already familiar with the approach may skip to the next section. Recognition is structured as a hierarchy of layered and concurrent *parameter transforms* [3]. Features that are structurally independent, for instance, planes and linear 3D edges, are extracted in concurrent paths of recognition. The transforms can be decomposed into hierarchical layers within a path. Each transform accumulates evidence for feature hypotheses in an associated parameter space, *compatibility relations* identify evidential links between hypotheses, both within and between spaces. A global interpretation of

138

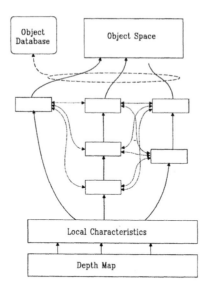

Figure 1. System architecture

the scene is achieved using *constraint satisfaction networks* [20] to fuse the evidence provided by these procedures.

The result is a highly parallel, uniform, and modular recognition system. The homogeneity allows different feature types (such as surfaces, curves, and surface/curve texture), and high and low-resolution images, or range and intensity images (e.g., a system such as [32]), to be easily integrated.

Due to the various noise sources and the correlated noise, created by the interaction of different features in the input [13], a large number of potential hypotheses about geometric features in the scene are generated. Many of these will be erroneous. The evidence for and against these hypotheses is integrated using a dynamically constructed constraint satisfaction network [20]. Each parameter space is instantiated as a subnetwork where units correspond to hypotheses concerning the existence of a feature. The links in the network are (1) bottom-up connections, representing support between input data points and the units, and (2) links between units themselves. The latter links can be inhibitory, in case the hypotheses are conflicting, or excitatory, in case of consistent hypotheses. For example, units generated by the same data points are connected by an inhibitory link, because they represent conflicting interpretations of the same data.

Each unit i computes an activation level u_i representing the confidence in the existence of the corresponding feature or object in the input. The activation level is updated as a function of support for the hypothesis and the presence of competing hypotheses. Quantitatively, for each iterative step t, the activation level of a node,

Figure 2. Real world scene

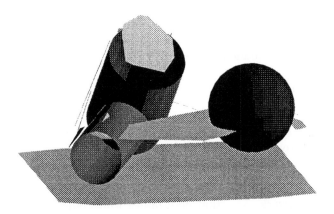

Figure 3. Reconstructed surfaces and curves

denoted by $u_i(t)$, is computed as

$$u_i(0) = 0 \tag{2}$$
$$u_i(t + 1) = u_i(t) + I_i + \sum_j w_{ij} u_j(t) - D_i \tag{3}$$

where I_i represents bottom-up reinforcement for unit i. That is, a measure of confidence that the corresponding hypothesis exists based only on data measurements (see [7]). The summation in Expression 2 embodies the collective inhibition and excitation of conflicting and cooperating hypotheses j on the hypothesis i. The weight factor w_{ij} is negative when hypothesis j conflicts with hypothesis i and positive when hypothesis j is supporting i. The term D_i is the decay term that suppresses spurious hypotheses.

If a unit in a space "survives" iteration, it creates or supports elements in higher-level spaces via the next parameter transform. Units also feed-back to their component features in lower-level spaces and to consistent hypotheses in parallel spaces. Thus, surviving interpretations form *stable coalitions* which represent globally consistent interpretations of the scene [20].

The top parameter space in Figure 1 represents object hypotheses. It is in this space where the best object hypotheses are found through iterative refinement (Expression 2) and matching features to object models. Active object models allow for more efficient matching of scene features to object hypotheses.

The system has been tested on numerous images. A recent example is a depth map taken from above from the scene in Figure 2. The surfaces shown in Figure 3 are extracted and the four objects successfully recognized. Detailed descriptions of the architecture can be found in [7, 8, 13].

3. Feature graphs

Our 3D object modeling scheme is integrated with our recognition network paradigm, described in the previous section. A model of a 3D object consists of two graph structures.

1. A relational graph, $\mathcal{M}_G$, that holds the geometric structure of the object.

2. An active network, $\mathcal{M}_A$, that determines which object features are active – i.e., which features are part of the search space.

The geometry of an object is represented by a feature graph, as the one in Figure 4. The graph is a pair $\mathcal{M}_G = (\mathcal{N}_M, \mathcal{R}_M) = (\mathcal{N}, \mathcal{R})$, a node $n_i \in \mathcal{N}$ in the graph represents a primitive feature of the object; currently, surfaces, their intersection curves, and surface texture. An arc $R_{ij} \in \mathcal{R}$ represents coordinate-free geometric relationships between feature n_i and feature n_j. Coordinate-free relationships do not depend on how an object is embedded in a 3D coordinate system, for example, R_{18} in Figure 4, indicates that the features n_1 and n_8 are related by the distance

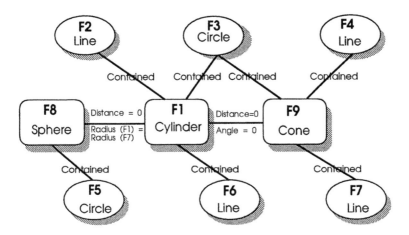

Figure 4. Feature graph of an object.

from the center of the sphere to the axis of the cylinder and by the relative size of their radii. For each pair of feature types, a set of such relationships is defined.

There are several advantages to the use of coordinate-free relationships.

- Matching a newly detected feature to an instance of an object can be done through pairwise comparison with previously matched features. When few features are matched, this is computationally efficient. No (partial) rigid body transformation between scene features and object features has to be computed, as is the case when the objects are modeled in an object-centered coordinate system.

- They allow for the representation of 3D features like the limbs of a cylinder. Such features, although they do not correspond to physical discontinuities of the object surface, are perceived as discontinuities in the image. For instance, in Figure 4, the circle n_5 is the limb of sphere n_8; the relation R_{58} indicates that the circle is contained in the sphere.

Of course, a disadvantage is that the positions of the features are not specified in an object-centered reference frame. When many features are matched to the models, this will be computationally more efficient; we are planning to incorporate an object-centered coordinate system as another module of an object model.

4. Active networks

The active network for an object $\mathcal{M}$ is a pair $\mathcal{M}_A = (\mathcal{N}_{\mathcal{M}}, \mathcal{F}_{\mathcal{M}}) = (\mathcal{N}, \mathcal{F})$. Here $\mathcal{N}$ is a set of K nodes corresponding to the features $\mathcal{N} = \{n_1, \ldots, n_K\}$ of the object

$\mathcal{M}$; $\mathcal{F}$ is a set of Boolean functions $\mathcal{F} = \{f_1, \ldots, f_K\}$. The network $\mathcal{M}_A$ can be in different states during the matching. Associated with each node, $n_i \in \mathcal{N}$, are two Boolean variables

$$m_i = \begin{cases} 1 & \text{if feature } n_i \text{ is matched "to" a scene feature} \\ 0 & \text{otherwise} \end{cases} \tag{4}$$

and

$$a_i = \begin{cases} 1 & \text{if feature } n_i \text{ is part of the search space} \\ 0 & \text{otherwise} \end{cases} \tag{5}$$

The activation a_i for each node n_i is a Boolean function of the matched features, i.e., $a_i = f_i(m_1, \ldots, m_K) \in \mathcal{F}$.

Let t be the iterative step in the iterative refinement of Expression 2. The set $\mathcal{N}'_{\mathcal{M}}(t) = \mathcal{N}'(t) \subset N$ are the activated features for an instantiated object model $\mathcal{M}$, the features that are part of the search space. Hence, it is the set of nodes (features) n_i for which a_i is 1. The objective is:

1. The size of the set $\mathcal{N}'(t) \subset \mathcal{N}$ is minimal for each instantiated model $\mathcal{M}$ at any time t during matching – i.e., the number of nodes n_i for which $a_i = 1$ is as small as possible.

2. The set $\mathcal{N}'(t) \subset \mathcal{N}$ is consistent in terms of the geometric relationships $\mathcal{R}$ of Section 3.

3. The set $\mathcal{N}'(\infty) \subset \mathcal{N}$ is both consistent in terms of the geometric relationships $\mathcal{R}$ and consistent in terms of the constraints.

The active models $\mathcal{M}_A$ hold the constraints on the scene-to-model matches. For each object, the active model is a multilevel feature network where the constraints are implemented as excitatory and inhibitory links. To form these networks, the nodes are divided up into clusters – referred to as *vsets* $\mathcal{V}_i$ (viewsets). If one central feature, $\alpha \in \mathcal{V}$, in a vset is visible from a viewpoint, then another feature $\beta \in \mathcal{V}$ in the vset is visible as well for most viewpoints. The vsets do not form a strict partitioning of the features, that is, a feature can be part of a number vsets at the same time. In the world of polyhedra, a vset is a planar surface patch with its delimiting edges. In richer visual worlds, using the relaxed definition, a vset is a bounded surface: a patch of object surface with its bounding curves (here a bounding curve can also be a surface limb).

For the moment, consider an object modeled by a collection of vsets $\{\mathcal{V}_i\}$ at a single resolution. The notion of object vsets allows us to define the *visibility constraints* between features of an object. Given three vsets, $\mathcal{V}_i$, $\mathcal{V}_j$, and $\mathcal{V}_k$, of an object, we distinguish two types of visibility constraints between features in these vsets. These visibility constraints are implemented by links in the active network, i.e.,

1. Excitatory links that allow features to become active based on viewpoint considerations. If $\alpha_i \in \mathcal{V}_i$ is matched, $\alpha_j \in \mathcal{V}_j$ should be active, that is, $m(\alpha_i) = 1 \rightarrow a(\alpha_j) = 1$.

2. Inhibitory links that express viewpoint inconsistencies between object features. If $\alpha_i \in V_i$ is matched $\alpha_k \in V_k$ should not be active, that is, $m(\alpha_i) = 1 \to a(\alpha_k) = 0$.

Two features in two vsets, V_i and V_j, of the object that contain a common feature (neighboring vsets) are connected by an excitatory link in the active model; the result is that if one feature in V_i is matched, all the features in V_j become active and part of the search space. For polygons, two contiguous faces are vsets that share a feature. For curved objects, vsets that share a feature have the property that if the central feature in one of the vsets is visible, on the average the central feature in the other vset is visible as well. These excitatory links in the active models ensure that scene features will only be matched to object features that are likely to be visible.

Next to excitatory links, we define links that inhibit certain scene-to-model matches. In this case, if two vsets V_i and V_k are viewpoint incompatible, then $m(\alpha_i \in V_i) = 1 \to a(\alpha_k \in V_k) = 0$. Or in words, if feature α_i in vset V_i is matched, all features α_k in vsets that are viewpoint incompatible with V_i are deactivated – $m(\alpha_i \in V_i) = 1 \to a(V) = 0$ for incompatible vsets. Clearly, if a vset V receives both excitation and inhibition, the features in the vset should be deactivated, i.e., $a(V) = 0$ (this is easily implemented with logical AND's) The result is that object features that become viewpoint incompatible during matching are dynamically hidden from the search. (In Appendix B we describe how visibility constraints can be generated from geometric object models.)

Active models provide a natural mechanism for incorporating *resolution information* in the 3D object representations. Each vset of an object may be grouped into an ordered set of resolution layers, $V = (V_1, V_2, \ldots)$. Here, V_1 are coarse resolution features, V_2 finer resolution features, and so on. Let us only look at two resolution layers in a vset, i.e., $V = (V_c, V_f)$, with V_c the coarse features and V_f the fine features. Again, activation of fine resolution features α' can be inhibited and excited through links:

1. Inhibition is guided through viewpoint inconsistency links. If V is deactivated, we have $a(V) = 0$ and, hence, $a(V_c) = 0$ and $a(V_f) = 0$.

2. Excitatory links that allow features to become active based on resolution considerations. If $\alpha \in V_c$ is matched, $\alpha' \in V_f$ should be active, that is, $m(\alpha) = 1 \to a(\alpha') = 1$.

Activation of model features based on resolution is very similar to activation based on viewpoint. In the former case, model features are activated that are expected to be found at a finer resolution, while in the latter, model features are activated that are expected to be found by a (small) change of viewpoint. Hence, in both cases, activation is spread through the model from matched features to features that become (better) visible by a change in viewing parameters – i.e., a change in the location of the observer or zooming in on the object.

Finally, we introduce *location* and *configuration constraints*. These type of constraints are different from viewpoint and resolution constraints, in that they can depend on the intrinsic geometry of objects. We define these constraints $a(\alpha)$ on a feature α as any Boolean function $a(\alpha) = f(m(\beta_1), m(\beta_2), \ldots)$, where the β's are other features associated with the same object. Imagine a die with six sides, $s_1, \ldots, s_6$, and six different patterns (features), $\alpha_1, \ldots, \alpha_6$. There are many examples of location and configuration constraints:

- If pattern α_1 has been found on a die hypothesis, the other sides contain different patterns. That is, $m(\alpha_1) = 1 \rightarrow a(\alpha_1) = 0$ and $a(\alpha_2) = 1, \ldots a(\alpha_6) = 1$.

- The patterns on opposite sides of a die add up to seven. This gives constraints like $m(\alpha_1) = 1 \rightarrow a(\alpha_1) = 0$ and $a(\alpha_6) = 0$; $m(\alpha_1) = 2 \rightarrow a(\alpha_2) = 0$ and $a(\alpha_5) = 0$; etcetera.

- If a pattern is detected, there probably is a side in the scene too, i.e., $m(\alpha_i) = 1 \rightarrow a(s_1) = 1$. Here, the patterns are loosely coupled to the sides.

Many other constraints are conceivable.

A few words about vnets and aspect graphs. The vset approach differs from the aspect graph approach [27–29] in that there is no requirement that a specific vset contains all the object features that are visible from a particular viewpoint. Rather, each aspect $\mathcal{A}$ of the object is formed by a union of several neighboring vsets. If an active model network is correctly built, the stable states of the network will correspond to aspects. On the other hand, a method to define inhibition links is to consider the aspects of an object. Typically, an aspect $\mathcal{A}$ of an object is the union $\bigcup \mathcal{V}_i$ of a collection of vsets. Inhibition can start when the viewing direction is completely established or partially established. That is, when it is known which aspect is viewed or when it is known that features in two (or more) neighboring aspects are viewed. In the former case, let the aspect be $\mathcal{A} = \bigcup \mathcal{V}_i$. Then if there is a feature $\alpha_i \in \mathcal{V}_i$, for which $m(\alpha_i) = 1$ for each vset $\mathcal{V}_i \in \mathcal{A}$, then the vsets $\mathcal{V}_k \notin \mathcal{A}$ should be deactivated. This can be made more involved by regarding "neighboring aspects" $\mathcal{A}_i$ and $\mathcal{A}_j$. The intersection $\mathcal{A}_i \cap \mathcal{A}_j$ is again a collection of vsets $\bigcup \mathcal{V}_i$. If in each of these vsets at least one model feature is matched to a scene feature, the vsets $\mathcal{V}_k \notin \mathcal{A}_i \cup \mathcal{A}_j$ should be inhibited. In Appendix A we show that an active model with viewpoint consistency and inconsistency links can be built with simple logic circuits. In this example, the inhibition links are based on aspects.

5. Matching to active models

Processing an input scene with the vision system, described in Section 2, results in a set of reconstructed features as shown in Figure 3. However, the features are not all reconstructed at the same time – different features are reconstructed at different rates by the iterative refinement of Expression 2. Object hypotheses are initiated by indexing into the database of objects. Here, features are used as indices in lookup tables; for this, approaches as in [16] or approaches similar to

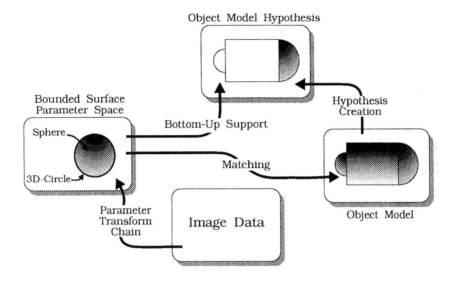

Figure 5. Creating object model hypothesis

Geometric Hashing [25] are under investigation. Hypotheses are instantiated in the top-level parameter space (Figure 1) which represent the matching of image features to model features. This process is summarized in Figure 5.

Once an object model is instantiated in the object parameter space, evidence for and against an object hypothesis i is accumulated through an activation level v_i in the constraint satisfaction of Expression 2. Supporting evidence comes from matched image features, and other consistent object hypotheses. Conflicting evidence arises from features which support more than one object hypothesis, and other inconsistent hypotheses.

For complex objects matching just the few features of an index is not enough to distinguish between the alternative interpretations; more features will have to be matched and their mutual relationships satisfied before the correct object hypothesis can gather enough activation. Thus unmatched features of existing hypotheses are also checked for matches to image features. If a match is found, the hypothesis is updated. Iteratively linking scene features with the instantiated hypotheses is equivalent to subgraph to graph matching and will pose some severe constraints on the tractability of the problem. Here the pruning of model features by active networks comes into play. Activation networks of instantiated models are queried to find the active features, and *only* these features are checked for potential matches. Through the connections of the active model networks, the object hypotheses will evolve toward stable states as more scene features find a suitable match.

Now, for example, let an object in the scene be a cube and let the object-model database contain several cubes – a simple cube and cubes with extra features on one or more of the faces of the cube. As the faces and edges of the cube are extracted, hypotheses representing all these cubes will be created. However, at a given scene resolution and observation viewpoint, the features needed to distinguish between these alternatives may not be reconstructed. Because of the active object network, it is known that features **should** be visible which would allow resolution of the conflict. Knowing these features, along with their relationships to matched features provides the information needed to change the viewing parameters – zooming in at certain locations in the scene [15] or viewing the object from a different viewpoint [1] – as needed to uniquely identify the correct cube.

6. Active networks: Dynamic behavior

The proposed model for active object models is characterized by the ability to dynamically reconfigure the number of model features that are matched against the features extracted from the image data. The total number of model features N_{Tot} evolve in time through a series of iterative stages $\{S(i)\}_{i=0}$. In the following we propose a dynamic model for this behavior.

At each iterative stage i, all reconstructed image features are matched only against the model features which are both *active* and *unmatched*. The number of such features in the model is $N_A(i) - N_M(i)$, where $N_M(i)$ is the number of matched model features, while $N_A(i)$ is the number of active model features. If a match is found, the corresponding active model feature has a transition from active to matched. After the matching, which, depending on $N_M(i)$ and $N_A(i)$ can require different completion times, the model feature network relaxes in fixed time to a new state by propagating the active-to-matched changes through the inhibitory and excitatory links. As a consequence some features will become active, while others will be inhibited and will no longer participate to the matching process. Let $N_I(i)$ indicate the number of inhibited features at iteration i. To extrapolate a general dynamical behavior for the active model networks, a number of assumption is required.

The first assumption deals with the probability that a given model feature in an active state is matched at iteration i. We will assume that this probability has a fixed value p_{AM} throughout the search. This implies that with a uniform probability density all visible features will be reconstructed from the image data. In fact, due to the iterative behavior of the extraction paradigm (see Section 2), the reconstructed features are approximately uniformly spread over time. (In the general case, where because of noise or occlusion not all features compatible with a given object pose are extracted, it is possible to set $p_{AM} = 0$ after a given number of features has been bound.) With these assumptions $N_M(i)$ evolves as:

$$N_M(i + 1) = N_M(i) + p_{AM}[N_A(i) - N_M(i)] \qquad (6)$$

since only active and unmatched features can become matched.

To write a similar iterative solution for the evolution of $N_A(i)$ and $N_I(i)$, a second assumption is required. We will assume that, on average, the number of excitatory and inhibitory output connections for each given model feature is approximately constant with values, L_E and L_I, respectively. In this case the iterative solution becomes:

$$
\begin{aligned}
N_A(i+1) &= N_A(i) + N_M(i)\, L_E\, P_{\bar{A}\bar{I}}(i) - N_M(i)\, L_I\, P_{A\bar{I}} \qquad &(7)\\
N_I(i+1) &= N_I(i) + N_M(i)\, L_I\, P_{I\bar{M}}(i) \qquad &(8)
\end{aligned}
$$

where

$$
N_M(i) = N_M(i) - N_M(i-1) \qquad (9)
$$

is the increment in the number of bounded model features at iteration i, and

$$
P_{\bar{A}\bar{I}} = \frac{N_{Tot} - N_A - N_I}{N_{Tot}} \; ; \; P_{\bar{A}\bar{I}} = \frac{N_A}{N_{Tot}} \; ; \; P_{I\bar{M}} = \frac{N_{Tot} - N_M - N_I}{N_{Tot}} \qquad (10)
$$

are the probability of a model feature at iteration i to be inactive and uninhibited, active and uninhibited, and uninhibited and unmatched, respectively. The normalizing terms in Expression 10 are required since:

1. Only inactive and uninhibited model features can become active.

2. Only uninhibited and active model features can have a transition from active to inhibited and thus inactive.

3. Only active and unmatched model features can become inactive.

The analytic solution for nonlinear systems of coupled finite difference equations, cannot be found in general. However, by recursively solving the system of equations in Expressions 6 and 7 with boundary conditions

$$
\begin{aligned}
N_M(0) &= 0, N_M(1) = 1 \qquad &(11)\\
N_A(0) &= N_{Tot}, N_A(1) = 1 \qquad &(12)\\
N_I(0) &= 0, N_I(1) = 0 \qquad &(13)
\end{aligned}
$$

for various values of the parameters, the following observations can be made.

1. The total number of iterations required for convergence ($N_M = N_A$) does not depend significantly on the number of features contained in the object model, this is shown in graphs of Figure 6, and Figure 7.

2. There is no significant modification of the evolutive behavior, except for faster or slower convergence, when the number of either excitatory or inhibitory output connections per feature is changed, see Figure 8. This supports our previous assumption to use average values for these parameters.

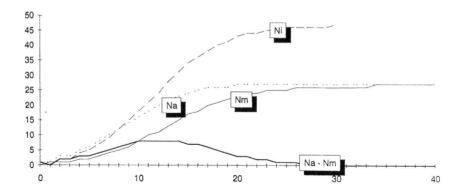

Figure 6. Evolution for a model with 400 features, 3 inhibitory and 3 excitatory links per feature

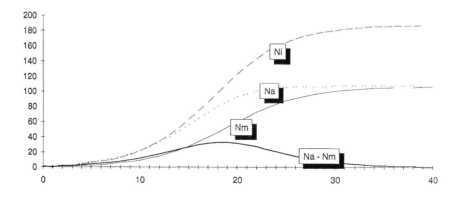

Figure 7. Evolution for a model with 400 features, 3 inhibitory and 3 excitatory links per feature

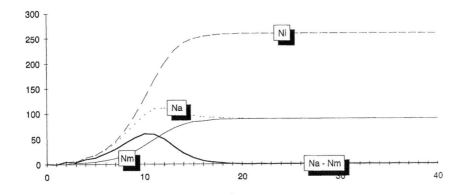

Figure 8. Evolution for a model with 200 features, 6 inhibitory and 6 excitatory links per feature

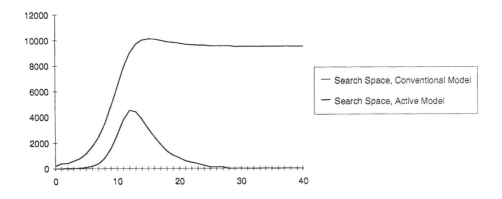

Figure 9. Search space size versus number of iterations

Let us now consider the number of active and unbound features that are used in the matching process at each iteration step. At each iteration step i, the number of matching operations required for each image feature, and hence the search space size, is proportional to

$$(N_A - N_M)N_M + N_M. \qquad (14)$$

In Figure 9 we show the behavior of the search space versus i for a model with 100 features and an average of 6 excitatory and 1 inhibitory connection per feature. This behavior is confronted with the same measure in the case of a traditional object model with all the unmatched features participating in the search. This behavior seems to be in very good agreement with the one shown by the experimental results of Section 7 and shows a significant computational advantage of active models.

7. Experiments

Figure 10 shows an active model for a die in a state of activation – the sides of the die are labeled with a square, a cross, etc. The features of the die are divided up into two resolution layers. Layer one contains the planar patches (side 1 through side 6); layer two contains the rest of the features, such as the edges. In Figure 10 side 1 and side 2 of the die are matched to planar surfaces reconstructed from the scene, indicated by the shaded left side of the nodes. This activates sides 5 and 6 (shaded right side of the nodes), and deactivates sides 3 and 4. This also activates and deactivates the appropriate features in the second resolution layer. That is, the patterns on side 1 and side 2 of the die are activated and those edges of the die that are formed by the intersection of sides that are either matched or activated.

To verify the analytical results, we modified the object matching parameter space (see Figure 1) to show the size of the search space. Using an actual object model (variations of the model of the die in Figure 10) and a complete set of features for the die, we plotted the size of the space searched by each successive feature matched to the model. The plots are averages of several runs, where image features attempt to match to the model in a random order.

The plots in Figure 11 show the number of active features at each match step. Without using active models, all features are initially active and the number decreases as each feature is bound. The plot also shows what happens when key features are used – key features are a small subset of features which must be matched before a model hypothesis is considered. Initially, just the key features are active but very quickly all other features are activated. Finally, a dramatic reduction in the number of active features is seen when a full active model is used. It is interesting to note what happens after 18 features are matched. Eighteen is the most features visible on the die at any one time. With the active model the number of active features has gone to zero, reflecting that no more features should be visible. The nonactive model is unaware of feature visibility, so even the last image feature to be matched to the model has several features to search through. Indeed, every subsequent image feature (features from other objects in the scene)

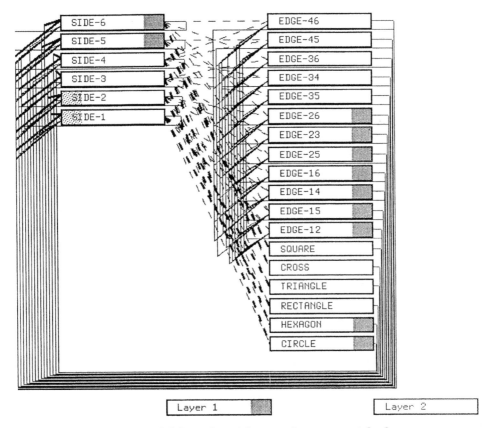

Figure 10. An active model for a die with some features matched

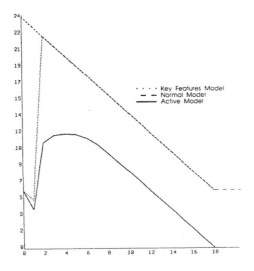

Figure 11. The number of active features.

152

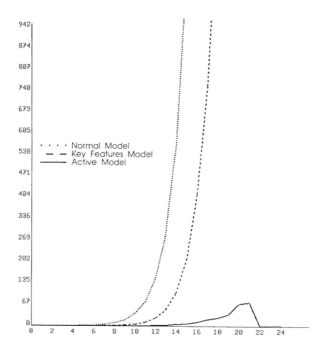

Figure 12. Work done matching in worst case.

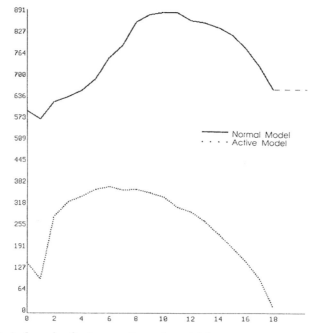

Figure 13. Work done by first commitment matching.

will have to search through these features as well, as their number will never decrease (demonstrated by the horizontal dashed line).

Figure 12 attempts to show the amount of computation done during each succeeding match in the worst case. Work includes verifying the unary constraints (feature type checking) between image and model features, as well as checking whether the relationships between the current image feature and previously matched features are satisfied. This is obviously approximate as there are several simplifying assumptions we had to make; for example, we do not know the exact amount of computation needed to verify a particular relationship, computation done by image features which do not succeed in matching to any model feature is not represented, etc.. Figure 12 compares the amount of computation needed to match scene features to the die both with and without active object networks. (Here because of various simplifying assumptions, the simulator was unable to recognize a completed model, and hence continued to attempt to match more than 18 features.) Key features delay the exponential explosion of computation, but do not avoid it. Active models avoid exponential behavior, even in the worst case.

The final set of graphs shows how active models curb search when used in combination with other search reduction techniques. Figure 13 shows the work done by features matching to a model using a maximal commitment approach. That is, as soon as a model feature is found whose relationships are satisfied, we commit to that match and stop our search. This relies on several assumptions such as complete information being contained in the relationships, and an optimized search order. With this technique, even the search of a static (nonactive) model is bounded, rather than exponential, however active models show a dramatic reduction in the amount of computation needed to match to this model. Again notice that the work done after 18 features are bound to this model will never decrease for subsequent image features (the horizontal dashed line), as the number of active features will not decrease. The active model is aware that the model is complete, so will not spend work attempting to bind future features to it.

8. Conclusions

We have introduced novel types of constraints on scene-to-model feature matching – viewpoint, resolution, location, and configuration constraints. These constraints dynamically inhibit model features from matching and depend on the model features that are already matched to scene features. This dynamic inhibition of model features is implemented as an active network where each stable state corresponds to a specific viewpoint, that is, a configuration where visible features are active and invisible features are nonactive. In the active network inhibition of features is expressed Boolean functions of matched model features. Hence, implementation of these constraints is computationally efficient.

These constraints on a scene-to-model match are a powerful addition to the constraints already in use for model matching, unary and binary constraints – constraints on single features and constraints on pairs of features. When a scene feature α_s is found, determining which model feature α_m to match to is a achieved

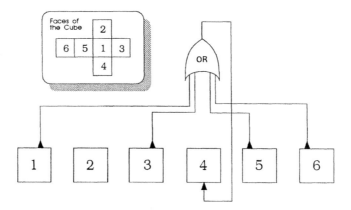

Figure 14. A matched plane activates neighboring views.

as sequence of tests. 1) The Boolean test if α_m active, e.g., is it likely to be visible. 2) Unary tests on α_m to determine if α_s is the right type, etcetera. 3) Binary tests involving previously matched features to determine the geometric correctness of the match.

Both an analysis of the dynamic behavior of the active model networks and experiments with the active models show very promising results.

Automatic model building is *not* the topic of this paper. Methods for building models from CAD/CAM representations can be found in the literature, e.g., [2, 21]; it is interesting to extend these methods to active object models.

A. An active network model for a cube

We show that active 3D models can be implemented using simple logic circuitry.

Let us consider a single layer of an object model of a cube, i.e., a given resolution scale. In the world of polyhedra, a vset corresponds to a face – a planar patch as central feature and its delimiting edges. Once a planar patch in a vset is matched, it can activate other vsets. Hence, the output of a vset is 1 if the planar patch in the vset is matched to a plane in the scene. An active model for a cube is instantiated when evidence for a cube is found in the scene.

Each edge is shared as a feature by two adjoining vsets. In the active model, vsets that share a feature – two contiguous planes of a polyhedron – are connected by weighted excitatory links. This is shown in Figure 14 for a cube where vset 4 is supported by four adjoining planes; using an OR gate, any of the neighboring planes of vset 4 can activate this vset. Hence, if any of these model planes is matched to a planar surface reconstructed from the data, vset 4 (the planar patch and its edges) will be activated and participate in the search space. By duplicating the OR gate circuit for each of the vsets, we obtain a circuit that activates any of the vsets if a neighboring plane is matched to a scene feature. Hence, the result of matching

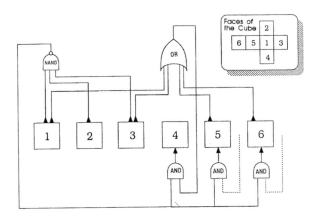

Figure 15. Only features in one viewpoint can be active.

a scene feature to plane 1 is that all the vsets of the cube, except vset 6, become active. The associated features are part of the search space.

However, these kind of excitatory connections are not sufficient. For example, vset 4 and vset 2 of the cube are both supported by plane 1. That is, if face 1 is matched to a plane in the scene, both vset 2 and vset 4 are activated and are part of the search space. But these vsets are viewpoint incompatible – they cannot be seen from the same viewpoint at the same time. The model network can be made more involved by adding inhibitory connections such that the previously described excitatory links are suppressed for conflicting viewpoints. An example of such a network is given in Figure 15, a largely incomplete viewpoint network is shown for the cube.

Note that the output of the NAND gate is zero when planes 1, 2, and 3 are matched and inhibits the lateral excitation of vset 4 (and vsets 5 and 6). The excitation of face 4 is caused by the fact that one of the adjoining planes of vset 4 is matched. Only three vsets that correspond to a viewpoint (or aspect) can be active and be part of the search space. If we complete the network by replicating the OR-AND sequence for each vset in the model and the NAND gating mechanism for each possible combination of planes that correspond to a viewpoint (the number of possible aspects), we obtain a network that has a finite number of stable and determined states. Each of these states corresponds to a set of active vsets consistent with a particular viewpoint. The excitatory and inhibitory links greatly reduce the search space during matching.

Using logic circuits, connection networks exhibiting different behaviors can be implemented.

B. Generating Visibility Constraints

Given a geometric model of an object, we can compute the visibility constraints.

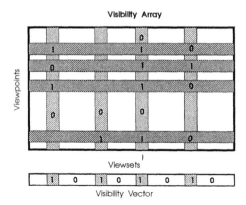

Figure 16. Computing the visibility vector.

The Gaussian viewing sphere is sampled at regular intervals, generating a fixed number m of viewpoints. At each of the viewpoints, visibility of the vsets is determined using geometric techniques such as ray tracing. For each model, a visibility array VA is generated with the viewpoints as rows and vsets as columns. Each entry VA_{ij} in the visibility array is set to 1 if $\mathcal{V}_j$ is visible from viewpoint i. Figure 16 shows a hypothetical visibility array.

For each vset $\mathcal{V}_j$, a visibility vector vv_j is computed by ORing all the rows that have 1 in column j. This vector has 1 in entry k if and only if $\mathcal{V}_k$ is visible with $\mathcal{V}_j$ from one or more viewpoints. This vector is complete, i.e., for all k, if $\mathcal{V}_k$ is visible with $\mathcal{V}_j$ from any viewpoint then $vv_j[k] = 1$, and if for any k, $vv_j[k] = 1$, then there is a viewpoint such that $\mathcal{V}_k$ and $\mathcal{V}_j$ are visible simultaneously. Given this completeness property of the visibility vector, the viewpoint incompatibility vector vi_j for $\mathcal{V}_j$ can be simply computed by taking the complement of the visibility vector. Element k of vi_j is 1 if and only if there is no viewpoint such that both $\mathcal{V}_j$ and $\mathcal{V}_k$ are visible simultaneously.

REFERENCES

1   J. Aloimonos and A. Bandyopadhyay, "Active vision," in *Proc. First Int. Conf. on Comp. Vision*, June 1987, pp. 35-55.
2   F. Arman and J.K. Aggarwal, "Object recognition in dense range images using a CAD system as a model base," in *Proc. IEEE Conf. on Robotics and Automation*, May 1990.
3   D.H. Ballard, "Parameter nets: A theory of low level vision," in *Proc. 7th Int. Joint Conf. on Artificial Intell.*, Aug. 1981, pp. 1068-1078.
4   D.H. Ballard and C.M. Brown, *Computer Vision*. New Jersey: Prentice-Hall,

Inc., 1982.

5   D.H. Ballard, "Reference frames for animate vision," in *Proc. 11th Int. Joint Conf. on Artificial Intell.*, Aug. 1989, pp. 1635-1641.

6   P.J. Besl, "Geometric modeling and computer vision," in *Proc. of the IEEE*, Vol. 76, No. 8, Aug. 1988, pp. 936-955.

7   R.M. Bolle, R. Kjeldsen, and D. Sabbah, "Primitive shape extraction from range data," in *Proc. IEEE Workshop on Comp. Vision*, Nov.-Dec. 1987, pp. 324-326.

8   R.M. Bolle, A. Califano, and R. Kjeldsen "A scalable and extendable approach to visual recognition," *IEEE Transactions on Pattern Analysis and Machine Intelligence*, Vol. 14, No. 5, May 1992, pp. 534–548.

9   R.M. Bolle and A. Califano, "A homogeneous approach to 3D recognition," in *Machine Vision, Acquiring Interpreting the 3D Scene*, H. Freeman (ed.). New York: Academic Press, 1990, pp 1–24.

10 R.C. Bolles and R.A. Cain, "Recognizing and locating partially visible objects: The local feature focus method," *Int. Journal of Robotics Research*, Vol. 1, No. 3, Fall 1982, pp. 57-82.

11 R.C. Bolles, P. Horaud, and M.J. Hannah, "3-DPO: A three-dimensional part orientation system," *Robotics Research: First Int. Symposium, 1984, pp. 413-424.*

12 A. Califano, "Feature recognition using correlated information contained in multiple neighborhoods," in *Proc. 7th Nat. Conf. on Artificial Intell.*, July 1988, pp. 831-836.

13 A. Califano, R.M. Bolle, and R.W. Taylor, "Generalized neighborhoods: A new approach to complex feature extraction," in *Proc. IEEE Conf. on Comp. Vision and Pattern Recognition*, June 1989, pp. 192-199.

14 A. Califano, Multiple window parameter transform for image feature extraction, U.S. Patent application, 1989.

15 A. Califano, R. Kjeldsen, and R.M. Bolle, "Data and model driven foveation," in *Proc. 10th Int. Conf. on Pattern Recognition,* June 1990.

16 A. Califano and R. Mohan, "Shape acquisition, recognition, and sub-part detection," in *Proc. AAAI*, May 1990.

17 C.H. Chen and A.C. Kak, " A robot vision system for recognizing 3-D objects in low-order polynomial time," *IEEE Trans. Systems, Man, and Cybernetics,* Vol. 19, No. 6, Nov./Dec. 1989, pp. 1535-1563.

18 G.J. Ettinger, "Large hierarchical object recognition using libraries of parameterized sub-parts," in *Proc. IEEE Conf. on Comp. Vision and Pattern Recognition*, June 1988, pp. 32–41.

19 S.E. Fahlman and G.E. Hinton, "Connectionist architectures for artificial intelligence," *IEEE Computer*, Jan. 1987, pp. 100-109.

20 J.A. Feldman and D.H. Ballard, "Connectionist models and their properties," *Cognitive Science*, Vol. 6, 1981, pp. 205-254.

21 P.J. Flynn and A.K. Jain, "CAD-Based computer vision: From CAD models to relational graphs," in *IEEE Transactions on Pattern Analysis and Machine Intelligence*, Vol. 13, No. 2, Feb. 1991, pp. 114–132.

22 W.E.L. Grimson and T. Lozano-Perez, "Model-based recognition and localization

from sparse range or tactile data," *The Int. Journal of Robotics Research*, Vol. 3, No. 3, Fall 1894, pp. 3-34.

23 W.E.L. Grimson, "The combinatorics of heuristic search termination for object recognition in cluttered environments," *IEEE Transactions on Pattern Analysis and Machine Intelligence*, Vol. 13, No. 9, Sept. 1991, pp. 920–935.

24 D.G. Hakala, R.C. Hillyard, P.F. Malraison, and B.F. Nource, "Natural quadrics in mechanical design," in *Proc. CAD/CAM VII*, Nov. 1980, pp. 363-378.

25 Y.C. Hecker and R.M. Bolle, "Is geometric hashing a Hough transform?" *IBM Tech. Rep.* RC 15202, Nov. 1989.

26 D.P. Huttenlocher and S. Ullman, "Object recognition using alignment," in *Proc. First Int. Conf. on Comp. Vision*, June 1987, pp. 102-111.

27 K. Ikeuchi, "Recognition of 3-D objects using the extended Gaussian image," in *Proc. 7th Int. Joint Conf. on Artificial Intell.*, Aug. 1981, pp. 595-600.

28 K. Ikeuchi and K. Sang Hong, "Determining linear shape change: Toward automatic generation of object recognition programs," in *Proc. IEEE Conf. on Comp. Vision and Pattern Recognition*, June 1989, pp. 450-457.

29 J. Koenderink and A. van Doorn, "The internal representation of solid shape with respect to vision," *Biological Cybernetics*, Vol. 32, 1979, pp. 211-216.

30 D.G. Lowe, *Perceptual Organization and Visual Recognition*. Boston, MA: Kluwer Academic Publishers, 1985.

31 D.G. Lowe, "Three-dimensional object recognition from single two-dimensional images, " *Artificial Intelligence*, Vol. 31, 1987, pp. 355-395.

32 R. Mohan and R. Nevatia, "Perceptual organization for scene segmentation and description," in *IEEE Transactions on Pattern Analysis and Machine Intelligence*, Vol. 14, No. 6, June 1992, pp. 616-652.

33 D. Sabbah, "Computing with connections in visual recognition of origami objects," *Cognitive Science*, Vol. 9, No. 1, Jan.-March 1985, pp. 25-50.

34 D. Sabbah and R.M. Bolle, "Extraction of surface parameters from depth maps viewing planes and quadrics of revolution," in *Proc. SPIE Conf. Intell. Robots and Comp. Vision*, Oct. 1986, pp. 222-232.

35 Technical Arts Corporation, *100X 3D Scanner: User's manual & application programming guide*. Redmond, WA: 1986.

36 A.J. Vayda and A.C. Kak, "Geometric reasoning for pose and size estimation of generic shape objects," in *Proc. IEEE Conf. on Robotics and Automation*, May 1990.

Three-Dimensional Object Recognition Systems
A.K. Jain and P.J. Flynn (Editors)
© 1993 Elsevier Science Publishers B.V. All rights reserved. 159

Image Prediction for Computer Vision

Octavia I. Camps [a] , Linda G. Shapiro [b] , and Robert M. Haralick [c]

[a] Dept. of Electrical and Computer Engineering
The Pennsylvania State University
University Park, Pennsylvania 16802 USA

[b] Dept. of Computer Science and Engineering, FR-35
University of Washington
Seattle, Washington 98195 USA

[c] Dept. of Electrical Engineering, FT-10
University of Washington
Seattle, Washington 98195 USA

Abstract

To recognize objects and to determine their poses in a scene we need to find correspondences between the features extracted from the image and those of the object models. Knowledge of the degree to which image features might break up or disappear under different lighting and viewing conditions is essential to automate the design of computer vision systems that can work with problems of practical complexity. In this paper we describe how the model-based vision system **PREMIO** (PREdiction in Matching Images to Objects) models some of the physical processes involved in the image formation and feature detection processes to predict and evaluate the features that can be expected to be detected in an image of an object in a semi-controlled environment. We also illustrate how these predictions can be used to successfully control the inherent combinatorial explosion of the relational matching approach commonly used in object recognition.

1. Introduction

To recognize objects and to determine their poses in a scene we need to find correspondences between the features extracted from the image and those of the object models. Most feature–based matching schemes assume that all the features that are potentially visible in a view of the object will appear with equal probability. The resultant matching algorithms have to allow for "errors" without really understanding what they mean, and usually get lost in the high combinatorics of the problem [12]. Thus, the application of these matching algorithms has been reduced to very simple tasks where only very few features are needed. Therefore the knowledge of the degree to which each feature might break up or disappear under different lighting and viewing conditions is essential to the design of computer vision systems that can work with problems of practical complexity. In this paper we describe how a new model-based vision system, **PREMIO** (PREdiction in Matching Images to Objects), models some of the physical processes involved in the image formation and feature detection processes to predict and evaluate the features that can be expected to be detected in an image of an object in a semi-controlled environment. We also illustrate how these predictions can be used to successfully control the inherent combinatorial explosion of the relational matching approach commonly used in object recognition.

2. PREMIO Overview

PREMIO uses CAD models, surface reflectance properties, light sources, sensor characteristics, and the performance of feature detectors to build a model called the **Vision Model**. The Vision Model is used to generate a model called the **Prediction Model** that will be used to automatically generate vision algorithms. The system is illustrated in Fig. 1. Our Vision Model is a more complete model of the world than the ones presented in the literature. It not only describes the object, light sources and camera geometries, but it also models their interactions.

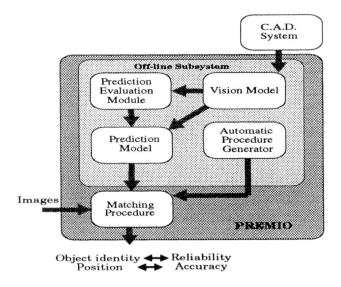

Figure 1. PREMIO: A Model-Based Vision System

The system has two major subsystems: an offline subsystem and a online subsystem. The offline subsystem, in turn, has three modules: a Vision Model generator, a feature predictor, and an automatic procedure generator. The Vision Model generator transforms the CAD models of the objects into topological models and incorporates them into the Vision Model. The feature predictor uses the Vision Model to predict and evaluate the features that can be expected to be detected in an image of an object, taken from a given viewpoint and under a given light source and sensor configuration. The output of the Prediction Module is organized as the Prediction Model. The automatic procedure generator takes as its input the Prediction Model and creates the matching procedure to be used for matching the image features against the object models. The online subsystem consists of the matching procedure generated by the offline subsystem. It uses the Vision Model, the Prediction Model, and the input images, first, to hypothesize the occurrence of an object and estimate the reliability of the hypotheses, and second, to determine the object position relative to the camera and estimate the accuracy of the calculated pose.

3. The Vision Model

A representation is a set of conventions about how to describe entities [25]. A *description* makes use of the conventions of a representation to describe some particular thing [25]. Finding an appropriate representation is a major part of any system-design effort, and in particular in the design of a machine vision system.

The vision model in a machine vision system is a *representation* of the world in which the system works. The entities that must be described by the representation of the world are the objects to be imaged and the characteristics that these images will have. In this paper we will limit ourselves to images that are formed when light is reflected off the surface of an object, such as photographs, but the concepts discussed here can be easily extended to other types of images such as range or X-ray images. The characteristics of an object image obtained by the usual optical means depends on several factors: the geometry of the object; the physical characteristics of the object surfaces; the position of the object with respect to the sensors, the light sources, and other objects; the characteristics of the light sources and the sensors; and ultimately on the characteristics of the processes that "observe" the image.

Following the previous discussion, we can divide the problem of designing the vision model into three subproblems: how to describe the objects, how to describe the light sources and sensors, and how to describe the processes that observe the images. Fig. 2 shows a block diagram of the PREMIO vision model.

3.1. The Object Model

An image of an object is a two-dimensional pattern of brightness. How this pattern is produced depends not only on the geometry of the object but also on the way that light interacts with the object surface and the sensors. Hence, we have divided the object model into two submodels: the topological object model and the physical object model. The topological object model describes the geometry of the objects in PREMIO's world. The physical object model describes how the light interacts with the object surfaces and the sensors according to the laws of physics.

3.1.1. The Topological Object Model

The topological object model describes the geometric characteristics of the objects in PREMIO's world, and the relations among their faces, edges and vertices. In many domains, all the possible objects to be recognized are known and can be, if they are not already, modeled using a CAD system. Hence, PREMIO assumes it

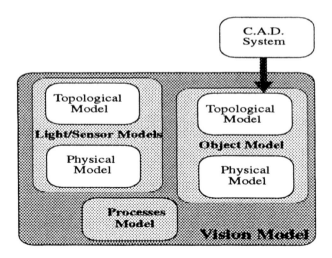

Figure 2. PREMIO's vision model.

has available CAD models of all the objects. A CAD system describes the geometry of the objects that it models, but not necessarily in the most adequate form for our application. CAD systems model 3D-objects and their component parts, providing view independent descriptions and interactive manipulation capabilities to observe and combine the different parts that can be automatically machined.

There are a number of commercial and experimental systems available for geometric modeling. A complete survey can be found in [21]. PREMIO uses the geometric modeler PADL2, designed by H. B. Voelcker and A. G. Requicha at the University of Rochester and distributed by Cornell University. This system is a constructive solid geometry (CSG) modeler, but has the ability to convert the CSG representation of any object to its boundary representation (BREP). Its primitives are: spheres, cylinders, cones, rectangular parallelepipeds, wedges and tori. Its main advantages are its capacity for fast wireframe drawing of the objects being designed that makes it very friendly to the user and its ability to convert any CSG representation to its boundary representation.

PREMIO's topological object model is a hierarchical, relational model similar to the one proposed by Shapiro and Haralick [22]. The object model is called a *topological object model* because it represents not only the geometry of the objects but also the relations among their faces, edges, and vertices. This information is redundant in the sense that it can be derived from the geometrical information, but it is included to speed up the system.

The model has five levels: a world level, an object level, a face level, a surface/boundary level, and an arc level. Fig. 3 shows a diagram of the topological object model. The world level at the top of the hierarchy is concerned with the arrangement of the different objects in the world . The object level is concerned

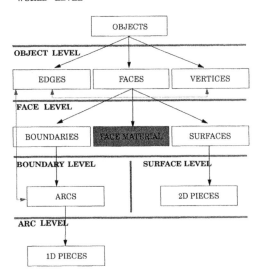

Figure 3. Topological Object Model Structure.

with the arrangement of the different faces, edges and vertices that form the objects. The face level describes a face in terms of its surfaces and its boundaries. The surface/boundary level specifies the elemental pieces that form those surfaces and the arcs that form the boundaries. Finally, the arc level specifies the elemental pieces that form the arcs.

To create the topological object model from the PADL2 model, we use the boundary file routines provided by PADL2. These routines give access to all the information concerning the face surfaces and the boundary arcs of the objects but do not provide a direct way to extract the boundary, topology and vertex information that we need. To find the boundaries of a face, its arcs must be grouped together to form closed loops. These loops are found using an algorithm developed by Welch [24] to find closed loops in an undirected graph. At the same time that the boundary information is obtained, the edge and the vertex information can be updated. The edge relation provides a way to relate two faces that have an arc in common, while the vertex relation relates all the arcs that have a vertex in common.

3.1.2. The Physical Object Model

Given the physical properties of a material it is possible to predict the properties of images of this material. PREMIO works with images produced by ordinary optical processes like photography or CCD cameras.

Buchanan [7] showed that the reflectance model presented by Cook and Torrance [10] is the most accurate when the light is completely unpolarized. However, polarized light is commonly used in vision tasks to suppress specular reflections from metal surfaces [3, 6]. Furthermore, monochromatic light is always polarized. Yi *et al.*'s proposed a light reflection model that is an extension of Cook and Torrance's model, but that incorporates the reflection effects [13] found when the incident light is partially or totally polarized, and it is the one used in PREMIO.

In Yi *et al.*'s model the light sources are assumed to be dimensionless point light sources that radiate in every direction. Let P be a point on a surface, N be the surface normal at P, and I_l^i be the incident light from light l. The reflection models predict the intensity of the reflected light at P, I^r, as the sum of two terms. The first term is the ambient component, and the second term contains the specular and diffuse components summed over the number of lights present:

$$I^r = I_a^i R_a f + \sum_l I_l^i (N \cdot L_l) dw_l^i (s R_s + d R_d) \tag{1}$$

where I^r is the intensity of the reflected light, I_a^i is the intensity of the incident ambient light, R_a is the ambient reflectance, f is the unblocked fraction of the hemisphere, I_l^i is the average intensity of the incident light l, N is the unit surface normal, L_l is the unit vector in the direction of the light l, dw_l^i is the solid angle of a beam of the incident light l, s is the fraction of reflectance that is specular, R_s is the specular bi-directional reflectance, d is the fraction of reflectance that is diffuse and R_d is the diffuse specular reflectance.

The model calculates the specular component by representing the object surface as a collection of many small planes which are called microfacets. The spectral reflectance function R_s is:

$$R_s = \frac{FDG}{\pi(N \cdot L)(N \cdot V)}$$

where F is the Fresnel term, D is the microfacet distribution, G is the geometric attenuation, N is the surface normal, L is the unit vector in the light direction and V is the unit vector in the viewer direction. Then

$$F = \rho_{\|} \cos^2 + \rho_{\perp} \sin^2$$

with

$$\tan^2 = \frac{I_\perp^i}{I_\|^i} \qquad \cos = N \cdot L$$

$$\rho_\perp = \frac{a^2 + b^2 - 2a\cos + \cos^2}{a^2 + b^2 + 2a\cos + \cos^2} \qquad \rho_{\|} = \rho_\perp \frac{a^2 + b^2 - 2a\sin\tan + \sin^2\cos^2}{a^2 + b^2 + 2a\sin\tan + \sin^2\cos^2}$$

$$a = \sqrt{\frac{\sqrt{c^2 + 4n^2k^2} + c^2}{2}} \qquad b = \sqrt{\frac{\sqrt{c^2 + 4n^2k^2} - c^2}{2}}$$

$$c = n^2 - k^2 - \sin^2$$

where n is the index of refraction and k is the extinction coefficient of the material; and

$$D = \frac{\exp^{-((\tan\delta)/m)^2}}{m^2 \cos^4 \delta}$$

where m is the root mean square slope of the facets, and δ is the angle between the surface normal N and the angular bisector between L and V, H. The parameter m

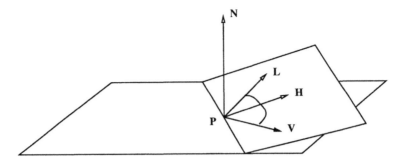

Figure 4. Light reflection on a microfacet.

is a scalar between 0 and 1 and is a measure of the roughness of the surface: the rougher the surface, the greater the value of m;

$$G = \min\left\{1, \frac{2(N \cdot H)(N \cdot V)}{(V \cdot H)}, \frac{2(N \cdot H)(N \cdot L)}{(V \cdot H)}\right\}$$

The diffuse component is modeled in both models as a constant denoted R_d. The diffuse component is independent of the angle of illumination, and hence it is taken to be equal to the specular reflection with the illumination on the surface normal $(L = V = N = H)$.

In summary, the physical object model consists of a set of constants for each of the object surfaces such that they completely specify the parameters needed in Yi *et al.*'s reflection model. These constants are:

R_a	ambient reflectance	R_d	diffuse reflectance
s	specular fraction	d	diffuse fraction
m	root mean square slope of microfacets	n	index of refraction
k	extinction coefficient		

3.2. The Light Sources and Sensors Model

Image formation occurs when a sensor registers radiation that has interacted with physical objects. The quality of an image is greatly affected by the sensor resolution and the scene illumination. In most industrial applications the imaging acquisition setup is known and controllable. That is, the type of light sources and sensors used as well as their relative geometric arrangement are known, if not fixed. This knowledge is incorporated into the vision model of PREMIO.

As with the object model, the light sources and sensors model can be divided into a topological and a physical model. The topological model describes the possible geometric or space configurations of the light sources and sensors with respect to the object being imaged. The physical light sources and sensors model describes their physical characteristics: the light wavelength and polarized components, the response of the sensor to the radiation input, and its resolution.

3.2.1. The Light Sources and Sensors Topological Model

The light sources and sensors topological model consists of all possible geometrical arrangements of the light sources and sensors with respect to the object reference frame. These configurations could be just a few or hundreds depending on the object and the particular application. In PREMIO the light sources and sensors are placed on spheres centered at the origin of the object coordinate system, called the *illumination* and *viewing* spheres. The radii of these spheres are large enough to contain the whole object. The points at the illumination and viewing spheres constitute a continuous space which is sampled uniformly.

3.2.2. The Light Sources and Sensors Physical Model

The light sources and sensors physical model describes properties and characteristics of the light sources and sensors that respond to the laws of physics and that are independent of their position in space. The light source physical model describes light characteristics such as the polarization and the wavelength distribution of the sources. The sensor physical model describes the magnification of the sensors and their transfer characteristics.

The Light Sources Physical Model

In PREMIO, the light sources are monochromatic point sources that irradiate partially or totally polarized light in all directions. Hence, the physical model of the light sources consists of the intensity values of the parallel and perpendicularly polarized components of the sources.

The Sensor Physical Model

The sensor physical model describes how images are formed. A gray-scale image is a two-dimensional pattern of brightness. To understand how this pattern is formed, we need to answer two questions: (1) *What determines where, on the image, some point on the object will appear?* and (2) *What determines how bright the image of some surface on the object will be?* The first question can be answered by making a first-order approximation of the camera as a "pinhole" camera. Using this approximation, images result from projecting scene points through a single point, the center of projection, onto an image plane at a distance f in front of the camera (See Fig. 5). Given a coordinate system, called the *camera coordinate system*, with its $z - axis$ parallel to the optic axis of the camera lens and such that the camera lens is at (x_0, y_0, z_0) and the image plane has equation $z = f + z_0$, the coordinates $(x', y', f + z_0)$ of the projection of a point with coordinates (x, y, z) onto the image plane are given by:

$$
\begin{aligned}
x' &= x^*/t^* \\
y' &= y^*/t^*
\end{aligned}
\tag{2}
$$

$$
\begin{pmatrix} x^* \\ y^* \\ t^* \end{pmatrix} = \begin{pmatrix} 1 & 0 & 0 & -x_0 \\ 0 & 1 & 0 & -y_0 \\ 0 & 0 & 1/f & -z_0/f \end{pmatrix} \cdot \begin{pmatrix} x \\ y \\ z \\ 1 \end{pmatrix}
\tag{3}
$$

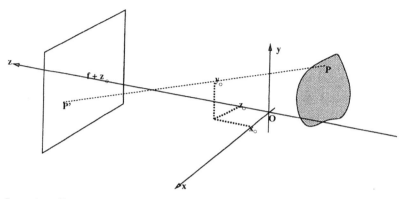

Figure 5. Imaging Geometry: Perspective Projection.

Next, we answer the second question of what determines how bright the image of some surface on the object will be. Different points on the object in front of the imaging system will have different intensity values on the image depending on the amount of the incident radiance, how they are illuminated, how they reflect light, how the reflected light is collected by the lens system of the imaging system, and how the sensor responds to the incoming light. The image intensity I is proportional to the scene radiance. The scene radiance, the amount of light emitted by the surface, depends on: (1) the amount of light falling on the surface; (2) the fraction of the incident light that is reflected; and (3) the geometry of the reflection, (*i.e.* the direction from which it is viewed as well as the direction from which it is illuminated). Mathematically, the image intensity of a given point P in a given object surface is given by [13]:

$$I = \int CS(\lambda)Q(\lambda)I^r(\lambda)d\lambda$$

where C is the lens collection factor, S is the wavelength dependent sensor responsivity, Q is the spectral distribution of the illumination source, λ is the wavelength of the light, and I^r is the reflected intensity at P given by equation 1. The lens collection factor, C, is the portion of the reflected light that comes through the lens system and affects the film. The lens collection C is given by [13]:

$$C = \frac{\pi}{4}\left(\frac{a}{f}\right)^2 \cos^4 \alpha$$

where f is the focal distance of the lens, a is the diameter of the lens, and α is the angle between the reflected ray from the object patch to the center of the lens. The sensor responsivity S, is in general a function of the wavelength of the incident light. However, for monochromatic sensors, it can be approximated by 1, independent of the wavelength of the incident light.

The *digital images* with which computers deal are arrays of integers. Hence, both the image domain and the image range have to be sampled. Sampling the domain involves the sampling interval and the pattern of the sampled points. The sampling interval determines how many sampled points (pixels) there are in the image. Television frames, for example, might be quantized into 450x560 pixels, while the resolution of general film is approximately 40 lines/mm, that is 1400x1400 pixels for a 35mm slide. The pattern into which the image range is sampled is called its tessellation. The tessellation pattern used in PREMIO, as with almost all machine vision systems, is a rectangular pattern. Sampling the range, corresponds to a *quatization* of the intensity values into a number of different *gray levels*. The number of levels determines the number of bits necessary to represent the intensity of a pixel. In PREMIO, images are quantized into 256 levels corresponding to 8-bit pixels.

In summary, a sensor physical model in PREMIO is specified by giving the following constants:

f	camera focal length	a	diameter of the camera lens
w	film width	h	film height
R	camera resolution		

3.3. The Processing Algorithms Model

An image is only as good as the processes used to extract the information that it contains. In the context of machine vision, these processes are the algorithms that are used in the feature extraction process. These algorithms and their performance must be included in the vision model of the system. Examples of processing algorithms are edge detectors, corner detectors, line finders, etc. The processing algorithms model in PREMIO consists of a set of gradient edge operator masks, a set of threshold values (used to decide when the gradient is large enough to assume that there is a local edge at a particular pixel), a thinning algorithm, a linking algorithm and a corner detection algorithm[13].

4. Feature Prediction Module

Given a vision model representing the world, the goal of the prediction module is threefold: (1) it must predict the features that will appear in an image taken of the object from a given viewpoint and under given lighting conditions; (2) it must evaluate the detectability of the predicted features; and (3) it must organize the data produced by (1) and (2) in an efficient and convenient way for later use. The visible features can be edges, corners, holes, or any complex relationship among these primitive features. The *detectability* of a feature for a given sensor and detector is the probability of finding the feature using that sensor. Additionally, the prediction module should evaluate the reliability and accuracy of the predicted features. The *reliability* of a feature is the probability of correctly matching the detected feature to the corresponding one in the model. The *accuracy* of a feature is a measure of the error or uncertainty propagated from the detected feature to a geometric property like the pose of the object.

4.1. Predicting Features

Given a three-dimensional object and the imaging geometry model, we want to determine which edges and surfaces are visible on the image of the object. There are two different approaches to the use of CAD-Vision models for feature prediction: synthetic-image-based prediction and model-based feature prediction.

Synthetic-image-based feature prediction consists of generating synthetic images and extracting their features by applying the same process that will be applied to the real images. Realistic image synthesis is a computer graphics area that has as its ultimate goal to produce synthetic images as realistic as photographs of real environments. Amanatides [1] surveyed different techniques used in realistic image generation. In general there is a tradeoff between processing time and realism. A particularly powerful technique used to achieve realism is ray casting: we cast a ray from the center of projection through each picture element and identify the visible surface as the surface intersecting the ray closest to the center of projection. Bhanu et al. [4] use ray casting to generate range images for their vision model. Other systems, [11, 15] use a different technique, called polygon mesh shading. This technique approximates the objects with polygon meshes, reducing considerably the required processing time while maintaining an appealing realism.

Model-based feature prediction uses models of the object, of the light sources and of the reflectance properties of the materials together with the laws of physics to analytically predict those features that will appear in the image for a given view without actually generating the gray tone images. Instead, only data structures are generated. This is a more difficult approach, but it provides a more computationally efficient framework suitable for deductive and inductive reasoning. This is the approach used by PREMIO.

4.1.1. Model-Based Feature Prediction

The model-based feature prediction task can be divided into three steps: The first step is to find the edges that would appear in the image, taking into account only the object geometry and the viewing specifications. The result is similar to a wireframe rendering of the object, with the hidden lines and surfaces removed. The second step is to use the material reflectance properties and the lighting knowledge to find the contrast values along the edges in a perspective projective image, and to predict any edges that may appear due to highlighted or shaded regions on the image. The third and last step is to interpret and group the predicted edges into more complex features such as line segments, triplets, corners, forks, holes, etc. A block diagram of the feature prediction module and its connections with the vision model is given in Fig. 6. Fig. 7 illustrates the model-based feature prediction process. Fig. 7(a) shows a raycasted image of an object. Fig. 7(b), (c) and (d) show the results of each of the steps.

Ponce and Chelberg [19] used this approach to predict features for generalized cylinder objects. However, they assumed an imaging system with orthographic projection, and they did not consider the effects of lighting or material properties.

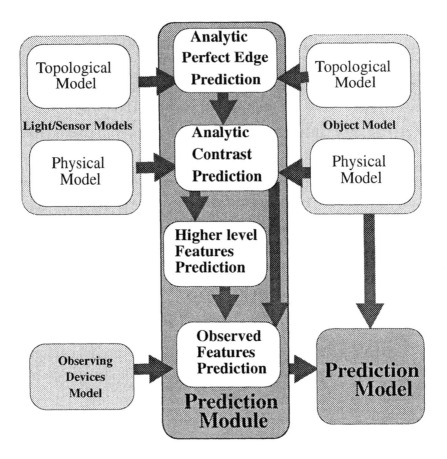

Figure 6. PREMIO's Feature Prediction Module.

4.1.2. Analytic Wireframe Prediction

The problem of determining which parts of an object should appear and which parts should be omitted is a well known problem in computer graphics. A complete survey of algorithms to solve the "Hidden-Line, Hidden-Surface" problem can be found in [23]. A particularly efficient way of solving this problem is using an analytical approach, by projecting the object surface and boundary equations onto the image plane and determining whether the resulting edges are visible or not. This approach obtains the edges as a whole, as opposed to the ray casting approach, which finds the edges pixel by pixel. The aim of the solution is to compute "exactly" what the image should be; it will be correct even if enlarged many times, while ray casting solutions are calculated for a given resolution. Hence this is the preferred method for our application.

In order to analytically predict a wireframe representation we need to introduce the following definitions [17]:

Def. 4.1 A **boundary** is a closed curve formed by points on the object where the surface normal is discontinuous.

Def. 4.2 A **limb** is a curve formed by points on the surface of the object where the line of sight is tangent to the surface, i.e. perpendicular to the surface normal.

Def. 4.3 A **contour** is the projection of a limb or a boundary onto the image plane.

Def. 4.4 A **T-junction** is a point where two contours intersect.

Def. 4.5 A **cusp point** is a limb point where the line of sight is aligned with the limb tangent.

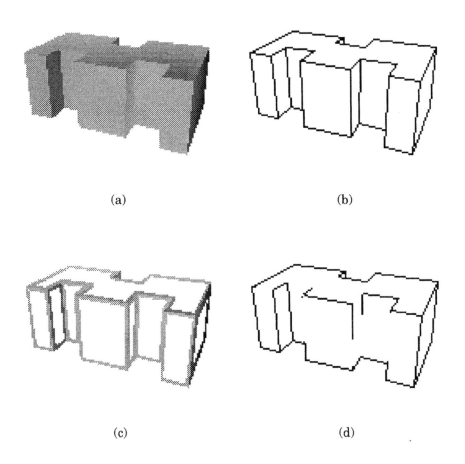

(a)   (b)

(c)   (d)

Figure 7. Model-Based Feature Prediction. (a) Raycasted Object. (b) Wireframe. (c) Intensity image along the edges. (d) Edges after applying an edge operator.

The edges in an image are a subset of the set of contours. A piece of a contour will not appear in the image if its corresponding boundary or limb is part of a surface that is partially or totally occluded by another surface closer to the point of view. Since the visibility of a contour only changes at a cusp point or a T-junction point, it follows that to find the edges on the image we have to take the following steps: (1) find all the limbs and cusp points, (2) project the boundaries and limbs to find the contours and all the T-junctions and (3) find for each cusp and T-junction point the object surface closest to the point of view.

Finding Limbs and Cusp Points

To find the analytical expressions for the limbs and cusp points we use an approach similar to the one used by Ponce and Chelberg [19], but designed for PADL2-modelable objects instead of generalized cylinders.

Let P_0 with object coordinates (X_0, Y_0, Z_0) be the projection center and let P with object coordinates (X, Y, Z) be a point on a limb on the surface S with implicit equation $f(X, Y, Z) = 0$. Then, the vector of sight $\vec{v}$ from P_0 to P is given by:

$$\vec{v} = (X - X_0, Y - Y_0, Z - Z_0) \tag{4}$$

and the normal $\vec{N}$ to the surface S is given by:

$$\vec{N} = \left(\frac{\partial f}{\partial X}, \frac{\partial f}{\partial Y}, \frac{\partial f}{\partial Z} \right). \tag{5}$$

In order for P to belong to the limb curve, P must be on the surface S and the line of sight must be perpendicular to the normal $\vec{N}$ at P. Hence the limb equations are given by:

$$\begin{cases} \vec{v} \cdot \vec{N} &= 0 \\ f(X, Y, Z) &= 0 \end{cases} . \tag{6}$$

Once the limb equations are solved, a limb can be expressed in a parametrized form:

$$\begin{cases} X &= X(t) \\ Y &= Y(t) \\ Z &= Z(t) \end{cases} \tag{7}$$

with $t_{min} \le t \le t_{max}$. Then, the tangent vector $\vec{T}$ to the limb is given by:

$$\vec{T} = \left(\frac{\partial X}{\partial t}, \frac{\partial Y}{\partial t}, \frac{\partial Z}{\partial t} \right) \tag{8}$$

Since a cusp point C is a limb point where the line of sight is aligned with the limb tangent, its coordinates must satisfy the following equations:

$$\begin{cases} \vec{T} \times \vec{v} &= 0 \\ X &= X(t) \\ Y &= Y(t) \\ Z &= Z(t) \end{cases} \tag{9}$$

with $t_{min} \le t \le t_{max}$.

The process of finding the limbs is performed in $O(s)$ time where s is the number of curved surfaces of the object.

Finding the contours and T-junctions

To find the contours we need to project the limbs and boundaries of the object onto the image plane; to find the T-junctions we need to intersect the resulting contours. The intersection detection problem for n planar objects has been extensively studied and it can be solved in $O(n \log n + s)$ time [20], where s is the number of intersections. In our case, the objects are the set of contours. The limb curves are either circles or straight lines, while the boundaries can be either straight lines, conics or more complex curves. Since the perspective projection of a straight line is another straight line, and the perspective projection of a conic is another conic, we can find a closed solution for the T-junctions between contours that result from projecting straight lines and conics [8]. To find the other T-junctions, a numerical approach must be used.

Determining Visibility

The next step is to determine the edges and surfaces that are hidden by occlusion. Appel [2], and Loutrel [17], have presented similar algorithms for analytical hidden line removal for line drawings. They define the *quantitative invisibility* of a point as the number of relevant faces that lie between the point and the camera. Then, the problem of hidden line removal reduces to computing the quantitative invisibility of every point on each relevant edge. The computational effort involved in this task is dramatically reduced by the fact that an object's visibility in the image can change only at a T-junction or at a cusp point. At such points, the quantitative invisibility increases or decreases by 1. This change can be determined by casting a ray through the point and ordering the corresponding object surfaces in a "toothpick" manner along the ray. Hence, if the invisibility of an initial vertex is known, the visibility of each segment can be calculated by summing the quantitative invisibility changes.

The quantitative invisibility of the initial vertex is determined by doing an exhaustive search of all relevant object faces in order to count how many faces hide the vertex. An object face is considered relevant if it "faces" the camera, i.e. its outside surface normal points towards the camera. A face hides a vertex if the line of sight to the vertex intersects the face surface and if the intersection point is inside the boundary of the face. To propagate the quantitative invisibility from one edge to another edge starting at its ending vertex, a correction must be applied to the quantitative invisibility of the starting point of the new edge. The complication arises from the fact that faces that intersect at the considered vertex may hide edges emanating from the vertex. This correction factor involves only those faces that intersect at the vertex. For an object with e edges, f faces, and with an average of 3 faces meeting at each vertex, the computational time needed to remove its hidden lines using this algorithm is $O(f + (2 \times 3 \times e))$.

4.1.3. Using Material and Lighting Knowledge

Once the wireframe of the object with the hidden lines removed has been obtained, the prediction module uses its knowledge about the sensor, the reflectance properties of the surfaces material, and the lighting conditions to predict the intensity of the reflected light in the neighborhood of the wireframe contours.

While the contours obtained in the previous stage are continuous, the digital images with which computers deal are discrete. Hence, the continuous contours have to be discretized according to the resolution of the sensor being modeled. This is accomplished by using the well known graphics algorithm due to Bresenham [5]. The intensity of the reflected light is computed in a neighborhood of the contour pixels by applying the reflectance surface model and the light model to the selected pixels. Furthermore, each of these pixels has associated one or more proximate contours and their corresponding 3D boundary or limb.

4.1.4. Using the Processing Algorithms Model

A line segment that is potentially visible in a set of views of an object may appear as a whole, disappear entirely, or break up into small segments under various lighting assumptions depending upon the contrast along the edges and the detector characteristics. The prediction module uses the processing algorithms model described in section 3 to predict which features can be detected. The module applies the modeled algorithms to the intensity pixels predicted in the previous stage while keeping track of their associated 3D features.

Simple features such as edges can be interpreted by themselves, or can be grouped to be considered as higher-level features. Matching *perceptual groupings* of features was suggested first by Lowe [18]. Henikoff and Shapiro [14] proposed using arrangements of triplets of line segments, called *interesting patterns*. Other useful high-level features are junctions and closed loops. In general a higher-level feature will be more useful than a lower-level feature to recognize and locate an object. Of course, there is a tradeoff between the amount of information that a feature represents and the cost of extracting that feature from the image.

4.2. Evaluating Predicted Features

After a feature is predicted, its potential utility must be evaluated. In order for a predicted feature to be useful, it has to be detectable in the image. For a given sensor and detector, the *detectability* of a feature is defined as the probability of finding the feature using that detector on an image taken with that sensor. We estimate the detectability of a feature by the frequency with which it shows up in a prediction, when the light and sensor positions are varied over the illumination and viewing spheres.

4.3. Output of the Predictor Module

For a given object, a configuration of light sources, and one or more sensors, the output of the predictor module is a hierarchical relational data structure similar to the one defined in section 2. This structure will be called a *prediction* of the object. Each prediction contains a set of image features, their attribute values such as detectability, and their corresponding three-dimensional features. The prediction also has five levels: the image level, an object level, a region level, a boundary level, and an arc level. The image level at the top of the hierarchy is concerned with the imaging conditions that generated the prediction, the general object position, the background information, etc. The object level is concerned with the different regions, edges and junctions that will appear on the image. The region level describes the regions in terms of their boundaries. The boundary level

 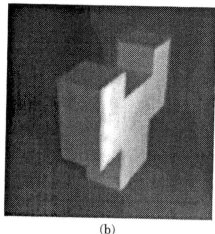

(a)   (b)

Figure 8. (a) Cube3Cut. (b) Fork.

specifies the arcs that form the boundaries of the regions. Finally, the arc level specifies the elemental pieces that form the arcs.

5. Illustrative Examples

In this section we will show the usefulness of the predictions obtained by PRE-MIO through some illustrative examples. Fig. 8 shows real images of *Cube3Cut* and *Fork*, two of the objects modeled in PREMIO. Fig. 9 shows the line segments obtained after processing the images with a sequence consisting of a Sobel edge operator, a connected shrinking, an edge linking, and a corner detection algorithms. As a result of the illumination conditions when the images were taken, and the use of "default" parameters in the image processing sequence, many segments are missing and others are broken.

The results shown in Fig. 9 are not surprising if we examine the predictions produced by PREMIO for similar viewing and illumination conditions. Fig. 10 shows some of the predictions generated for views within the same view aspect of the ones shown in Fig. 8 and with the light at approximately the same location. Even though the predicted views are different from the real ones being considered, they show which segments are more prone to disappear or appear fragmented.

PREMIO summarizes hundreds of predictions of an object into a single model M. The model consists of a set of features (segments), a set of relational tuples of features (junctions and chains of segments), and an attribute mapping for the features (midpoint coordinates, length, and orientation statistics of the segments). The features and the relational tuples are ranked in decreasing order of detectability. Fig. 11 shows visualizations of the sets of features of the models of Cube3Cut

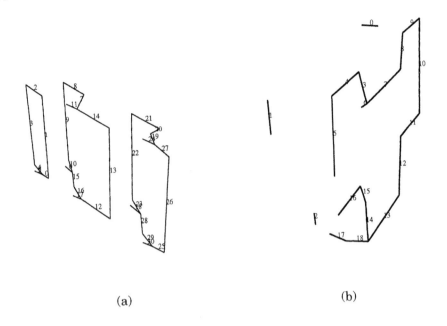

(a)

(b)

Figure 9. (a) Cube3Cut segments. (b) Fork segments.

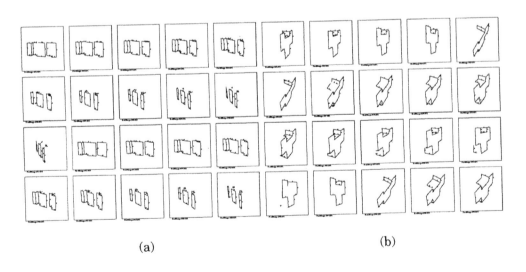

(a)

(b)

Figure 10. (a) Cube3Cut: predicted images. (b) Fork: predicted images.

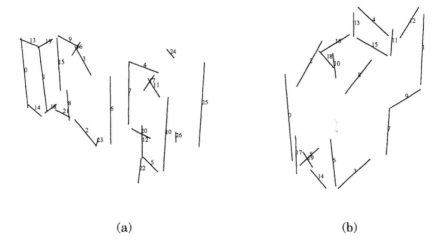

Figure 11. Cube3Cut and Fork models (segments). (a) Cube3Cut. (b) Fork.

and Fork respectively. The features are drawn as segments with their mean attribute values. The numbers shown by the segments are the feature ID's indicating the relative detectability, with the lower the number, the higher the detectability.

An example of how these predictions can be used to reduce the combinatorial explosion of the relational matching problem can be found in [8, 9]. There, the relational matching problem is framed as the Bayesian problem of finding a set of correspondences h between features of a model M and features of an image I, such that the *a posteriori* probability $P(M, h|I)$ that the given image I is an observation of M under the transformation h is maximized. Furthermore, it is shown that this problem is equivalent to maximizing the joint probability $P(M, h, I)$, since the maximization does not involve I.

In order to solve this problem, we need first to estimate the probability distribution $P(M, h, I)$ and then find the set of correspondences h such that the probability is maximum. Given a model M, the predicted images produced by PREMIO are observations of the model, each one having associated a different set of correspondences. Thus, these predictions correspond to samples of the probability distribution $P(M, h, I)$ and they can be used to estimate the parameters of the probability distribution (See [8, 9].). A set of correspondences h that maximizes the $P(M, h, I)$ probability, can be found by using an iterative-deepening-A* (IDA*) search [16] in a tree where each node represents a model feature, the branches represent the possible image feature assignments, and a path from the root to a leaf represents a possible set of feature correspondences. The IDA* algorithm consists of a sequence of depth-first searches. It starts with an initial threshold value equal to the estimated probability for a path starting at the root. In each iteration, the algorithm is a pure depth-first search, cutting off any branch that has an estimated probability smaller than the current threshold value. If a solution is expanded, the algorithm is finished. Otherwise, a new threshold value is set to the maximum

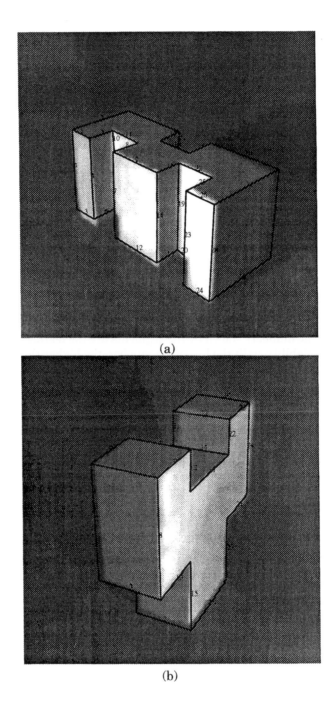

(a)

(b)

Figure 12. Cube3Cut and Fork pose estimation. (a) Cube3Cut. (b) Fork model.

estimated probability that was below the previous threshold, and another depth-first search is begun from scratch. As in the well known A* heuristic search, if the estimated probability is an overestimate of the real probability, IDA* finds the optimal solution with time complexity equal to an exponential function of the error of the estimate. The advantage of IDA* over A* is that since each iteration of the algorithm is a depth-first search, the memory complexity is a linear function of the depth of the solution, instead of being exponential. The number of nodes opened by IDA* is asymptotically the number of nodes opened by A*, provided that the tree grows exponentially. In practice, IDA* runs faster than A*, since its overhead per node is less than the overhead for A*.

It was found that the use of the predictions resulted in a pruning ratio, defined as the percentage of pruned paths relative to the total number of opened paths in the tree, between 40% and 60%.

Fig. 12 shows the estimated wireframes obtained by using 11 and 10 feature correspondences superimposed on the images of Cube3Cut and Fork respectively. For Cube3Cut, only 51 paths were opened and 29 of these were pruned, resulting on a pruning ratio of 56.86% and an execution time of 1.75 sec in a SPARC station 2. For Fork, 196 paths were opened and 126 of them were pruned, resulting in a pruning ratio of 64.29% and an execution time of 2.9 sec.

6. Conclusion

The CAD-based vision system PREMIO combines techniques of analytic graphics, CAD models of 3D objects and knowledge of surface reflectance properties, light sources, sensors characteristics, and the performance of feature detectors to predict and evaluate the features that can be expected to be detected in an image. The predictions generated in this way are powerful tools for object recognition. These predictions can be used to dramatically reduce the search space in a feature-based object recognition algorithm.

REFERENCES

1   J. Amanatides. Realism in computer graphics: A survey. *IEEE Computer Graphics and Applications*, 7:44–56, January 1987.

2   A. Appel. The notion of quantitative invisibility. In *Proc. ACM National Conference*, pages 387–393, 1967.

3   B. Batchelor, D. Hill, and D. Hodgson. *Automated Visual Inspection*. IFS Publications Ltd., Bedford, UK, 1985.

4   B. Bhanu, T.Henderson, and S.Thomas. 3-D model building using CAGD techniques. In *Proc. IEEE Computer Vision and Pattern Recognition*, pages 234–239, June 1985.

5   J. Bresenham. Algorithm for computer control of digital plotter. *IBM Syst. J.*, 4(1):25–30, 1965.

6   A. Browne and L. Norton-Wayne. *Vision and Information Procesing for Automation*. Plenum Press, New York, N.Y., 1986.

7 C. Buchanan. Determining surface orientation from specular highlights. Master's thesis, Dep. Comp. Sc., Univ. of Toronto, Toronto, Ontario, Canada, 1986.

8 O. I. Camps. *PREMIO: The Use of Prediction in a CAD-Model-Based Vision System*. PhD thesis, Department of Electrical Engineering, University of Washington, Seattle, Washington, 1992.

9 O. I. Camps, L. G. Shapiro, and R. M. Haralick. Object Recognition Using Prediction and Probabilistic Matching. In *Proc. of the IEEE/RSJ International Conference on Intelligent Robots and Systems*, pages 1044–1052, Raleigh, North Carolina, July 1992.

10 R. Cook and K. Torrance. A reflectance model for computer graphics. *ACM Trans. on Graphics*, 1(1):7–24, January 1982.

11 P. J. Flynn. *CAD-Based Computer Vision: Modeling and Recognition Strategies*. PhD thesis, Michigan State University, 1990.

12 W. E. L. Grimson. The combinatorics of object recognition in cluttered environments using constrained search. In *Proc. of the International Conference on Computer Vision*, pages 218–227, 1988.

13 R. Haralick and L. Shapiro. *Computer and Robot Vision*. Addison-Wesley, 1992.

14 J. Henikoff and L. Shapiro. Interesting patterns for model–based matching. In *ICCV*, 1990.

15 P. Horaud and R. Bolles. 3DPO: A system for matching 3-D objects in range data. In A. Pentland, editor, *From Pixels to Predicates*, pages 359–370. Ablex Publishing Corporation, Norwood, New Jersey, 1986.

16 R. E. Korf. Search: A survey of recent results. In H. E. Shrobe and The American Association for Artificial Intelligence, editors, *Exploring Artificial Intelligence*, chapter 6, pages 197–237. Morgan Kaufmann Publishers, Inc., 1988.

17 P. P. Loutrel. A solution to the hidden-line problem for computer-drawn polyhedra. *IEEE Trans. on Computers*, 19(3):205–210, March 1970.

18 D. G. Lowe. Three–dimensional object recognition from single two–dimensional images. *Artificial Intelligence*, 31:355–395, 1987.

19 J. Ponce and D. Chelberg. Finding the limbs and cusps of generalized cylinders. *Int. J. Comp. Vision*, April 1987.

20 F. P. Preparata and M. I. Shamos. *Computational Geometry: An Introduction*. Springer-Verlag New York Inc., 1985.

21 A. A. G. Requicha and H. B. Voelcker. Solid modeling: A historical summary and contemporary assesment. *IEEE Computer Graphics and Applications*, pages 9–24, March 1982.

22 L. Shapiro and R. Haralick. A hierarchical relational model for automated inspection tasks. In *Proc. 1st IEEE Int. Conf. on Robotics*, Atlanta, March 1984.

23 I. E. Sutherland, R. F. Sproull, and R. A. Schumacker. A characterization of ten hidden-surface algorithms. *Computing Surveys*, 6(1), March 1974.

24 J. T. J. Welch. A mechanical analysis of the cyclic structure of undirected linear graphs. *Journal of the Association for Computing Machinery*, 3(2):205–210, April 1966.

25 P. Winston. *Artificial Intelligence*. Addison-Wesley Publishing Company, second edition, July 1984.

Three-Dimensional Object Recognition Systems
A.K. Jain and P.J. Flynn (Editors)
© 1993 Elsevier Science Publishers B.V. All rights reserved.

Tools for 3D Object Location from Geometrical Features by Monocular Vision

Michel Dhome, Jean-Thierry Lapresté, Marc Richetin, and Gérard Rives [a]

[a]Electronics Laboratory
Blaise Pascal University of Clermont-Ferrand
63177 Aubiere Cedex
France

1. Introduction

Computer Vision may be used to give a robot the ability of recognizing and locating objects situated in the field of view of optical sensors. When using video cameras, the basic available data are 2D brightness images from which it is then necessary to infer 3D information about the shape and the spatial attitude of these objects.

All the shape-from-X methods are to solve this problem. They differ from the stereovision approach in the sense that the later is related with binocular or trinocular vision, and the former with monocular vision. This paper is focused on the shape-from-contour paradigm. The model of the projection of the 3D world on the image plane of a video camera being the perspective transform in the most general case, this paper brings some answers to the following central question in monocular vision: given the perspective projection of some curves on the object surface, is it possible to deduce the spatial location of the object?

A precise answer can be given when the object model is known a priori and for some geometrical primitives extracted from the contour image: straight segments, quadratic curves, zero-curvature points, angular points or in the case of objects of revolution paired points of the limb projection. For polyhedral objects, it is demonstrated that the knowledge of the projection of only three straight ridges allows the determination of the three translations and rotations which define a spatial attitude with a good accuracy and with little ambiguity. An experiment with a polyhedral real object is provided thereafter which shows how this method can be incorporated in a prediction-verification procedure.

Circular or elliptical edges generally give elliptical contours when viewed under perspective projection. For a given radius or elongation and eccentricity, it is shown that there are only two space planes which support the given circle or ellipse, such that its perspective projection is compatible with the elliptical contour detected in the image. Consequently, there are four possible space attitudes for the object supporting the given circle or ellipse. A real experiment is presented which indicates the efficiency of this method.

The third case concerns space or planar curves. It is known that the perspective projection of a zero-curvature point of a space curve is also a zero-curvature point, the converse being true only for planar curves. This property is exploited in two experiments. One deals with a straight homogeneous generalized cylinder, and the other with an object of revolution. It is shown that they can be located from their contours under the hypothesis that their scaling function has at least one zero-curvature point.

At last, the previous method is generalized for any pair of limb points of an object of revolution which belong to some cross section. An experiment is also presented.

This chapter is mainly focused on the results obtained for the inverse perspective problems presented above. For each of them, theoretical details can be found in provided references.

2. Interpretation of Three Contour Segments

2.1. Principle

We presented in [2] a method to determine with a good accuracy the spatial attitude of a polyhedral object in the viewer coordinate system. The principle is based on the interpretation of a triplet of any image lines as the perspective projection of a triplet of linear ridges of the object model, and on the search of the model space attitude consistent with these projections.

In short, the principle of the method is :

- to write down the equations expressing that each of the three projections of the straight lines supporting the model edges are collinear to the corresponding image segments,

- to choose a peculiar reference system to simplify the equations and to solve them by reducing them to an eight degree equation in one unknown.

Recently many people have worked on this problem [8, 1, 13, 7] but they provide no general solution as they assume restrictions such as orthographic projection or orthogonality between model ridges. Horaud has developed a new approach [6] without any restriction but the object attitude is obtained via an hypothesis accumulation process (looking for maxima in a big accumulator array).

Our method gives all the spatial attitudes of the object model for which the projections of the triplet of model ridges are collinear with the supporting lines of the matched image segments. The number of solutions (up to 16) may seem very important, but simple logical rules permits the retention of only the realistic ones (at most 1 or 2) among them.

These pruning rules can be detailed as follows:

- first, all the solutions having a negative value of translation along the z-axis must be suppressed since this axis is the viewing axis of the camera and the corresponding location of the model would be situated behind the camera.

- second, since only visible lines can be selected in the images, all solutions implying a model attitude such that one of the selected lines is not visible must be discarded. Being visible means that a projected model line is in the image field and belongs to the visible part of the object.

- last, in the mathematical formulation used in this method, the equations written only imply the collinearity of the model lines projections with the corresponding image lines. When the model is projected onto the image for a given admissible spatial attitude it is possible that no overlapping occurs between image lines and model lines projection. Thus all the solutions which produce such a situation must also be suppressed.

An example can be seen in figure 1. Figure 1.a presents a synthetic image and the triplet of selected image lines. The upper left box in figure 1.b corresponds to the model attitude used for the image synthesis and the other boxes show the model spatial attitudes compatible with the supporting lines of the image segments. Only solution #1 can be retained as a realistic one.

In practice the attitude determination has to be included in a general prediction-verification procedure in order to find the correct matching sequentially and to calculate the precise location of the objects at last.

Lowe's iterative method [10] can give one solution when convergence is obtained with at least three lines. But this solution may be wrong depending on the initial values provided for the rotations and the translations. Thus it is not possible to say that the matching is not correct when a solution is found and reveals to be wrong after verification. With a great number of lines Lowe's algorithm would probably converge to the correct and accurate solution for a correct matching, but in this case the complexity of the combinatorial process involved in the matching would be great.

It appears also that it is highly desirable to have good initial conditions at the very beginning of the matching by prediction-verification.

As all the possible solutions can be quickly calculated with any triplet of lines, we propose to start the matching with the admissible ones found with this method and to use, when enough lines have been matched, Lowe's algorithm with the corresponding initial conditions to obtain a more precise object location.

2.1.1. Experiment

An experiment from the image of a real object is presented. Figure 2.a is the image of a tape drive. Figure 2.b shows the admissible attitude of the model calculated from three manually selected lines. A more precise object location is obtained in an iterative way. At each step, new image lines are selected with a prediction-verification process and submitted to Lowe's algorithm. The iterations end when the position of the model is stable. Figures 2.c and 2.d show respectively selected image lines and Lowe's result at the final step. The attitude parameter values for these two steps are given in table 1. It is noticeable that the small difference between the initial and final values however involves great difference between model projections. In this experiment the exact spatial attitude of the object was not known a priori, and the tape drive is not really a polyhedron since its ridges are rather smooth. But the final result (figure 2.d) proves that the calculated spatial attitude in the viewer coordinate system can be very close to the exact one.

184

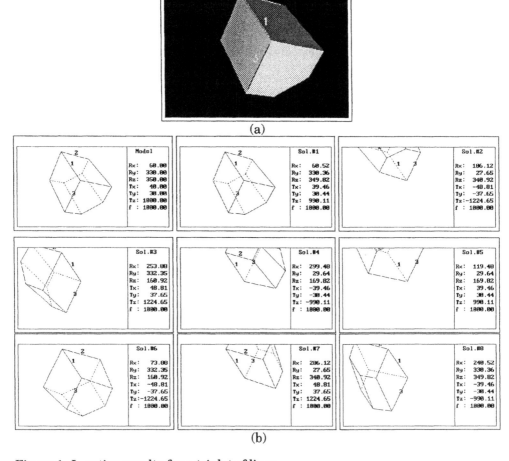

(a)

(b)

Figure 1. Location results for a triplet of lines.
(a) - synthetical image with detected and selected edges.
(b) - real and calculated model attitudes.

	value after 3-lines matching	final value with Lowe's algorithm
Rx	355.4	357.0
Ry	324.7	328.6
Rz	3.7	3.5
Tx	71.8	82.4
Ty	−72.7	−80.0
Tz	721.7	766.4

Table 1
Attitude parameter values.

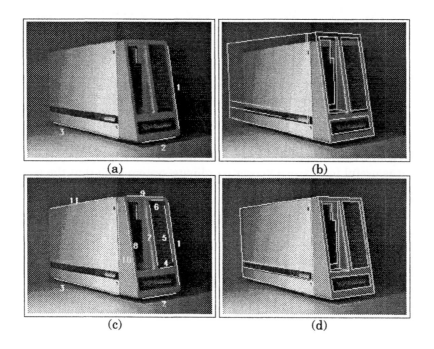

Figure 2. Three lines matching and iterative improvement
(a) - Real image with detected and selected edges.
(b) - Real image and calculated model attitude.
(c) - Selected edges submitted to Lowe's algorithm.
(d) - Final Lowe's solution.

3. Interpretation of Elliptical Contours

3.1. The cyclic plane problem

Let us consider an object having a circular ridge, for which the support plane, the center and the radius are known in an object coordinate system. Let us suppose also that the perspective projection of this circle is viewed in the image. From this projection, which is an ellipse in the general case, it is possible to recover the spatial attitude of the object.

This inverse perspective problem can be decomposed in two parts [3]:

3.1.1. Determining the cyclic planes.

Let C be the cone defined by the optical center of the camera and by the viewing directions of all the points of the elliptical contour.

The first step consists in determining the orientation of the families of planes whose intersection with C is a circle. This problem is known as the "cyclic plane determination" and its analytical solution involves a symmetrical matrix diagonalization and generally leads to two families of planes.

If the equation of the ellipse is given by the following equation :

$$a_0 x^2 + b_0 xy + c_0 y^2 + d_0 x + e_0 y + 1 = 0,$$

the equation of the cone defined by the optical center of the camera and by the viewing directions of all the point of the ellipse is :

$$a_0 x^2 + b_0 xy + c_0 y^2 + \frac{d_0}{f} xz + \frac{e_0}{f} yz + \frac{1}{f^2} z^2 = 0$$

Let us get further in the computation details. In order to simplify this determination, it is useful to choose a viewer coordinate system such that the equation of the cone is reduced to :

$$\frac{x_1^2}{a_1^2} + \frac{y_1^2}{b_1^2} - \frac{z_1^2}{f^2} = 0 \text{ with } \frac{1}{a_1^2} \geq \frac{1}{b_1^2}.$$

This new coordinate system is obtained after a rotation R_D of the original one which makes diagonal the following matrix :

$$\begin{pmatrix} a_0 & \frac{b_0}{2} & \frac{d_0}{2f} \\ \frac{b_0}{2} & c_0 & \frac{e_0}{2f} \\ \frac{d_0}{2f} & \frac{e_0}{2f} & \frac{1}{f^2} \end{pmatrix}$$

In this coordinate system, the great axis of the ellipse is always horizontal (x-axis). To solve the cyclic plane problem, then it is sufficient to find the rotation R_α of angle α around the horizontal axis for which the intersection of the cone by a vertical plane is a circle.

After the application of rotation R_α , the equation of the cone becomes :

$$\frac{x_1^2}{a_1^2} + \left(\frac{\cos^2 \alpha}{b_1^2} - \frac{\sin^2 \alpha}{f^2} \right) y_1^2$$
$$+ \left(\frac{\sin^2 \alpha}{b_1^2} - \frac{\cos^2 \alpha}{f^2} \right) z_1^2 + 2 \left(\frac{\sin \alpha \cos \alpha}{b_1^2} - \frac{\sin \alpha \cos \alpha}{f^2} \right) y_1 z_1 = 0$$

The intersection of the cone by a vertical plane (constant z_1) is a circle if and only if the coefficients of x_1^2 and y_1^2 are equal. This gives equation :

$$\frac{1}{a_1^2} = \frac{\cos^2 \alpha}{b_1^2} - \frac{\sin^2 \alpha}{f^2}$$

which implies the following determinations for $\cos \alpha$:

$$\cos^2 \alpha = \frac{b_1^2(a_1^2 + f^2)}{a_1^2(b_1^2 + f^2)}$$

and a multiple determination for angle α :

$$\alpha_1, \; \alpha_2 = \pi - \alpha_1, \; \alpha_3 = \pi + \alpha_1, \; \alpha_4 = -\alpha_1,$$

It is easy to see that the angles α_1 and α_3 (respectively α_2 and α_4) define the same plane family.

3.1.2. Finding the object circular ridge pose.

To solve the localization problem completely, the plane situated in front of the camera and for which the radius of the obtained circle is equal to the radius of the object circular ridge must be selected, from the two families of cyclic planes.

Thus we obtain two particular cyclic planes. To each of these plane correspond two poses for the object as it can be placed in each of the two half spaces delimited by each cyclic plane, and that, with no further consideration it is no possible to know to which half-space the object does belong.

3.2. Experiment

The preceding method permits the computation of the four spatial attitudes of a circle with given radius, defined by the position of its center and by the normal to its support plane, which are compatible with an elliptical contour. Knowing the position of a circular edge on the surface of an object, it is then straightforward to determine the rigid transform to apply to the object model to bring it in a position such that the projection of the circular edge will coincide with the observed elliptical contour.

Figure 3.a shows the detected contours in a brightness image of a vase. One of the detected ellipses appears in figure 3.b. Figure 4 presents the four attitudes of the object model which have been determined from the interpretation of this ellipse. It is clear that three of them are not valid. The quality of the matching of the external contours of the projected model image with the contours of figure 3.a can be evaluated [12]. A score between 0 and 1 is calculated (see figure 4) and the selected solution has the highest score[1]. Both the brightness image and the projected model image for the selected solution are presented in figure 5.a.

A localization error occurs in the area of the top of the vase. It is due to an inaccurate determination of the ellipse parameters. But an iterative adjustment

---

[1]This score corresponds to the percentage of the external contour of the model projection (for a given attitude) overlapping the image contour.

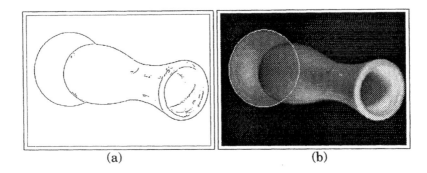

Figure 3. Location using elliptic contours
(a) Detected contours. (b) Selected detected ellipse.

procedure can be used to refine the attitude. It involves matching of straight segments (straight contours after polygonal segmentation of the contour image and projections of the external edges of the object polyhedral model), a bucketing technique implemented to increase the speed of this matching, and the Lowe's algorithm [10] to calculate a new attitude after the matching. At the end of this procedure, a better attitude is obtained (figure 5.b).

4. Interpretation of Zero-Curvature Contour Points

In this section, we recall how zero-curvature points of the contour of an object and especially of a straight homogeneous generalized cylinder (SHGC) can be used to find its precise pose.

For the family of SHGCs, a partial answer to this problem is proposed by Ponce [11] when orthographic projection is assumed. The presented approach extends Ponce's results in two ways, first by using the perspective projection, and second by solving the localization problem completely. These results rely on the two following theorems:

Theorem 1: *If P is a zero-curvature point of a space curve (C), then its image is a zero-curvature point of the perspective projection of (C).*

Theorem 2: *The points of the limbs of a straight homogeneous generalized cylinder which correspond to zero-curvature points of the scaling function, are viewed under perspective projection as zero-curvature points of a contour.*

4.1 Localization using Theorem 1

The method presented in section 2 and theorem 1 permits the determination of the localization of any kind of object providing curves with at least two zero-curvature points belonging to its surface are visible in its image.

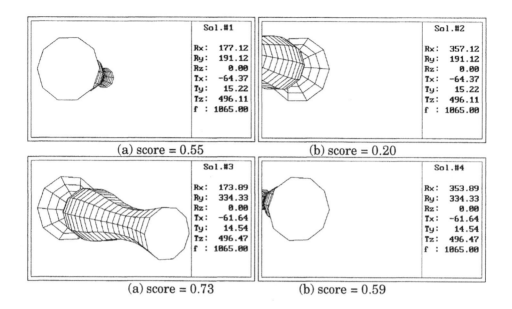

Figure 4. The four compatible attitudes and the corresponding scores.

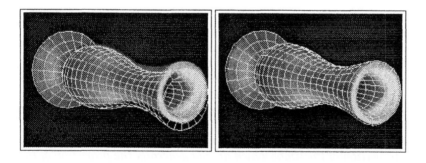

Figure 5. Location with elliptic contour and iterative improvement
(a) Superimposition of the brightness image and
of the projection of the model in the initial selected attitude.
(b) Superimposition of the brightness image and
of the projection of the model in its final calculated attitude.

Indeed, two zero-curvature points of a contour and the tangents at these points define three straight lines in the image. If we suppose that a correspondence is established between these three image lines and a triplet of straight lines in the model reference system, the method described in section 2 yields all the localizations of the model triplet compatible with the image triplet.

In the related experiments, we have used a salt-shaker corresponding to a straight homogeneous generalized cylinder whose cross section is nearly a regular hexagon, and whose scaling function has zero-curvature points. The studied images (see figure 6.a) are taken from a point of view such that the contours of the object contain the projection of the angular ridges corresponding to the scaling function (i.e. not too close to the axis).

The selection in the image of two zero-curvature points on the external contours of the object (supposed to belong to the same cross section of the model) and the determination of the contour tangent at these points define a triangle (see figure 6.b).

Let C be the cone defined by this cross section and the tangents to the scaling function at the points of the cross section.

Using theorem 1, the triangle previously defined is the projection of an other one made of the vertex of C and any two vertices of the polygonal basis of C. In the case of the salt-shaker, since its cross section is approximated by a regular hexagon (central symmetry), only five triangles among all such triangles can be matched with the image triangle.

For each matching, the method presented in section 2 gives all the possible spatial attitudes of the cone triangle and the related position of the object model (see figure 7). The correct spatial attitude is selected by verification of contour localization. The quality of the covering of the image contours and the external contours of the projected model is scored as in the case of localization by elliptical contour.

The selected attitude has the highest score and can be seen in figure 8.

4.2. Localization using Theorem 2

From theorem 2, if the scaling function of a SHGC has at least one zero-curvature point and if the point of view is such that the limbs go through at least one section corresponding to a zero-curvature point of this scaling function, then the external contours of the object image contain at least one pair of zero-curvature points. The two points of a pair and the tangents to the contours at these points define a triangle.

Remark: *For any point of view, the contours of a SHGC image mainly coincide with the projection of the limbs which are not generally the projection of the curves defined by the scaling function.*

Nevertheless, the method presented in section 4.1 can be used since we know the model cross section corresponding to the pair of zero-curvature point selected in the image. To obtain accurate results, it would be necessary to approximate precisely this cross section by a polygon. In this case the great number of vertices of the polygon could lead to a highly combinatorial resolution.

In case of solids of revolution, a non combinatorial solution can be obtained. It has

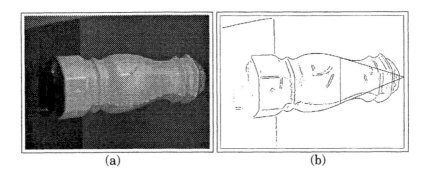

Figure 6. Location using zero-curvature points
(a) Grey level image of a salt-shaker.
(b) Detected contours, two zero-curvature points selected and their tangents.

been shown [12] that the triangle defined in the image by a pair of zero-curvature points and their tangent to the contour, is the perspective projection of a cone of revolution. This cone is defined by the cross section of the model corresponding to the zero-curvature point of the scaling function and by the tangents to the surface at each point of this section. It has also been demonstrated that there are in general two spatial attitudes of the cone of revolution compatible with the observed triangle.

This approach has been applied on the image of figure 3.a. Figure 9 shows the selected pair of zero-curvature points on the external contours of the vase, and the tangents to the contours at these two points. The two possible attitudes of the cone of revolution are visible on figure 10. A superimposition score is calculated in a same way as in the former section. The selected attitude which corresponds with the highest score leads to figure 11 on which one can see the good fit between the projection of the polyhedral model and of the brightness image.

The result obtained with this approach is better than the one derived from the interpretation of an ellipse. This could be surprising as it is known that it is quite difficult to locate zero-curvature points accurately. But the triangle which is interpreted is made of two segments which are the tangents to the contours at these points. Thus their orientation is very precise and the triangle is not so badly defined.

5. Interpretation of Limb Projection from Paired Points

5.1. Overview

The method presented in this section and detailed in [3] is a generalization of the preceding one in the case of objects of revolution. Using a procedure described

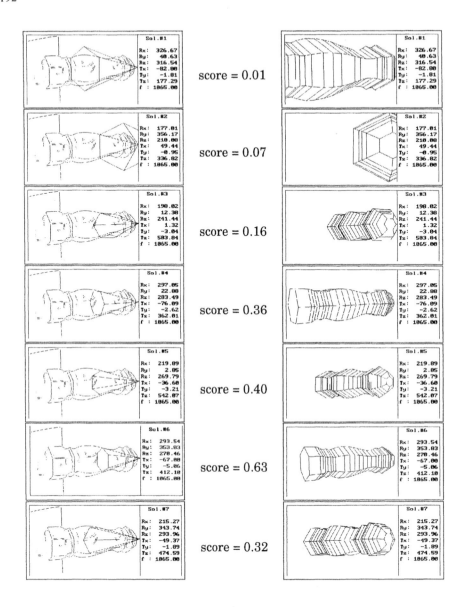

Figure 7. All the possible spatial attitudes of the polyhedral cone, the same salt-shaker model ones, and the corresponding scores.

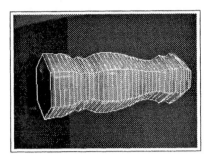

Figure 8. Retained attitude of the salt-shaker model.

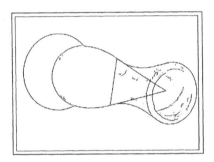

Figure 9. The triangle defined by a pair of zero-curvature contour points and the tangents to the contours at these points.

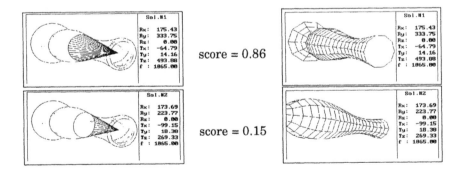

Figure 10. The two spatial attitudes of the cone of revolution and of the object model.

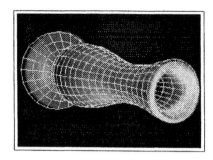

Figure 11. The superimposition of the projection of the model
in its calculated attitude and of the brightness image.

in the next section, two points of the projection of the limbs which belong to the
projection of a same circular section of the object are selected. As before, these
two points and the tangents to the contours at these points define a triangle. But
here the position of this section in the object coordinate system is assumed not to
be known. So it is necessary to scan all the possible sections of the object and to
calculate all the possible attitudes compatible with this triangle. In order to select
the best one, a superimposition score is calculated.

5.2. Choice of a judicious viewer coordinate system

At first, it must be noted the plane going through the optical center of the camera
and the axis of revolution of the object is a plane of symmetry for the object. Thus
the limbs which are situated on either sides of this plane are symmetrical curves
on the surface of the object with respect to this plane. The perspective projection
of these curves are symmetrical if and only if the optical axis intersects the axis of
revolution of the object.

If this is the case then the problem is much simpler since two symmetrical contour
points in the image belong to the same section of the object. Fortunately it is always
possible to be in this situation by use of a transform which permits the computation
of a virtual image in which the external contours are symmetrical.

We have developed a robust algorithm [4] to compute the accurate location in the
image plane of the projection of the spatial axis of an object of revolution.

This method looks up for a pair of contour points belonging to the projection
of a same cross-section. It is based on a prediction-verification scheme using the
invariance of the cross-ratio of aligned points under perspective projection. When
this step is performed (see 12.a), it is easy to apply a rotation R_0 to the viewer
coordinate system in order to make the limb projection symmetrical with respect
to the vertical axis of the image (see figure 12.b).

5.3. Compatible attitudes and evaluation of the quality of an attitude

Let us choose a pair of symmetrical contour points (point P_0 in figure 12.b) in the new coordinate system. These points and their tangents to the contour define a triangle. For each hypothesis about the corresponding cross section of the model it is possible to find (result of section 4.2) two model attitudes compatible with this image triangle. At this stage in order to locate the object it is necessary to have a criterion to select the correct attitude among the possible ones.

For the evaluation of the quality of an attitude, we choose some control points P_i (see P_1, P_2, P_3, P_4 in figure 12.b) along the contours. For each point P_i and for a given attitude A_j we compute the distance between P_i and a particular point P_{ij} belonging to the projection of the model limbs in the spatial attitude defined by A_j. By definition the points P_{ij} must correspond to points P_i if the given attitude A_j is the correct one.

To select the good cross section we used the criterion

$$C(Z) = \min_{j=1}^{m}(\sum_{i=1}^{n} ||\overrightarrow{p_i p_{ij}}||)$$

where m is the number of real possible attitudes for a given section which is defined in the model reference system by its coordinate Z. The retained attitude is the one for which $C(Z)$ is minimal.

6. Experiment

We also experimented this method on the vase image presented in figure 3.a. Figure 12 shows the result of the axis projection detection. After the application of R_0, symmetrical contours are obtained as shown in figure 12.b. On this figure, the four control points used in this experiment are also visible.

The height of the vase is nearly 250 mm. The Z axis was sampled each millimeter. The value of the criterion $C(Z)$ is drawn in figure 13. Obviously this curve has two relative minima $C(158) = 21$ and $C(60) = 1341$. The corresponding attitudes for the object model in the virtual image are presented in figure 14, in which two other attitudes are also shown for $Z = 153$ mm and $Z = 161$ mm.

When the best solution is selected, it is easy to compute the corresponding model attitude in the original viewer coordinate system. The superimposition of the projected object model and of the brightness image is presented in figure 15. The quality of the mutual covering of the two images is quite good. The only noticeable differences appear to be due to the polyhedral approximation.

A very accurate pose of an object of revolution can be obtained from the interpretation of all the limbs points and not only of a few ones as it is done above [5]. The proposed method requires a rough estimate of the pose and involves the minimization of a "distance" between the projection of the limbs of the object detected in the image and of the projection of the limbs of the model for the estimated attitude.

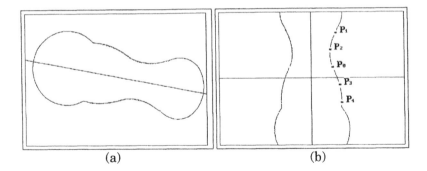

(a) (b)

Figure 12. Perspective projection of a vase
(a) Revolution axis projection.
(b) External contours in the virtual image.

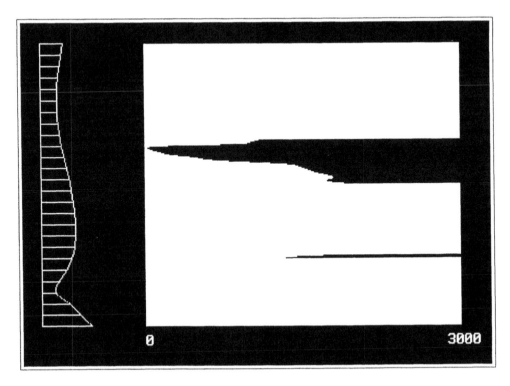

Figure 13. Values of criterion $C(Z)$.

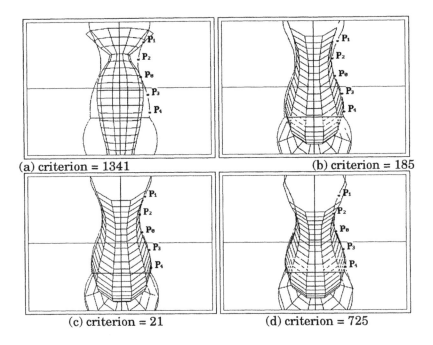

(a) criterion = 1341

(b) criterion = 185

(c) criterion = 21

(d) criterion = 725

Figure 14. Vase attitudes for the control point p_0
for $Z = 60, 153, 158$ et 161
(from left to top and from top to bottom)

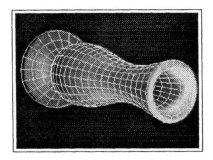

Figure 15. Superimposition of the projected model and of the brightness image.

7. Conclusion

Recovering the spatial attitude of an object in the viewer coordinate system from a single perspective image is still a challenge because there is no general solution to the inverse perspective problem.

Using a model-based approach, which is not a restriction in robotics as the environment is generally known, we have demonstrated that some geometrical features extracted from the brightness perspective image of an object and coming from the projection of lines or points situated on the surface of the object, can be used to find the spatial attitude of this object in the viewer coordinate system.

The various methods presented in this paper involve different kinds of geometrical features and are complementary. So, this proves that much information about a 3D world can be obtained from a single image when the geometrical model of the objects which are present in the scene are known.

Lastly, some of the tools presented in sections 3 and 5 have been successfully applied to the geometrical modelization of objects of revolution from a few number of perspective images, without any knowledge of the location of the various view points [9].

REFERENCES

1   S. T. BARNARD *Choosing a Basis for Perceptual Space* Computer Vision Graphics and Image Processing, 29, #1, 1985, pp 87-99.

2   M. DHOME, M. RICHETIN, J. T. LAPRESTÉ, G. RIVES *Determination of the Attitude of 3D Objects from a Single Perspective View.* IEEE Trans. on PAMI, 11 #12, pp 1265-1278, December 1989.

3   M. DHOME, J. T. LAPRESTÉ, G. RIVES, M. RICHETIN *Spatial Localization of Modelled Objects of Revolution in Monocular Perspective Vision.* Proc. of First European Conf. on Computer Vision, Antibes, April 1990, pp 475-488.

4   R. GLACHET, M. DHOME, J. T. LAPRESTÉ *Finding the perspective projection of an axis of revolution.* Pattern Recognition Letters, vol 12, pp 693-699, November 1991.

5   R. GLACHET, M. DHOME, J. T. LAPRESTÉ *Finding the pose of an object of revolution.* Proc. of Second European Conf. on Computer Vision, Genova, May 1992, pp 681-686.

6   R. HORAUD *New Methods for Matching 3-D Objects with Single Perspective View.* IEEE Trans. on PAMI, vol PAMI-9, #3, May 1987, pp 401-412.

7   D. P. HUTTENLOCHER & S. ULLMAN *Object Recognition using Alignment.* Proc. of the First Int. Conf. on Computer Vision, London, June 1987, pp 102-111.

8   T. KANADE *Recovery of the Three Dimensional Shape of an Object from a Single View.* Artificial Intelligence, Special Volume on Computer Vision, Vol 17, #1-3, August 1981.

9   J.M. LAVEST, R. GLACHET, M. DHOME, J. T. LAPRESTÉ *Modelling Solids of Revolution by Monocular Vision.* Proc. of IEEE Computer Vision and Pattern Recognition '91 Conf. Lahaina, Hawaii,

June 1991, pp 690-691.

10 D. G. LOWE *Perceptual Organization and Visual Recognition*. Kluwer, Boston, 1985, chapter 7.

11 J. PONCE & D. CHELBERG *Finding the Limbs and Cusps of Generalized Cylinders*. Int. J. Computer Vision, vol 1, #3, 195-210, October 1987.

12 M. RICHETIN, M. DHOME, J. T. LAPRESTÉ, G. RIVES. *Inverse Perspective Transform Using Zero-Curvature Curve Points: Application to the Localization of Some Generalized Cylinders from a Single View*. In IEEE Trans. on PAMI vol 13 #2 pp 185-192, February 1991.

13 T. SHAKUNAGA & H. KANEKO *Perspective Angle Transform and Its Application to 3-D Configuration Recovery* Proc. of Int. Conf. on Computer Vision and Pattern Recognition, Miami Beach, Florida, June 1986, pp 594-601.

Three-Dimensional Object Recognition Systems
A.K. Jain and P.J. Flynn (Editors)
© 1993 Elsevier Science Publishers B.V. All rights reserved.

Part-Based Modeling and Qualitative Recognition

Sven J. Dickinson[a]

[a]Department of Computer Science
University of Toronto
6 King's College Road
Toronto, Ontario, M5S 1A4
Canada

1. Introduction

A multitude of object recognition paradigms have been proposed, each differing in their primitive (feature) extraction, matching, model representation, verification, or overall control strategies. Despite the tremendous variety of approaches, however, there is a very powerful metric that can be used to compare them. By examining the *indexing primitives* (image structures that are matched to object models) used in the various approaches, we can draw some powerful conclusions about building object recognition systems. In addition, we will see that the selection of indexing primitives not only affects system performance, but constrains the design of other recognition system modules.

A comparison of object recognition systems according to their indexing primitives is given in Figure 1. In the left column are various indexing primitives ranging in complexity[1] from low (e.g., 2-D points) to high (e.g., 3-D volumes), as depicted by the width of the leftmost bar (bar 1). Some of the indexing primitives are two-dimensional, while others are three-dimensional, often reflecting the type of input as intensity or range image data. Accompanying each indexing primitive is a reference to an example system that employs that primitive. Note that this list of indexing primitives is not complete; it is meant only to exemplify the range in complexity of possible indexing primitives.

Working from left to right in Figure 1, we see that as the complexity of indexing primitives increases, the number of primitives making up the object models decreases (bar 2), since an object can be described by a few complex parts or by many simple parts. This, in turn, implies that the search complexity, i.e., the number of hypothesized matches between image and model primitives, decreases with increasing primitive complexity (bar 3). The high search complexity involving simple indexing primitives is compounded by large object databases. As a result, most systems using simple indexing primitives, e.g., Lowe [19], Huttenlocher and Ullman [15], Thompson and Mundy [24], and Lamdan et al. [17], are applied to small databases typically containing only a few objects.

---

[1]By complexity, we mean a primitive's descriptive power, typically proportional to the number of bits used to represent the primitive.

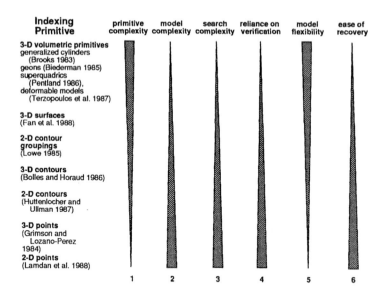

Figure 1. A comparison of object recognition systems according to their indexing primitives (reprinted from [7], ©1992 Academic Press)

Since simple indexing primitives imply a more ambiguous interpretation of the image data (e.g., a few corners in the image may correspond to many corner triples on many objects), systems that employ simple primitives must rely heavily on a top-down verification step to disambiguate the data (bar 4). In this manner, the burden of recognition is shifted from the recovery of complex, discriminatory indexing primitives to the model-based verification of simple indexing primitives. Since many different objects may be composed of the same simple features, these systems are faced with the difficult task of deciding which object to use in the verification step. However, there is a more fundamental problem with simple indexing features.

Reliance on verification to group or interpret simple indexing primitives has two profound effects on the design of recognition systems. First, verifying the position or orientation of simple indexing primitives such as points or lines requires an accurate determination of the object's pose with respect to the image. If the pose is incorrect, the search of a local vicinity of the image for some model feature may fail. Needless to say, accurately solving for the object's pose can be computationally complex, particularly when a perspective projection camera model is used, e.g., Lowe [19].

Relying on verification also affects object modeling. Specifically, the resulting object models must specify the exact geometry of the object, and are not invariant to minor changes in the shape of the object (bar 5). Consider, for example, a polyhedral model of a chair. If we stretch the legs, broaden the seat, or raise the back, we would require a new model if our verification procedure were checking the position of points and lines in the image. Excellent work has been done to extend this approach to certain types of parameterized models, e.g., Grimson [10], Huttenlocher [13], and Lowe [18]. However, by nature of the indexing primitives, these models do not explicitly represent the gross structure of the object, and therefore cannot easily accommodate certain types of shape changes.

So far, bars 1 through 5 in Figure 1 clearly indicate the advantages of using complex indexing features over simple ones. What is the trade-off? Why are most 3-D from 2-D recognition systems using simple indexing primitives?[2] First of all, in certain domains, e.g., typical CAD-based recognition, in which the object database is very small, object models are constructed from simple primitives, object shape is fixed, and exact pose determination is required, simple indexing primitives have proven to be quite successful. However, more importantly, the reliable recovery of more complex features, particularly from a single 2-D image, is a very difficult problem (bar 6), particularly in the presence of noise and occlusion. Clearly, the major obstacle in the path of any effort to build a recognition system based on complex indexing primitives will be the reliable recovery of those primitives. This chapter addresses this challenge. From a single 2-D image, we present an approach to the recovery and recognition of 3-D objects using 3-D volumetric indexing primitives.

2. Object Modeling

2.1. Choosing the 3-D Primitives

Given a database of object models representing the domain of a recognition task, we seek a set of three-dimensional volumetric primitives that, when assembled together, can be used to construct the object models. In addition, the chosen primitives should be qualitatively defined so that the object models are invariant to minor changes in shape of the primitives. To demonstrate our approach, we have selected an object representation similar to that used in Biederman's Recognition by Components (RBC) theory [2]. RBC suggests that from nonaccidental relations in the image, a set of contrastive dichotomous (e.g., straight vs. curved axis) and trichotomous (e.g., constant vs. tapering vs. expanding/contracting cross-sectional sweep) 3-D primitive properties can be determined. The Cartesian product of the values of these properties gives rise to a set of volumetric primitives called *geons*.

Biederman's geons constitute only one possible selection of qualitatively defined volumetric primitives; the general approach of applying the Cartesian product to a set of contrastive primitive properties can be used to generate many different volumetric primitive representations. For our investigation, we have chosen three properties including cross-section shape, axis shape, and cross-section sweep. The

---

[2]Many of the more complex indexing primitives, e.g., 3-D surface patches, deformable models, and superquadrics are typically recovered from range data images.

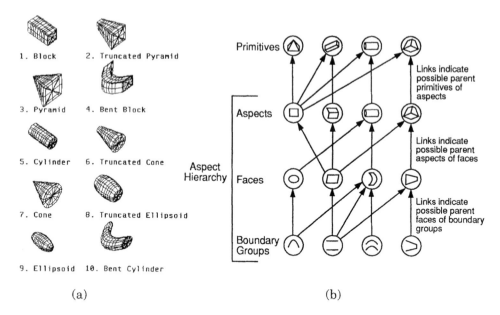

Figure 2. (a) The ten modeling primitives (reprinted from [8], ©1992 IEEE), (b) The aspect hierarchy (reprinted from [8], ©1992 IEEE)

values of these properties give rise to a set of ten qualitative volumetric primitives.[3] To construct objects, the primitives are simply attached to one another with the restriction that any junction of two primitives involves exactly one attachment surface from each primitive, i.e. an attachment cannot lie on a surface discontinuity.

In our system, these ten primitives were modeled using Pentland's SuperSketch 3-D modeling tool [21], as illustrated in Figure 2(a).[4] We believe that this taxonomy of volumetric primitives is sufficient to model a large number of objects; however, nothing in our approach is specialized for superquadrics or geons. If necessary, our approach can easily accommodate other sets of volumetric primitives bearing little resemblance to geons or superquadrics.

2.2. Defining the 2-D Aspects

Traditional aspect graph representations of 3-D objects model an entire object with a set of aspects, each defining a topologically distinct view of an object in terms of its visible surfaces (Koenderink and van Doorn [16]). Our approach differs

---

[3]The Cartesian product of the values of these properties results in a set of 20 primitives; however, to simplify the investigation in terms of generating the conditional probability tables described in the next section, we have chosen a subset of 10 primitives which we believe to be a good basis for modeling a wide range of objects.
[4]SuperSketch models each primitive with a superquadric surface that is subjected to bending, tapering, and pinching deformations.

in that *we use aspects to represent a (typically small) set of volumetric primitives from which each object in our database is constructed, rather than representing an entire object directly.* Consequently, our goal is to use aspects to recover the 3-D primitives that make up the object in order to carry out a recognition-by-parts procedure, rather than attempting to use aspects to recognize entire objects.

To minimize the number of aspects needed to represent the primitives, we constrain the aspects to be invariant to minor changes in primitive shape. By encoding only region topology and qualitative region shape, a particular aspect of a primitive becomes invariant to changes in primitive size, curvature, taper, etc. As a result, a small set of qualitatively different aspects describes a small set of qualitatively different volumetric primitives; each primitive, in turn, describes an enormous range of 3-D shape. The advantage of this approach is that since the number of qualitatively different primitives used to build objects is generally small, the number of possible aspects is limited and, more important, *independent* of the number of objects in the database. In contrast, the number of aspects required to model complete objects grows with the size of the database, and is further compounded when objects are articulated.

The disadvantage is that if a primitive is occluded from a given 3-D viewpoint, its projected aspect in the image will also be occluded. Clearly, we must accommodate the matching of occluded aspects, which we accomplish by introducing a hierarchical aspect representation we call the *aspect hierarchy*. The aspect hierarchy consists of three levels, based on the faces appearing in the aspect set; Figure 2(b) illustrates a portion of the aspect hierarchy.

- *Aspects* constitute the top level of the aspect hierarchy and represent all possible views of the primitives in terms of visible faces. Identification of the aspects can allow identification of the visible primitives. However, due to occlusion, some of the faces in an aspect may be partially or completely missing. When this occurs, we may need to analyze the arrangement of the remaining faces, and so we introduce the second level of the aspect hierarchy.

- *Faces* constitute the second level of the aspect hierarchy and represent all possible component faces making up the aspects. Reasoning about the type and arrangement of visible faces can allow identification of an aspect even when it is partially occluded. However, again due to occlusion, some of the contours that make up a face may be partially or completely missing. When this occurs, we may need to analyze the arrangement of the remaining contours bounding the face, and so we introduce the lowest level of the aspect hierarchy.

- *Boundary Groups* constitute the third and lowest level of the aspect hierarchy and represent all subsets of the faces' bounding contours. The boundary groups provide a mechanism for identifying the face type even when the face is partially occluded.

2.3. Relating the 2-D Aspects to the 3-D Primitives

A given boundary group may be common to a number of faces. Similarly, a given face may be a component of a number of aspects, while a given aspect may be the projection of a number of primitives. To capture these ambiguities, we have created a matrix representation that describes conditional probabilities associated with the mappings from boundary groups to faces, faces to aspects, and aspects to primitives. To generate these conditional probabilities, we first model our 3-D volumetric primitives using the SuperSketch modeling tool [21], as shown in Figure 2(a). The next step in generating the probability tables involves rotating each primitive about its internal x, y, and z axes in 10° intervals. The resulting quantization of the viewing sphere gives rise to 648 views per primitive.[5] For each view, we project the primitive onto the image plane, and note the appearance of each feature (boundary group, face, and aspect) and its parent. The resulting frequency distribution gives rise to the three conditional probability matrices (which can be found in [6]).

This procedure implicitly assumes that all primitives are equally likely to appear in the image, and that all spatial orientations of the primitives are equally likely. In practice, this is a strong assumption since both the frequency of occurrence and spatial orientation distribution of a primitive is governed by the contents of the object database. A more effective approach would be to preprocess the object database, counting the number of times each primitive appears in each object and noting the primitive's orientation. The resulting set of *a priori* probabilities of occurrence and orientation could then be easily incorporated into the analysis, providing a set of tables that more accurately reflect the contents of the object database.

It should be emphasized that these results offer only a rough approximation to the true probabilities. A more thorough analysis would use a finer quantization of both the primitives' parameters and the viewing sphere, and would measure the conditional probabilities directly from image data. The resulting explosion of views would require an automated tool to perform the analysis and generate the probabilities; much of the current analysis is performed manually. Nevertheless, the computation of the aspect hierarchy is performed off-line and is *independent* of the contents of the object database.

3. Primitive Recovery

The aspect hierarchy effectively *prunes* the mapping from boundary groups to primitives by introducing topological and probabilistic constraints on the boundary group to face, face to aspect, and aspect to primitive mappings. An analysis of the conditional probabilities [8] suggests that for 3-D modeling primitives which resemble the commonly used generalized cylinders, superquadrics, or geons, the most appropriate image features for recognition appear to be image regions, or faces. Moreover, the utility of a face description can be improved by grouping the

---

[5]Rotating the superquadric about its z axis in 10° intervals results in 36 views for a given x-y orientation. If we consider 18 x-y orientations by fixing either the x or the y orientation and varying the other at 10° intervals, we can effectively cover the viewing sphere with 648 views.

faces into the more complex aspects, thus obtaining a less ambiguous mapping to the primitives and further constraining their orientation. Only when a face's shape is altered due to primitive occlusion or intersection should we descend to analysis at the contour or boundary group level. Our approach, therefore, first segments the input image into regions and then determines the possible face labels for each region. Next, we assign aspect labels to the faces, effectively grouping the faces into aspects. Finally, we map the aspects to primitives and extract primitive connectivity.

3.1. Extracting Faces

The first step in extracting faces consists of extracting the bounding contours of image regions. We begin by applying Canny's edge detector [5] to the image followed by Beymer's algorithm [1] to fill gaps in the detected edges; minimal cycles in the resulting edge map correspond to the bounding contours of regions (or faces) in the image. The next step is to partition the contours at significant curvature discontinuities. We apply Saint-Marc and Medioni's scale-space adaptive smoothing algorithm [22] to come up with a partitioned contour at low, medium, and high scales.

Once the faces have been extracted, we must classify each face according to the faces in the aspect hierarchy. Both an image face and an aspect hierarchy face are represented by a graph in which nodes represent the face's bounding (partitioned) contours, and arcs represent relations between contours. Contours are characterized as straight or curved depending on how well a straight line can be fitted to them; furthermore, curves are characterized as convex or concave. Two non-coterminating lines are considered parallel if the angle between their fitted lines is small, while two non-coterminating curves are considered parallel if one is convex, one is concave, and the angle between their directions is small.[6] Two two non-coterminating, non-parallel lines are considered symmetric if there is sufficient overlap when one line is projected onto the other.[7]

Each of the three scales in the curve partitioning step gives rise to a graph representing an image face. Consequently, the classification of an image face consists of comparing its three graphs to those graphs representing the faces in the aspect hierarchy. We begin with the graph corresponding to the low scale in the curve partitioning step, often representing an oversegmentation of the face's bounding contour. If there is an exact match, as shown in Figure 3, then we immediately generate a *face hypothesis* for that image face, identifying the label of the face. If for any reason (e.g., occlusion, segmentation errors, noise, etc.), there is no match, we must descend to the boundary group level of the aspect hierarchy, as shown in

---

[6]The direction of a curve is computed as the vector whose head is defined by the midpoint of the line joining the two endpoints of the curve, and whose tail is definied by the point on the curve whose distance to the line joining the endpoints is greatest.
[7]Two non-parallel vectors will have an intersection point. When one vector is rotated about that point it can be brought into correspondence with the other. If the resulting overlap of the two lines is a large portion of the smaller of the two lines, the lines are said to be symmetric.

208

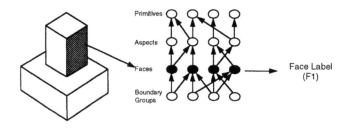

Figure 3. Labeling an unoccluded face (reprinted from [7], ©1992 Academic Press)

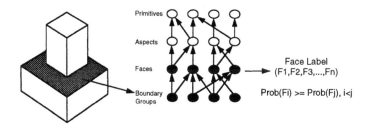

Figure 4. Labeling an occluded face (reprinted from [7], ©1992 Academic Press)

Figure 4. We then compare *subgraphs* of the graph representing the image face to those graphs at the boundary group level of the aspect hierarchy. For each subgraph that matches, we generate a face hypothesis with a probability determined by the appropriate entry in the conditional probability matrix mapping boundary groups to faces. This process is repeated for each of the other two graphs (medium and high scales). If, for any scale, a graph matches an aspect hierarchy face, the graphs representing all other scales are discarded; otherwise, we collect together the face hypotheses generated at all three scales. Each face hypothesis defines one or more *seed contour sets* representing those bounding contours of the image face (entire face or specific boundary groups) which define the label of the face.[8]

---

[8]In the case of a face hypothesis whose image face exactly matches an aspect hierarchy face (defined by the label of the face hypothesis), there will be a single seed contour set containing all the bounding contours of the image face. However, in the case of a face hypothesis whose image face does not match an aspect hierarchy face, there will be one seed contour set for every boundary group supporting the face hypothesis.

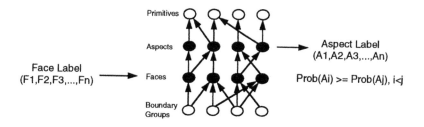

Figure 5. Generating the aspect labels of a face (reprinted from [7], ©1992 Academic Press)

3.2. Extracting Aspects
3.2.1. Problem Definition

The image can now be represented by a *face graph*, in which nodes represent faces (or regions) and arcs represent face adjacencies. Furthermore, following face extraction, each node (face) in the face graph has one or more face hypotheses associated with it. We can formulate the problem of extracting aspects as follows: Given a face graph and a set of face hypotheses at each face, find a covering of the face graph using aspects in the aspect hierarchy, an *aspect covering*, such that no face is left uncovered and each face is covered by only one aspect. Or, more formally: Given an input face graph, FG, partition the vertices (faces) of FG into disjoint sets, $S_1, S_2, S_3, \ldots, S_k$, such that the graph induced by each set, S_i, is isomorphic to the graph representing some aspect, A_j, from a fixed set of aspects, $A_1, A_2, A_3, \ldots, A_n$.

There is no known polynomial time algorithm to solve this problem (see [8] for a discussion on the problem's computational complexity); however, the conditional probability matrices provide a powerful constraint that can make the problem tractable. After the previous steps, each face in the face graph has a number of associated face hypotheses. For each face hypothesis, we can use the face to aspect mapping to generate the possible *aspect hypotheses* that might encompass that face, as shown in Figure 5; the face hypothesis becomes the *seed face hypothesis* of each of the resulting aspect hypotheses. The probability of an aspect hypothesis is the product of the face to aspect mapping and the probability of its seed face hypothesis. At each face, we collect all the aspect hypotheses (corresponding to all face hypotheses) and rank them in decreasing order of probability.

3.2.2. Aspect Instantiation

Each aspect hypothesis is merely an informed guess as to the aspect label of its seed face hypothesis. The process of verifying the hypothesized aspect label is called *aspect instantiation*. For an aspect to be instantiated from an aspect hypothesis, the relations between the seed face hypothesis and neighboring face hypotheses must be consistent with the definition of the aspect. More formally, there must exist a set of faces, S, including the face corresponding to the seed face hypothesis, such that the face subgraph induced by S is isomorphic to the graph representing the aspect. Since there may be multiple sets of faces which satisfy this criteria, there may be multiple aspects instantiated from a single aspect hypothesis. Hence, the process of aspect instantiation produces a (possibly empty) set of instantiated aspects for a given aspect hypothesis.

Let us explore the aspect instantiation process in more detail. Consider a face graph, FG, and an aspect hypothesis, ah, with label t, seeded at face f in FG. The aspect hierarchy aspect corresponding to label t, herein called the aspect definition, specifies that the aspect contains k faces, each with specified label and adjacency relations. We first collect together all neighboring faces of f (including f itself) in FG. Next, we generate all face subsets of size $\leq k$ from this collection; recall that there is an upper bound on k which is fixed (specified by the aspect hierarchy) and independent of the size of FG. For each subset, we check to see if the subgraph of FG (i.e., face subgraph) induced by the face subset is isomorphic to the aspect definition. For each matching subset, we instantiate an aspect; the result is a (possibly empty) list of instantiated aspects.

An aspect can be instantiated from an aspect hypothesis and a face subgraph if and only if the following conditions are satisfied:

- For each face in the face subgraph, there must exist, among its list of face hypotheses, a hypothesis whose label agrees with the label of its matching face in the aspect definition; if such a face hypothesis is found, it is *assigned* to the face in the subgraph.

- For each arc (or face adjacency relation) in the face subgraph, there must exist a corresponding arc in the aspect definition. Similarly, for each arc in the aspect definition, there must exist a corresponding arc in the face subgraph.

- For each arc in the face subgraph involving two faces, A and B, there must exist a seed contour set belonging to the face hypothesis assigned to A, and a seed contour set belonging to the face hypothesis assigned to B, such that each of the two seed contour sets includes the contours shared by A and B. Or, more intuitively, the contour(s) shared by two faces must be seed contours of both faces.

- For each face in the face subgraph, there must exist at least one seed contour set belonging to its assigned face hypothesis that satisfies all face adjacency relations involving that face.

If an aspect with k faces cannot be instantiated from an aspect hypothesis, it may be due to the fact that the aspect is occluded in the image. In this case, our goal is to find subsets of image faces that match portions of the aspect definition. Consider the set S of all subsets of image faces such that for each s in S, $|s| < k$ and the face subgraph induced by s matches some portion of the aspect definition (according to the above set of conditions). In addition, according to the partial match, let the valid seed contour sets at face i in s be $SC_1^i, SC_2^i, \ldots, SC_w^i$. Finally, let r represent the faces in the aspect definition not included in s (presumably occluded). We instantiate the aspect encompassing s provided the following conditions are satisfied:

- There exists no other subset t in S such that s is a proper subset of t and the aspect encompassing t has been instantiated. Or, more intuitively, if a set of faces satisfies an aspect, we ignore its subsets (which may also satisfy the aspect).

- For each arc in the face graph involving a face, fs, in s, and a face, fr, in r, there must exist a valid seed contour set SC_j^{fs} belonging to the face hypothesis assigned to fs such that the contours shared by fs and fr do not appear in the seed contour set. Or, more intuitively, if face A is occluded by face B, then the contours shared by faces A and B (which belong to face B) should not be seed contours of face A.

The above restrictions have a significant impact upon the selection of boundary groups. If we have a weak (i.e., low probability) face hypothesis, then it is likely that each of its seed contour sets represents a small fraction of the contours comprising the face. Consequently, instantiation of an aspect including such a face hypothesis may fail since it is likely that the required neighboring faces do not border at seed contours. However, with the lack of seed contours, smaller subgraphs may match the aspect definition since it is likely that neighboring faces do not border at seed contours. We conclude that there is a trade-off between selecting only the best boundary groups and exhaustively selecting all boundary groups. In the former case, a strong face hypothesis supported by strong boundary groups will likely match few aspect definitions, pruning out many interpretations of the face. However, if weaker boundary groups are not included in the face hypothesis, a correct interpretation may be impossible. Conversely, the presence of weak boundary groups allows occluded aspects to be instantiated. Although this may guarantee a solution, the increased number of interpretations may lengthen the search for a solution, and may result in less likely solutions being prematurely generated.

3.2.3. Algorithm

We can now reformulate our problem as a search through the space of aspect labelings of the faces in our face graph.[9] In other words, we wish to choose one aspect hypothesis from the list at each face, such that the instantiated aspects completely cover the face graph. Figure 6 illustrates the correct aspect covering of the face graph representing a scene containing an object composed of two blocks; one aspect label, in this case, A27 (see [8] for a description of all aspects), is selected from the list at each face to completely cover the graph. There may be many labelings which satisfy this constraint. Since we cannot guarantee that a given aspect covering represents a correct interpretation of the scene, we must be able to enumerate, in decreasing order of likelihood, all aspect coverings until the objects in the scene are recognized.

---

[9]The size of this space is $O(A^F)$, where A is the number of aspects in the aspect hierarchy, and F is the number of faces in the image.

212

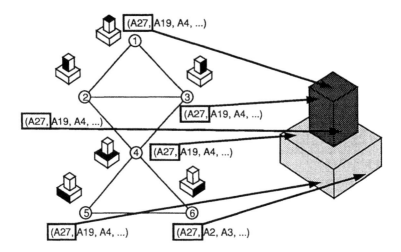

Figure 6. Covering the face graph with aspects (the aspect labeling problem) (reprinted from [7], ©1992 Academic Press)

For our search through the possible aspect labelings of the face graph, we employ Algorithm A (Nilsson [20]) with a heuristic based on the probability estimates for the aspect hypotheses. The different labelings are ordered in the open list according to the heuristic. At each iteration, a labeling, or state, is removed from the open list and checked (using a depth-first search) to see if it represents a solution (a covering). The successor states are then generated, evaluated, and added to the open list. The actual instantiation of aspects is performed during successor generation. The algorithm continues until all possible solutions are found, i.e. all labelings are checked. However, it should be pointed out that in an object recognition framework, once a solution is found, the search is only continued if the recovered shapes (inferred from the aspect covering) can not be recognized.

Before adding a successor state to the open list, it is evaluated using a heuristic function. The function has been designed to meet three objectives. First, we favor selections of aspects instantiated from higher probability aspect hypotheses. Second, we favor selections whose aspects have fewer occluded faces, since we are more sure of their labels. Finally, we favor those aspects covering more faces in the image; we seek the minimal aspect covering of the face graph. These three objectives have been combined to form an algorithm for evaluating a state, as shown in Figure 7; note that a node consists of a number of *indices*, one per face, with each *index* referring to a particular aspect hypothesis for that face.[10]

---

[10]The values of c_1 and c_2 were empirically chosen to be 0.25 and 0.50, respectively.

input: *node*
output: *value*
value = 0
for each *index* **in** *node* **do**
    aspect set = set of instantiated aspects pointed to by *index*
    aspect hypothesis = aspect hypothesis from which *aspect set* was instantiated
    value = *value* -
          probability(*aspect hypothesis*) -

$$(c_1 * \frac{\text{maximum number of visible faces of any aspect in } aspect\ set}{\text{number of faces in } aspect\ hypothesis \text{ definition}}) -$$

          (c_2 * maximum number of visible faces of any aspect in *aspect set*)
return *value*

Figure 7. Heuristic for Evaluating a State (reprinted from [8], ©1992 IEEE)

3.3. Extracting Primitives

We can represent an aspect covering by a graph in which nodes represent aspects and arcs represent aspect adjacencies. For each aspect in the aspect covering, we can use the aspect to primitive mapping to hypothesize a set of *primitives*, as illustrated in Figure 8.[11] As in the case of aspect hypotheses generated from face hypotheses, we can rank the primitives in decreasing order of probability. A selection of primitives, one per aspect, represents a 3-D interpretation of the aspect covering; we call such a selection a *primitive covering*.[12] Since we cannot guarantee that a given primitive covering represents a correct interpretation of the scene, we must be able to enumerate, in decreasing order of likelihood, all primitive coverings until the objects in the scene are recognized. To enumerate the selections, we employ a variation on the search algorithm used to enumerate the aspect coverings. The heuristic function simply negates the sum of the probabilities of the primitive, thereby favoring higher probability interpretations.

A primitive covering, represented by a graph in which nodes represent primitives and arcs represent primitive adjacencies, is then compared to the object database during the recognition process. If two aspects are not connected in the aspect covering, their corresponding primitives are not connected in the primitive covering. However, if two aspects are connected in the aspect covering, this does not mean that their corresponding primitives are necessarily connected in 3-D; one primitive may be occluding the other without being attached to it. A primitive connection

---

[11] In addition, the aspect hierarchy defines a mapping from the faces in an aspect to the attachment surfaces of a primitive.
[12] The number of possible primitive coverings is $O(P^A)$, where P is the number of primitives in the aspect hierarchy, and A is the number of aspects in the aspect covering.

214

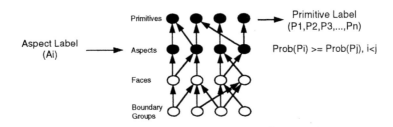

Figure 8. Generating the primitive labels of an aspect (reprinted from [7], ©1992 Academic Press)

between primitives P_1 and P_2 is said to be visible if the following condition is satisfied:

- There exists a pair of faces, F_1 and F_2, such that F_1 belongs to the aspect corresponding to P_1 and F_2 belongs to the aspect corresponding to P_2, F_1 and F_2 are adjacent in the face graph, and F_1 and F_2 share a contour.

Therefore, we define two types of primitive connectivity based on connection visibility:

- Two primitives are said to be *strongly* connected if their corresponding aspects are adjacent in the aspect covering, and the primitive connection is visible; in this case, we assume that the primitives are attached.

- Two primitives are said to be *weakly* connected if their corresponding aspects are adjacent in the aspect covering, and the primitive connection is *not* visible; in this case, one primitive occludes the other and it is not known whether or not they are attached.

A strong primitive connection strongly suggests the existence of a connection between two primitives. We can enhance the indexing power of a strongly connected subgraph if the attachment surfaces involved in each connection are hypothesized. Although it is impossible to define a set of domain independent rules which will, for any given set of primitives, correctly specify the attachment surfaces involved in a connection, we can define a set of heuristics which will specify a set of likely candidates. If a strongly connected subgraph is common to two object models, these heuristics can then be used to rank order the candidates for verification.

Hypothesizing the attachment surfaces proceeds as follows. Let S_1 be the set of faces belonging to the aspect corresponding to P_1 which are adjacent to a face belonging to the aspect corresponding to P_2. Similarly, let S_2 be the set of faces belonging to the aspect corresponding to P_2 which are adjacent to a face belonging to the aspect corresponding to P_1. There are three cases to consider:

1. *Sets S_1 and S_2 each contain a single face.* The attachment surface for P_1 is among the set of attachment surfaces that are adjacent to, and including, the surface representing the face in S_1. The attachment surface for P_2 is among the set of attachment surfaces that are adjacent to, and including, the surface representing the face in S_2. More intuitively, we believe that the attachment surface is in the local vicinity (on the primitive) of the attachment surface corresponding to the single visible face.

2. *Set S_1 contains a single face and set S_2 contains multiple faces.* In this case, the attachment surface for P_1 is the surface that the face in S_1 maps to. The attachment surface for P_2 is among the set of surfaces that are adjacent to, but not included in, the surfaces representing the faces in S_2. More intuitively, we believe that P_2 penetrates P_1; since the connection is visible, the attachment surface for S_1 is therefore attached to an occluded surface of P_2. (The same holds true when set S_2 contains a single face and set S_1 contains multiple faces.)

3. *Sets S_1 and S_2 both contain multiple faces.* In this case, the attachment surface for P_1 is among the set of surfaces that are adjacent to, but not included in, the set of surfaces representing the faces in S_1. The attachment surface for P_2 is among the set of surfaces that are adjacent to, but not included in, the set of surfaces representing the faces in S_2. More intuitively, we believe that although the attachment of P_1 and P_2 is visible, both their attachment surfaces are occluded.

4. Object Recognition

Given a primitive covering representation of the scene, in which nodes represent 3-D volumetric primitives and arcs represent strong or weak connections between the primitives, the final task is to identify the object(s) in the scene. There are two cases to consider. In an *unexpected* object recognition domain, we have no a priori knowledge of the contents of the scene. In this case, the recognition task consists of two steps: 1) identifying possible candidate models that might be present in the scene (model indexing), and 2) verifying that these models actually appear in the scene. In an *expected* object recognition domain, we search the image for one or more instances of a particular object.

4.1. Unexpected Object Recognition

The simplest unexpected object recognition strategy is to compare the entire primitive covering to each model in the object database, i.e., verify each object model in the image. If the graph representing the primitive covering is isomorphic to the graph (or subgraph) representing an object in the database, then the object in the scene has been identified. However, there are two major problems with this naive approach. First, for large object databases, the cost of verification may be prohibitive, as was shown by Grimson [11]. Second, this approach assumes that a primitive covering represents a single object. If the scene contains multiple occluded objects, the primitive covering will not match a single object in the database. Thus, we are left with two problems: 1) How do we avoid matching the recovered primitives to each object in the database?; and 2) What portions of the primitive covering likely belong to a single object, and should hence participate in the matching process?

4.1.1. Model Indexing

An alternative to sequentially matching the recovered primitives to each model object is provided by *hashing* techniques. A hash table is a precomputed data structure each of whose entries (in our case) map some recovered image feature(s) to a list of object models that contain that feature. The mapping between a recovered image feature and a location in the table is provided by a *hash function*. Once an image feature is "hashed" to an entry in the hash table, each of the objects referenced in the table entry must be verified. The advantage of hashing is that by preprocessing off-line the models in the object database, considerable on-line search can be avoided.

The hash table alone does not solve the problem, for if the recovered image features are simple, e.g., points, lines, or corners, they will be present in every object. The resulting hash table will have few entries (corresponding to a few simple indexing primitives), with each entry pointing to every object. Unfortunately, such a hash table leads us back to a sequential search of the database. Clearly, the goal in designing the hash table is to increase the size of the table, so that there are more entries, each having fewer pointers to object models.

As stated in Section 1, our goal has been to recover from the image richer, more complex primitives whose combination offers a more discriminating index into the object database. Unlike simple features such as lines, points, or corners which are abundant in every object, a particular collection of 3-D volumetric primitives is unlikely to be common to many objects. Our solution, therefore, is to index using a collection of recovered primitives. Since we have a variety of primitives which can be connected in a variety of ways, the size of our hash table will be larger than if we index using simple primitives. However, our second problem still remains: What collection of recovered primitives do we use as an index?

From a primitive covering of the input scene, we would like to index using a collection of recovered primitives that belongs to the same object. In Section 3.3, we hypothesized that if a connection between two primitives was *visible*, i.e., a *strong* connection, the two primitives were connected in 3-D. Our model indexing strategy therefore consists of identifying all the *strongly* connected components in the primitive graph, each hypothesizing a set of object candidates according to the two-level hash function described in Figure 9(a).

At the first level, we hash on the basis of a string formed from the labels of the primitives in the strongly connected component. Each entry in this table points to a separate hash table at the second level which encodes primitive connections; each hash table at the second level corresponds to objects that contain a particular set of primitives (number and type). Once at the second level, we hash on the basis of a string formed from the connections in the strongly connected component. It is

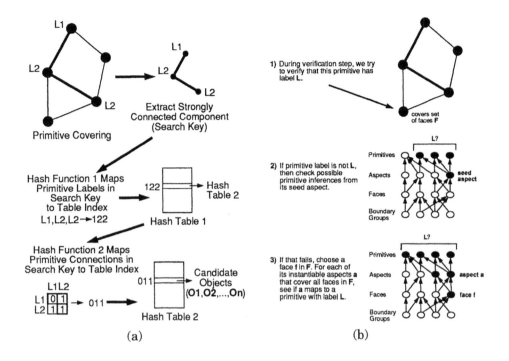

Figure 9. Object recognition strategies: (a) Unexpected object recognition (reprinted from [7], ©1992 Academic Press), (b) Expected object recognition (reprinted from [7], ©1992 Academic Press)

at this level that we further discriminate the objects on the basis of their primitive connections. The entire process is repeated for each strongly connected component in the primitive covering, resulting in a set of candidate object hypotheses.

4.1.2. Model Verification

Given a set of candidate models, each containing a given strongly connected component (or part thereof), the final step is to evaluate how well each model fits the scene (primitive covering). This process is known as hypothesis verification and consists of two stages. The first stage finds maximal correspondences between recovered primitives and model primitives.

Since our hash function ignores the connections in the strongly connected subgraph, the first step is to check that the strong connections in the strongly connected subgraph exist between the corresponding primitives in each candidate model; this may result in some candidate models being discarded. At this point, for each remaining candidate model, the strongly connected subgraph in the primitive graph is isomorphic to a subgraph of the candidate model. We then grow this correspondence according to the following steps:

goodness of fit = 0
for each *image primitive* **in** *correspondence* **do**
    goodness of fit = *goodness of fit* + probability(*image primitive*)
for each *arc* **in** *primitive subgraph* **do**
    if *arc* is weak **then**
      goodness of fit = *goodness of fit* + c_1
    else (*arc* is strong)
      if *arc* correctly specifies attachment surfaces **then**
        goodness of fit = *goodness of fit* + c_2
      else
        goodness of fit = *goodness of fit* + c_3

Figure 10. Goodness of fit algorithm for model candidates (reprinted from [7], ©1992 Academic Press)

1. Given a correspondence between a primitive (covering) subgraph PS and a model subgraph MS, we first choose a model primitive M_i that is not contained in the model subgraph, but is connected to a primitive M_j in the model subgraph. In the primitive subgraph, let the primitive corresponding to M_j be P_j.

2. Among the neighbors (through strong or weak connections) of P_j in the primitive covering which are not contained in PS, select those whose label matches that of M_i. If more than one such neighbor exists, we create a new correspondence for each neighbor.

We repeat this sequence of steps for each correspondence until its size stabilizes. The entire process is then repeated for each strongly connected subgraph in the primitive covering. The final result is a list of correspondences, each mapping a subgraph of the primitive covering to a model subgraph.

The final step ranks the correspondences according to a goodness of fit measure defined by the algorithm shown in Figure 10. The goodness of fit measure is a function of the size of the correspondence, the probability of the recovered primitive hypotheses, the visibility of the primitive connections, and the degree to which the connections are correctly specified. The input to the algorithm is a *correspondence*, consisting of a *primitive subgraph* whose nodes represent *image primitives*, and a *model subgraph* whose nodes represent *model primitives*.[13]

Once the correspondences are ranked according to the goodness of fit measure, we choose the best correspondence and remove those aspects from the image that cor-

---

[13] For the experiments described in Section 5, the values of c_1, c_2, and c_3 were chosen to be 1.0, 3.0, and 2.0, respectively.

respond to the recognized primitives. From the remaining aspects, forming a new aspect covering, we repeat the entire process. We first apply the primitive covering algorithm, establish primitive connectivity, extract strongly connected components, determine candidate models, grow and rank the correspondences, and select the most likely correspondence. The process is repeated until no aspects remain in the image. At any stage, a primitive covering may not yield any recognizable objects, i.e., candidate models. In this case, we generate a new primitive covering from the current aspect covering and repeat the process. Only when all primitive coverings are exhausted do we generate a new aspect covering.

4.2. Expected Object Recognition

In the domain of expected object recognition, the image is searched for one or more instances of a particular object. In this case, there is no need for a complicated indexing step since we know to what object the image features will be matched. Instead, we are faced with the question: What features of the object do we search for in the image? Our approach is to start with a primitive covering and then constrain further primitive and aspect covering generation by exploiting knowledge of the object. The assumption here is that the first aspect and primitive coverings of the scene represent a correct interpretation for much of the scene, and provide a good starting point for object search.

Figure 9(b) illustrates our approach to expected object recognition. The first step is the generation of the first primitive covering given the first aspect covering; this represents the most likely interpretation of the scene in terms of recovered shape. Next, as in the case of our approach to unexpected object recognition, we extract the strongly connected components. For each strongly connected component, the indexing step returns a list of model candidates that contain the component. Since we are looking for a particular object, we can discard all candidates but the object we are searching for (if it exists as a candidate). As before, we grow the correspondence between image and model primitives. However, it is during this last step that our approach differs.

In the unexpected object recognition algorithm, we attempt to completely recognize the entire primitive covering. Only when recognition fails do we generate another primitive covering. Furthermore, only when all primitive coverings are exhausted do we generate another aspect covering. Our expected object recognition approach attempts to integrate the three processes based on knowledge of which object part we are searching for.

During the correspondence growing step, some primitive in the primitive covering is checked to see if it matches some primitive in the model. If not, the unexpected object recognition approach does not include that primitive in the correspondence, i.e., growth of the correspondence is discontinued through that primitive. However, there may be an alternative primitive interpretation of that primitive's seed aspect (aspect used to infer primitive) that matches the expected model primitive. Recall that from the aspect covering, each aspect was used to infer a list of primitives in decreasing order of probability. By checking the various primitive hypotheses (in which the current primitive is included), we may find that the primitive label we are searching for is among the possible primitive interpretations of the seed aspect. If so, we choose the alternate interpretation and add it to the correspondence. If not, we can probe deeper for the correct interpretation.

Descending one more level of the aspect hierarchy, there may be other aspect interpretations of the faces belonging to the seed aspect. Furthermore, one of these interpretations may be used to infer the primitive we are searching for. Therefore, our strategy consists of searching the aspect labels of the faces belonging to the seed aspect for an aspect which not only can be verified, but whose mapping to the desired primitive has a nonzero conditional probability. If such an aspect is found, the desired primitive is inferred and added to the correspondence. If the search is unsuccessful, the correspondence will not include the faces belonging to the seed aspect.

5. Results

We have built a system to demonstrate our approach to 3-D object recognition. The system is called OPTICA (**O**bject recognition using **P**robabilistic **T**hree-dimensional **I**nterpretation of **C**omponent **A**spects), and has been implemented in Common Lisp on a Sun $4/330^{TM}$ workstation. The image preprocessing which takes an image and returns a gap-filled skeletonized image is performed in the KBVision environment. In this section we apply OPTICA to the six images presented in Figure 11; images (c) and (d) are real images of objects constructed out of clay, while (a), (b), (e), and (f) were generated using Pentland's Thingworld modeling tool.

Figures 12(a) through (d) present the results of applying OPTICA to the images in Figures 11(a) through (d). There are four windows in each figure. At the top, the image window contains the contours extracted from the image, along with the face numbers. To the left is the diagnostic window describing the recovered primitives (primitive covering). The mnemonics PN, PL, PP, and PS, refer to primitive number (simply an enumeration of the primitives in the covering), primitive label (see Figure 2(a)), and primitive probability, respectively. The mnemonics AN, AL, AP, and AS refer to the aspect number (an enumeration), aspect label (see [8]), aspect probability, and aspect score (how well aspect was verified), respectively. The mnemonics FN, FL, FP, and PS refer to face number (in image window), face label (see [8]), face probability, and corresponding primitive attachment surface (see [8]), respectively, for each component face of the aspect.

To the right are the "Recognized Objects" and "Primitive Connections" windows. The "Recognized Objects" window indicates the aspect covering iteration and primitive covering iteration (given the aspect covering). In addition, this box lists all objects currently (at the above iterations) identified in the image, including their corresponding primitive numbers (PN). The "Primitive Connections" window indicates the primitive connections by primitive number (PN); if two primitives are strongly connected, a list of probable attachment surfaces appears in parentheses next to the primitive number. This list is not exclusive, but rather a list of likely candidates.

In each case ((a) through (d)), the first aspect and primitive coverings represent

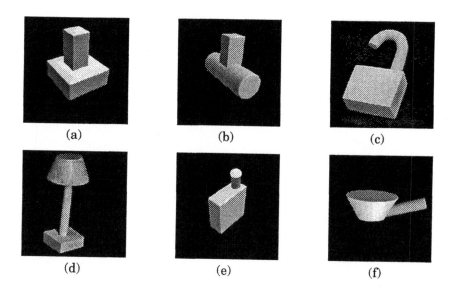

Figure 11. The six images input to OPTICA ((c) and (d) are real images, while (a), (b), (c), (d) were generated using Pentland's Thingworld modeling tool) ((c) and (d) reprinted from [8], ©1992 IEEE)

the correct interpretation of the scene. Total execution time, excluding the time required to extract the contours (90 seconds), was approximately 60 seconds in each case; the majority of that time was spent in building a graph representation of an image face, i.e., connected components analysis, curve partitioning, curve classification, and symmetry and parallelism detection.

Figure 13(a) presents the results of applying OPTICA to the image in Figure 11(e). In this case, the first aspect and primitive coverings do not represent the correct interpretation of the scene (arm of lock is misinterpreted as a cylinder). If we let the algorithm continue, we arrive at the correct interpretation with the second primitive covering given the first aspect covering, as shown in Figure 13(b). When we apply the expected object recognition algorithm (searching for a lock), the search for the bent cylinder arm is constrained to the area covered by the cylinder faces (first covering). In this case, the correct primitive interpretation was found by descending only one level of the aspect hierarchy, i.e. the aspect was correct but the primitive inference was wrong.

Similarly, Figure 13(c) presents the results of applying OPTICA to the image in Figure 11(f). Again, the first aspect and primitive coverings do not represent the

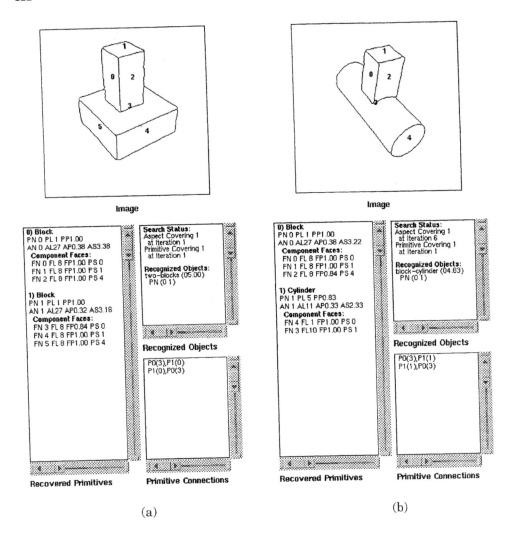

(a)

(b)

Figure 12. Results of applying OPTICA to images in Figures 11(a) and (b)

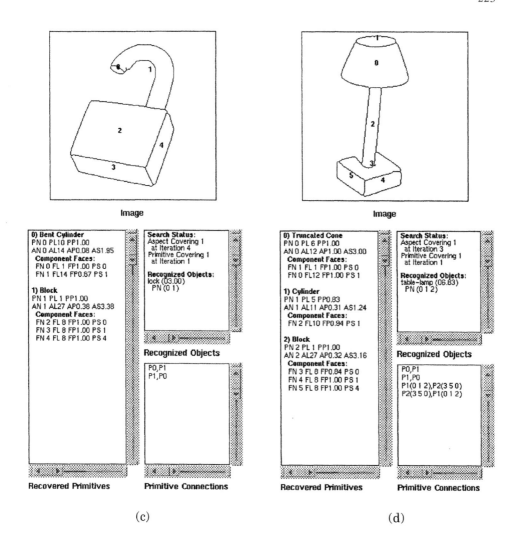

Figure 12. Results of applying OPTICA to images in Figures 11(c) and (d)

224

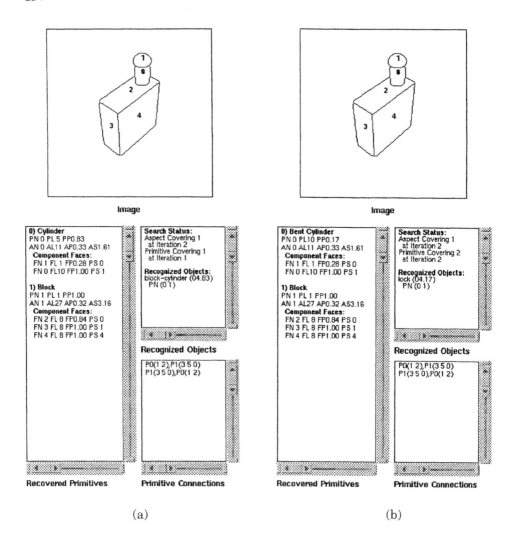

| Image | | Image |

0) Cylinder	**Search Status:**	**0) Bent Cylinder**	**Search Status:**
PN 0 PL 5 PP0.83	Aspect Covering 1	PN 0 PL10 PP0.17	Aspect Covering 1
AN 0 AL11 AP0.33 AS1.61	at Iteration 2	AN 0 AL11 AP0.33 AS1.61	at Iteration 2
Component Faces:	Primitive Covering 1	**Component Faces:**	Primitive Covering 2
FN 1 FL 1 FP0.28 PS 0	at Iteration 1	FN 1 FL 1 FP0.28 PS 0	at Iteration 2
FN 0 FL10 FP1.00 PS 1		FN 0 FL10 FP1.00 PS 1	
	Recognized Objects:		**Recognized Objects:**
1) Block	block–cylinder (04.83)	**1) Block**	lock (04.17)
PN 1 PL 1 PP1.00	PN (0 1)	PN 1 PL 1 PP1.00	PN (0 1)
AN 1 AL27 AP0.32 AS3.16		AN 1 AL27 AP0.32 AS3.16	
Component Faces:		**Component Faces:**	
FN 2 FL 8 FP0.84 PS 0		FN 2 FL 8 FP0.84 PS 0	
FN 3 FL 8 FP1.00 PS 1		FN 3 FL 8 FP1.00 PS 1	
FN 4 FL 8 FP1.00 PS 4		FN 4 FL 8 FP1.00 PS 4	

| | **Recognized Objects** | | **Recognized Objects** |

| | P0(1 2),P1(3 5 0) | | P0(1 2),P1(3 5 0) |
| | P1(3 5 0),P0(1 2) | | P1(3 5 0),P0(1 2) |

| **Recovered Primitives** | **Primitive Connections** | **Recovered Primitives** | **Primitive Connections** |

| (a) | (b) |

Figure 13. Results of applying OPTICA to image in Figure 11(e)

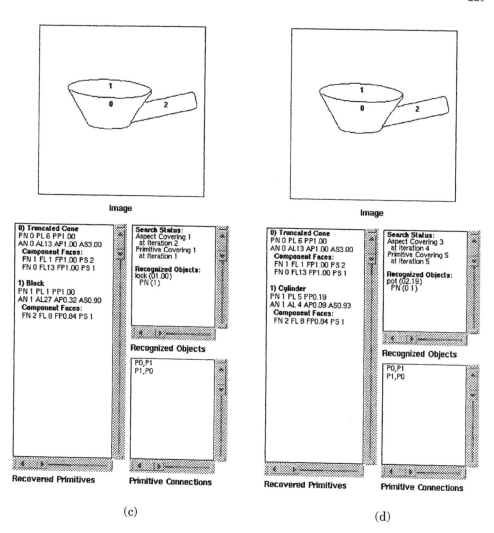

(c)

(d)

Figure 13. Results of applying OPTICA to image in Figure 11(f)

correct interpretation of the scene (handle of pot is misinterpreted as a block). If we let the algorithm continue, we eventually arrive at the correct interpretation with the fifth primitive covering given the third aspect covering, as shown in Figure 13(b). When we apply the expected object recognition algorithm (searching for a pot), the search for the cylinder handle is constrained to the area covered by the block faces (first covering). In this case, the correct primitive interpretation was found by descending two levels of the aspect hierarchy, i.e. a new aspect was recovered before the correct primitive inference could be made.

6. Conclusions

The inefficiency of most 3-D object recognition systems is reflected in the relatively small number of objects in their databases; in many cases, algorithms are demonstrated on a single object model. The major problem is that these systems terminate the bottom-up primitive extraction phase very early, resulting in simple primitives such as lines, corners, and inflections. These primitives do not provide very discriminating indices into a large database, resulting in a large number of hypothesized matches. Consequently, the burden of recognition falls on top-down verification, which for simple geometric image features requires both accurate estimates of the object's pose and prior knowledge of the object's geometry.

We instead index into the model database with more discriminating primitives, i.e., ones that do not require precise knowledge of model geometry or accurate estimates of pose. An appropriate choice for higher-order indexing primitives is the class of volumetric primitives which capture the intuitive notion of an object's parts. In this approach, object models are constructed from object-centered 3-D volumetric primitives. The primitives, in turn, are represented in the image by a set of viewer-centered aspects.

Unlike typical aspect-based recognition systems which model each entire object in a database using a set of aspects, we use aspects to model a finite number of volumetric parts used to construct the objects. The size of the resulting aspect set is fixed and, more important, *independent* of the contents of the object database. To accommodate the representation of occluded aspects arising from occluded primitives, we introduce a hierarchical aspect structure, called the *aspect hierarchy*, based on the faces appearing in the aspect set. The ambiguous mappings between levels of the aspect hierarchy are captured by a set of conditional probabilities resulting from a statistical analysis of the aspects. The aspect hierarchy is precomputed *once* off-line and remains fixed while objects are added or removed from the database.

We have demonstrated our approach using a vocabulary of primitives resembling Biederman's geons [2]; however, our approach is *not* dependent on geons as object modeling primitives. Although any selection of volumetric primitives can be mapped to a set of aspects, our hierarchical aspect representation is particularly appropriate for primitives with distinct surfaces, i.e., primitives whose aspects contain distinct faces. The use of a face-based aspect hierarchy is the backbone of our approach, allowing us to obtain probabilistic rules for inferring more complex features from less complex features, and for merging oversegmented contours and regions [8]. Although the individual features represented in our aspect hierarchy may change when using other types of volumetric primitives, the concepts of representing a set of 3-D volumetric primitives as a probabilistic hierarchy of image features, and exploiting these probabilities during recovery and recognition are applicable to any object representation that models objects using volumetric parts.

The cost of extracting more complex primitives from the image is the difficulty of grouping less complex features into more complex features; the number of possible groupings is enormous. Our recovery algorithm uses a statistical analysis of the aspects (explicitly represented in the aspect hierarchy) to rank-order the possible groupings. The result is a heuristic that has been demonstrated to quickly arrive at the correct interpretation. Note, however, that our approach will, if need be, enumerate *all* possible interpretations (or groupings); the correct interpretation of any scene, no matter how ambiguous or unlikely, will eventually be generated.

We have presented a database indexing scheme that maps a group of recovered primitives and their connections to a hash table location whose contents specify those models containing a similar primitive structure. The candidate hypotheses are then topologically verified and ranked based on the strengths of the hypothesized primitives. We show that for both the problems of unexpected and expected object recognition, the same recognition engine can be used.

7. Acknowledgements

I would like to thank Azriel Rosenfeld, Sandy Pentland, and more recently, John Tsotsos for their generous support and guidance throughout this work. I would also like to thank Larry Davis, Göran Olofsson, Lars Olsson, Jan-Olof Eklundh, and especially Suzanne Stevenson for many insightful discussions, and for reviewing earlier drafts of this work. Finally, I would like to thank Gérard Medioni for supplying the curve partitioning software, and Ken Shih for implementing the Beymer gap-filling algorithm.

REFERENCES

1 D. Beymer. Finding junctions using the image gradient. Memo 1266, Artificial Intelligence Laboratory, Massachusetts Institute of Technology, 1991.
2 I. Biederman. Human image understanding: Recent research and a theory. *Computer Vision, Graphics, and Image Processing*, 32:29–73, 1985.
3 R. Bolles and P. Horaud. 3dpo: A three-dimensional part orientation system. *The International Journal of Robotics Research*, 5(3):3–26, 1986.
4 R. Brooks. Model-based 3-D interpretations of 2-D images. *IEEE Transactions on Pattern Analysis and Machine Intelligence*, 5(2):140–150, 1983.
5 J. Canny. A computational approach to edge detection. *IEEE Transactions on Pattern Analysis and Machine Intelligence*, 8(6):679–698, 1986.
6 S. Dickinson, A. Pentland, and A. Rosenfeld. The recovery and recognition of three-dimensional objects using part-based aspect matching. Technical Report CAR-TR-572, Center for Automation Research, University of Maryland, 1991.
7 S. Dickinson, A. Pentland, and A. Rosenfeld. From volumes to views: An approach to 3-D object recognition. *Computer Vision, Graphics, and Image Processing: Image Understanding*, 55(2), 1992.

8   S. Dickinson, A. Pentland, and A. Rosenfeld. 3-D shape recovery using distributed aspect matching. *IEEE Transactions on Pattern Analysis and Machine Intelligence*, 14(2):174–198, 1992.

9   T. Fan, G. Medioni, and R. Nevatia. Recognizing 3-D objects using surface descriptions. In *Proceedings, Second International Conference on Computer Vision*, pages 474–481, Tampa, FL, 1988.

10 W. Grimson. Recognition of object families using parameterized models. In *Proceedings, First International Conference on Computer Vision*, pages 93–100, London, UK, 1987.

11 W. Grimson. The effect of indexing on the complexity of object recognition. Memo 1226, Artificial Intelligence Laboratory, Massachusetts Institute of Technology, 1990.

12 W. Grimson and T. Lozano-Pérez. Model-based recognition and localization from sparse range or tactile data. *International Journal of Robotics Research*, 3(3):3–35, 1984.

13 D. Huttenlocher. Three-dimensional recognition of solid objects from a two-dimensional image. Technical Report 1045, Artificial Intelligence Laboratory, Massachusetts Institute of Technology, 1988.

14 D. Huttenlocher and S. Ullman. Object recognition using alignment. In *Proceedings, First International Conference on Computer Vision*, pages 102–111, London, UK, 1987.

15 D. Huttenlocher and S. Ullman. Recognizing solid objects by alignment with an image. *International Journal of Computer Vision*, 5(2):195–212, 1990.

16 J. Koenderink and A. van Doorn. The internal representation of solid shape with respect to vision. *Biological Cybernetics*, 32:211–216, 1979.

17 Y. Lamdan, J. Schwartz, and H. Wolfson. On recognition of 3-D objects from 2-D images. In *Proceedings, IEEE International Conference on Robotics and Automation*, pages 1407–1413, Philadelphia, PA, 1988.

18 D. Lowe. Fitting parameterized three-dimensional models to images. *IEEE Transactions on Pattern Analysis and Machine Intelligence*, 13(5):441–450, 1991.

19 David Lowe. *Perceptual Organization and Visual Recognition*. Kluwer Academic Publishers, Norwell, MA, 1985.

20 N. Nilsson. *Principles of Artificial Intelligence*, chapter 2. Morgan Kaufmann Publishers, Inc., Los Altos, CA, 1980.

21 A. Pentland. Perceptual organization and the representation of natural form. *Artificial Intelligence*, 28:293–331, 1986.

22 P. Saint-Marc and G. Medioni. Adaptive smoothing: A general tool for early vision. *IEEE Transactions on Pattern Analysis and Machine Intelligence*, 13(6):514–529, 1991.

23 D. Terzopoulos, A. Witkin, and M. Kass. Symmetry-seeking models and 3d object recovery. *International Journal of Computer Vision*, 1:211–221, 1987.

24 D. Thompson and J. Mundy. Model-directed object recognition on the connection machine. In *Proceedings, DARPA Image Understanding Workshop*, pages 93–106, Los Angeles, CA, 1987.

Three-Dimensional Object Recognition Systems
A.K. Jain and P.J. Flynn (Editors)
© 1993 Elsevier Science Publishers B.V. All rights reserved.

Appearance-Based Vision and the Automatic Generation of Object Recognition Programs[1]

Keith D. Gremban and Katsushi Ikeuchi[a]

[a] School of Computer Science, Carnegie Mellon University, 5000 Forbes Avenue, Pittsburgh, Pennsylvania 15213-3890

Abstract

The generation of recognition programs by hand is a time-consuming, labor-intensive task that typically results in a special purpose program for the recognition of a single object or a small set of objects. Recent work in automatic code generation has demonstrated the feasibility of automatically generating object recognition programs from CAD-based descriptions of objects. Many of the programs which perform automatic code generation employ a common paradigm of utilizing explicit object and sensor models to predict object appearances; we refer to the paradigm as appearance-based vision, and refer to the programs as vision algorithm compilers (VACs). In this paper, we discuss the paradigm of appearance-based vision and present in detail two specific VACs: one that computes feature values analytically, and a second that utilizes an appearance simulator to synthesize sample images.

1. Introduction

The generation of object recognition programs by hand is a time-consuming, labor-intensive process that typically results in a special purpose program for the recognition of a single object or a small set of objects. The reason for this lies in the design methodology: on the basis of a representative set of sample images, the designer experiments on the images, selects a set of image features, and specifies the procedure to be used in matching image features to object features. The entire process requires a highly skilled and motivated designer.

Recently, research has been conducted towards the goal of automatically generating recognition code from a CAD-based description of an object. Most current industrial parts are designed and manufactured using computer-aided tools, so CAD descriptions exist for most parts; automatic generation of recognition code from the same model information used for design and manufacture would be an efficient, cost-effective approach. A number of programs for automatic generation of object recognition code have been written, and many of these programs employ

[1]This research was sponsored by the Avionics Laboratory, Wright Research and Development Center, Aeronautical Systems Division (AFSC), U. S. Air Force, Wright-Patterson AFB, Ohio 45433-6543 under Contract F33615-90-C-1465, ARPA Order No. 7597.

a common paradigm in which explicit object and sensor models are used to predict object appearances. We refer to the paradigm as *appearance-based vision*, and programs which generate object recognition programs are called *vision algorithm compilers*, or *VACs*.

Appearance-based vision represents an extension to the familiar paradigm of model-based vision. Model-based vision defines an execution-time strategy of matching observed image features to model features, but does not address the issue of defining a strategy for selecting either features or matching procedures. Appearance-based vision systems employ model-based matching during the execution-time recognition phase, but also employ a characteristic methodology during the off-line compilation phase, during which features are selected and processing strategies determined.

In principle, a CAD-like object model, augmented with sensor-specific information like surface color, roughness, and reflectance, can be used in conjunction with a sensor model to predict the appearance of the object under any specified set of viewing conditions. VACs can predict object appearances in two different ways: analytically, or synthetically.

In relatively simple domains, feature values can be analytically determined from model information. As objects and their properties grow in complexity, however, effects such as self-shadowing and inter-reflection become more important, but are difficult to incorporate into an analysis. Analytic prediction of appearances is therefore impractical for some domains. An alternative to analytic prediction of appearances is the use of an appearance simulator. An appearance simulator generates synthetic images of objects under specific viewing conditions, with respect to a given sensor. An appearance simulator can be used to generate a representative collection of sample images, which can then be automatically processed and analyzed to extract the feature values that characterize object appearances.

In this paper, we discuss the paradigm of appearance-based vision. In the next section, we review the state-of-the-art in appearance-based vision and the automatic generation of object recognition programs. In the course of the review, the defining characteristics of appearance-based vision systems will be noted. Following the review, we will present in detail two VACs which typify appearance-based systems. In section 3, we present a VAC that employs analytic prediction of appearances, and the advantages and limitations of this approach are discussed. Then, in section 4, we present a VAC that utilizes an appearance simulator. A brief summary concludes the paper.

2. The Paradigm of Appearance-Based Vision

The history of computer vision research has largely been a study in making vision systems work. Little attention has been paid to the study of how to design and build application systems. For example, the dominant paradigm in computer vision is that of *model-based* vision. Briefly, the model-based paradigm can be characterized as *hypothesize-predict-verify*: given a collection of image features, *hypothesize* a match of an image feature to a model feature; use the hypothesized match to *predict* the image locations of other model features; *verify* the predictions and update the hypothesis. The paradigm does not specify how to select the features to use, or how to perform the matching. The paradigm defines an approach to execution-time processing, rather than an approach to system building.

Because every computer vision system is essentially a custom solution to a specific problem, a typical computer vision system is expensive to develop and install, and is capable of recognizing only a single part or a small number of parts under very special conditions. Modifications to existing systems are difficult to make, and the cost to develop a new system is as high as that of the first system. Clearly this is an unacceptable situation.

Appearance-based vision addresses the problem of building cost-effective computer vision systems; it specifies a methodology for the automatic generation of object recognition programs. Appearance-based vision is an extension of the model-based paradigm that formalizes and automates the design process. Appearance-based vision can be characterized as an automated process of analyzing the appearances of objects under specified observation conditions, followed by the automatic generation of model-based object recognition programs based on the preceding analysis, and ending with with repeated execution of the generated program.

One characteristic of appearance-based vision is that both objects and sensors are explicitly modeled, and therefore exchangeable. Hence, a given VAC can generate object recognition code for many different objects using the same sensor model, or the set of objects can be fixed and the sensor models varied.

A VAC incorporates a two phase approach to object recognition. The first phase is performed off-line and consists of analysis of predicted object appearances and the generation of object recognition code. The second phase takes place on-line, and consists of applying the previously generated code to input images. The first phase consists ideally of a single program execution for a given object recognition task, and can be relatively expensive. The second phase consists of many executions of the output object recognition program, which must be both fast and cost-effective. The high cost of the first phase is amortized over a large number of executions of the second phase.

In the next subsection, we present an historical overview of research on appearance-based vision. Then, building on the historical perspective, we enumerate and elaborate on the commonalities between the systems; it is this set of common characteristics that define the paradigm of appearance-based vision.

2.1. Historical Perspective

Goad [10] presented an early version of a VAC. In Goad's system, an object is described by a list of edges and a set of visibility conditions for each edge. Visibility is determined by checking visibility at a representative number of viewpoints obtained by tessellating the viewing sphere. Object recognition is performed by a process of iteratively matching object and image edges until either a satisfactory match is found, or the algorithm fails. The sequence of matchings is compiled during the off-line analysis phase. Goad's system was not completely automatic, however. Goad selected edges as the features to be used for recognition, and the order of edge matching was specified by hand.

The 3DPO system of Bolles and Horaud [4] was built with the intended goal of using off-line analysis to produce the fastest, most efficient on-line object recognition program possible. 3DPO utilized the local-feature-focus method, in which a prominent *focus* feature is initially identified, and then secondary features predicted from the focus feature are used to fine-tune the localization result. The system was not fully automatic, as the feature matching strategies were determined interactively.

Ikeuchi and Kanade [15] [17] first pointed out the importance of modeling sensors as well as objects in order to predict appearances, and noted that the features that are useful for recognition depend on the sensor being used. Their system predicts object appearances at a representative set of viewpoints obtained by tessellating the viewing sphere. The appearances are grouped into equivalence classes with respect to the visible features; the equivalence classes are called *aspects*. A recognition strategy is generated from the aspects and their predicted feature values, and is represented as an interpretation tree. Each interpretation tree specifies the sequence of operations required to precisely localize an object.

Hansen and Henderson [11] demonstrated a system that analyzed 3D geometric properties of objects and generated a recognition strategy. The system was developed to make use of a range sensor for recognition. The system examines object appearances at a representative set of viewpoints obtained by tessellating the viewing sphere. Geometric features at each viewpoint are examined, and the properties of robustness, completeness, consistency, cost, and uniqueness are evaluated in order to select a complete and consistent set of features. For each model, a strategy tree is constructed, which describes the search strategy used to recognize and localize objects in a scene.

The system of Arman and Aggarwal [1] was designed to be capable of selecting the proper sensor for a given task. Starting with a CAD model of an object, the system builds up a tree in which the root node represents the object, and the leaves represent features (where features are dependent upon the sensor selected), and a path from the root to a leaf passes through nodes representing increasing specificity. Each arc in the tree is weighted by a "reward potential" that represents the likely gain from traversing that link. At run time, the system traverses the tree from the root to the leaves, choosing the branch with the highest weight at each level, and backtracking when necessary.

The PREMIO system of Camps et al [5] predicts object appearances under various conditions of lighting, viewpoint, sensor, and image processing operators. Unlike other systems, PREMIO also evaluates the utility of each feature by analyzing the detectability, reliability, and accuracy. The predictions are then used by a probabilistic matching algorithm that performs the on-line process of identification and localization.

The BONSAI system of Flynn and Jain [8] identifies and localizes 3D objects in range images by comparing relational graphs extracted from CAD models to relational graphs constructed from range image segmentation. The system constructs the relational graphs off-line using two techniques: first, view-independent features are calculated directly from a CAD model; second, synthetic images are constructed for a representative set of viewpoints obtained by tessellating the viewing sphere, and the predicted areas of patches are determined and stored as an attribute of the appropriate relational graph node. During the on-line recognition phase, an interpretation tree is constructed which represents all possible matchings of the graph constructed from a range image, and the stored model graph. Recognition is performed by heuristic search of the interpretation tree.

Sato, et al [21] demonstrated a system for recognition of specular objects. During an off-line phase, the system generates synthetic images from a representative set of viewpoints. Specularities are extracted from each image, and the images are grouped into aspects according to shared specularities, and each specularity is evaluated in terms of its detectability and reliability. At execution time, an input image is classified into a few possible aspects using a continuous classification procedure based on Dempster-Shafer theory. Final verification and localization is performed using deformable template matching.

2.2. The Common Threads

After reviewing the different appearance-based systems that have been constructed, it is useful to go back and point out the common processing steps.

2.2.1 Two Phases of Processing

Each of the systems discussed employ two distinct phases of processing. The first phase, variously called off-line, compilation, or analysis, consists of analyzing object appearances and constructing recognition strategies. The second phase, called on-line, run-time, or execution, consists of applying the strategies generated in the first phase to an actual recognition task. In general, computational efficiency is not a concern in the first phase, since it does not directly affect the actual time or effort required to perform object recognition, and the cost is only incurred once. In contrast, the second phase is expected to execute many times as part of an application, and consequently must be efficient. In effect, the time spent during strategy generation can be amortized over the number of executions of the resultant strategy.

2.2.2. Explicit Object and Sensor Models

Any object recognition system must match the appearance of an object with respect to some sensor, to a model of the object. Consequently, to automatically generate a program for object recognition, it is necessary to predict and analyze object appearances. Objects appear differently to different sensors, so in order to predict object appearances, sensors must be modeled as well. An appearance-based system therefore includes both object and sensor models.

The early appearance-based systems only made use of explicit object models and utilized implicit sensor models, although the need for different types of models to represent different types of detectable features was acknowledged. All recent systems have emphasized the fact that appearance depends upon the sensor and include explicit models of both objects and sensors. Models may be exchanged so that the same VAC can generate object recognition programs for a variety of objects and sensors.

2.2.3. Appearance Prediction and Analysis

In general, there are two approaches to predicting object appearances: analytic and synthetic. The analytic approach uses the information stored in object and sensor models to analytically predict the appearance of an object from various viewpoints. Alternatively, it is possible to generate images of objects under specific sensor conditions and analyze the synthetic images. Both techniques have advantages and disadvantages that are discussed more completely in the next two sections.

The appearance of an object with respect to a sensor is characterized by means of the features that can be extracted from the sensor image. Hence, each sensor model includes a feature set, and a collection of image processing operators that are used to extract the features.

The appearance of an object also varies with respect to the relative geometry between sensor and object, which can be referred to as the viewpoint. Potentially, there are six degrees of freedom in viewpoint, each of which spans an infinite number of parameter values. Clearly, exhaustive computation of all possible appearances is impossible. To make the set of possible appearances manageable, similar appearances are grouped into sets called aspects. Formally, an aspect is a class of topologically equivalent views of an object [18]. However, since different sensors detect different features, the formal definition of an aspect is usually relaxed to be a class of appearances that are equivalent with respect to a feature set.

A substantial amount of work has been performed on deriving methods for analytically determining the collection of aspects of an object. Representative examples of work in this area include Plantinga and Dyer [20], Kriegman and Ponce [19], and Chen and Freeman [6].

An alternative to the exact analytic computation of aspects is the exhaustive approach, in which viewpoints are sampled uniformly throughout the space of possible viewpoints, and then similar viewpoints are grouped together. An approach of this sort was used by Ikeuchi and Kanade [17], Hansen and Henderson [11], Flynn and Jain [8], and Sato, et al [21]. In each of these systems, the space of possible viewpoints is uniformly tessellated, and the object appearance is predicted from each viewpoint corresponding to the center of a tessel. The fidelity of the sampling can be increased by subdividing each tessel. This approach is more general than the analytic approach, since the same procedure can be used independently of the sensor or feature set.

2.2.4. Generation of a Recognition Strategy

The result of the off-line, compilation phase of an appearance-based system is a strategy for object recognition. The strategy is often represented in the form of a tree that represents the sequence of operations to perform at each step of the recognition process. Since the generation of a strategy is performed off-line, it is possible to perform relatively expensive optimization.

There are many different computational approaches that can be employed for object recognition. Suetens, et al [23] is a recent survey of the range of approaches.

Other surveys of object recognition include Besl and Jain [3], and Chin and Dyer [7]. A VAC can be constructed for any given approach. Consequently, there is no standard form for the recognition strategy output by a VAC; all that can be said is that the strategy consists of executable code.

3. A VAC Utilizing Analytic Feature Prediction

The VAC discussed in this section was designed to generate an object localization (pose determination) program for a bin-picking task, and was initially presented in [14], at which time the system was not fully automatic. Further research has led to complete automation of program generation [16], as well as optimization of the resulting code [13]. This section presents the complete system.The inputs to the VAC consist of an object model, specifying geometric and photometric characteristics of the object, and a sensor model, specifying the sensor characteristics necessary for predicting object appearances and feature variations. The output consists of a recognition strategy in the form of an interpretation tree.

The VAC presented below utilizes a two-stage approach to object localization. The first stage, *aspect classification*, classifies an input image into an instance of an aspect. This stage is equivalent to rough localization, since the range of possible object poses is constrained to be consistent with the observed aspect. The second stage, *linear shape change determination* performs more accurate localization by analyzing the shape change within an aspect.

3.1 Explicit Object and Sensor Models

Computer vision systems can use many different types of sensors. A sensor can be considered to be a transducer that transforms object features into image features, and different sensors yield different image features. For example, a laser range finder detects the range and orientation of object surfaces, while edge-based binocular stereo yields the range computed by triangulation on detected edges. A sensor model must specify the features detectable by the sensor.

The list of features describes the qualitative characteristics of a sensor. The quantitative characteristics are given by the *detectability* and *reliability* of each feature. Detectability specifies the conditions under which a given feature can be detected. Reliability specifies the expected error in the feature value. Both characteristics depend on the configuration of the object feature and the sensor.

The detectability of a feature by a given sensor depends on factors such as range, relative attitude, reflectivity and so on. In many applications, such as industrial workstations, many of the factors can be fixed, and relative attitude becomes the dominant factor. To consider relative attitude, fix the sensor coordinate system, and consider the relationship of a feature coordinate system with respect to it. The feature coordinate system is defined such that the z-axis is aligned with the surface normal; x and y axes are assumed to be defined arbitrarily. There are three degrees of freedom in relative attitude. The *feature configuration space*, described below, provides a convenient representation for relative attitude that considerably simplifies many of the operations involved in predicting both detectability and reliability.

Consider a solid unit sphere, called the *orientation sphere*, or *o-sphere*, in which each relative orientation of the feature coordinate system corresponds to a point. The direction from the center of the sphere to the point defines the orientation of the feature z-axis. Each point on the surface of the o-sphere represents the unique orientation of feature coordinates obtained by rotation about an axis perpendicular to the plane formed by the north pole, the center of the sphere, and the surface point itself. Interior points represent rotation about the (new) z-axis defined by the ray from the center of the sphere to the point. The north pole is taken to be the case in which feature coordinates and sensor coordinates are aligned. One o-sphere can be defined for each object feature, and is referred to as the feature configuration space.

As an example, consider the case of a feature coordinate system aligned with sensor coordinates; this is represented as the point $(0, 0, 1)$ on the o-sphere. If the feature system in now rotated by $+90$ deg about the y-axis, the new orientation is represented by the point $(1, 0, 0)$ on the o-sphere. An additional rotation of $+90$ deg about the *new* z-axis moves the point inward to $(0.75, 0, 0)$. Figure 1 further illustrates the concept.

A sensor system consists of two components: an illuminant, and a detector. For a feature to be detected, it must be visible to both components. For a given feature, a separate configuration space can be defined for each sensor component. Within each configuration space, the configurations for which the feature is detectable can be easily defined by geometric constraints. The detectability of a feature with respect to a given sensor is then the intersection of the detectable regions of the configuration spaces of each of the sensor components. An example of the detectability computation for a light-stripe range finder is shown in Figure 2.

For a given viewpoint, the appearance of an object with respect to a particular sensor can be defined by a list of the detectable object features, along with the values of parameters extracted from those features. A viewpoint corresponds to a single point in the configuration space of each feature of the object. A feature is detectable if and only if the point representing the sensor viewpoint lies in the detectable region of the feature configuration space, and if no other part of the object occludes the feature. These conditions can easily be checked using a geometric modeler.

3.2. Predict and Analyze Appearances

The techniques presented above make it possible to analytically determine the detectability of any feature from any viewpoint. The combination of features, along with predicted feature values, defines the appearance of the object. Next, the capability to predict appearances must be used to determine the aspects of the object.

The system employs an exhaustive approach to predicting appearances. The exhaustive approach is relatively easy to implement, especially for cases in which the distance between sensor and object is assumed fixed. Then, all possible viewpoints can be represented as points on the surface of a sphere centered on the object.

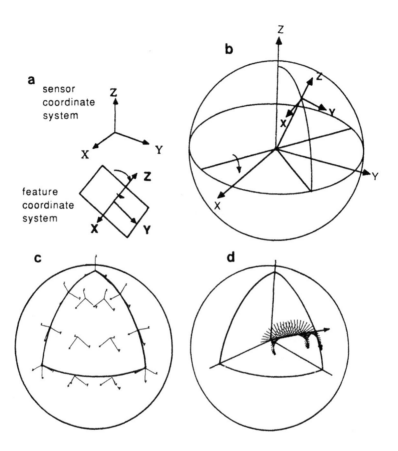

Figure 1. Feature Configuration Space. (a) Relationship between sensor coordinates and feature coordinates. (b) Feature coordinates as points on the o-sphere.
The bottom left drawing depicts the coordinates corresponding to points on the surface of the o-sphere, while the bottom right drawing depicts the coordinates along
one axis of the o-sphere.

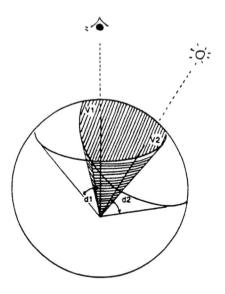

Figure 2. Detectability of a Face for a Light-stripe Range Finder. The detectable region is the intersection of the detectability of the illuminant and the detectability of the sensor.

A tessellation of the sphere using a geodesic dome which divides the sphere into many small spherical triangles yields a nearly uniform sampling of viewpoints. The triangles can be subdivided repeatedly to yield any desired level of sampling resolution. At the center of each spherical triangle, the detectability of each feature can be computed as outlined above.

Aspects can be selected in many different ways, depending upon the features being considered. For example, aspects can be defined as collections of viewpoints for which the same set of features are visible. Alternatively, as used here, aspects can be defined on the basis of detectable faces. Consider an object with N faces (planar or curved) $S_1, S_2, \ldots, S_N$ and define the face-label $X = (X_1, X_2, \ldots, X_N)$, where $X_i = 1$ or 0 according to whether or not face S_i is detectable. Viewpoints with identical face labels are grouped together into aspects. For each aspect, a representative attitude is selected and used to calculate representative feature values. Each aspect can be characterized by the feature values of the representative attitude. Further, the ranges of feature values can be obtained by examining the range of values of the features of each of the viewpoints constituting an aspect. Figure 3 illustrates the process of view generation and aspect selection.

3.3. Generation of a Recognition Strategy

The generation of a recognition strategy depends to some extent upon the sensor used, or at least upon the features used. In this section, results are presented for

239

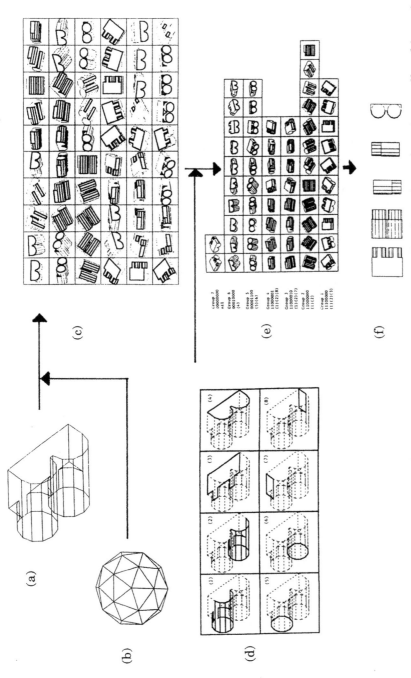

Figure 3. Extraction of Aspects. (a) Geometric model of an object. (b) The Gaussian sphere is tessellated into sixty triangles to represent viewpoints sampled. (c) Sixty computed appearances. Faces surrounded with bold lines are detectable by photometric stereo. (d) Eight component faces to be used for shape labeling. (e) The five aspects obtained through classification by shape label. Regions smaller than a certain threshold are regarded as undetectable. (f) Representative attitudes, one for each detectable aspect.

sensors which produce dense range maps.

3.3.1. Aspect Classification

Aspect classification is the process of classifying an input image into an instance of an aspect. Since an aspect represents a contiguous set of viewpoints, aspect classification is equivalent to rough localization. The parameters of object pose determined through aspect classification also provide good starting parameters for the stage of linear shape change determination that follows.

One way of performing aspect classification is to extract feature values from the input image and compare this set of values to the stored value ranges that characterize aspects. This approach may be very inefficient, however, since only a few of the features may be needed to perform classification, yet all are computed. A more cost-effective approach is to determine the computational cost of each feature, and then determine a discriminating set of features that minimizes the expected cost of classification.

A *classification tree*, or *decision tree*, is a tree in which each node represents a collection of classes and an associated test, and arcs represent the possible results of a test. Leaf nodes represent the final results of classification. Using a classification tree, a classification is performed by traversing the tree from the root to a leaf. A classification tree provides a convenient framework for optimizing the aspect classification process, since using a classification tree permits tests (and corresponding computations of feature values) to be performed sequentially.

Aspect classification trees can be used to optimize the classification process in the following way. In the off-line stage of processing, the entire set of possible aspect classification trees is examined systematically, and the minimum cost classification tree is identified and saved; this tree stores feature identifiers and test values at each node.

A path from the root of a classification tree to a leaf represents a complete classification operation. Computing the cost of a single path is straightforward. Each test requires a feature to be computed, and each such computation incurs a computational cost. Each node in the classification tree is assigned the cost of computing the feature needed for the corresponding test. The cost of a path is the sum of the costs of the intermediate nodes.

A classification tree contains many paths from the root to the leaf nodes, and different paths may be taken with different frequencies. Therefore, the cost of a classification tree is defined to be the expected cost of a classification; that is, the average cost taken over all possible inputs. The expected cost can be computed by weighting the cost at each node by the proportion of the sample population that will pass through the node. The cost of every node in the tree is summed and divided by the population size to yield the overall cost.

Finding a minimum cost classification tree can be formalized as a search problem over another kind of tree, called a *strategy tree*. A strategy tree for aspect classification is a tree in which each path from the root to a leaf represents a complete strategy for classification; that is, each path in a strategy tree can be expanded into a classification tree. Therefore, finding the least cost classification tree is

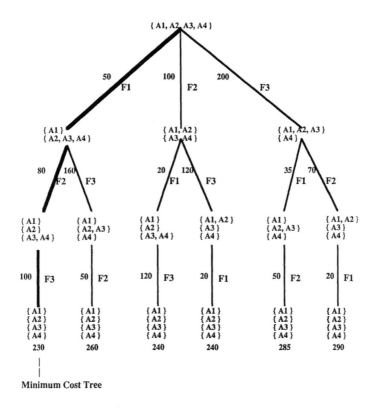

Figure 4. Strategy Tree. Each path from the root to a leaf represents a complete strategy for aspect classification. Finding the minimum cost classification strategy is equivalent to finding the minimum cost path from the root to a leaf.

equivalent to finding the least cost path from root to a leaf in the strategy tree.

In a strategy tree, each node contains the results of applying a test using a given feature, while arcs represent the tests. Each arc is labeled with the product of the cost of the feature and the expected number of samples to which the test will be applied. An arc is present when a feature can be used to break up a set into smaller sets. At the leaf nodes, all the constituent sets should be singletons, unless the feature set is incapable of distinguishing some of the input classes.

Figure 4 illustrates a strategy tree for a simple case consisting of 4 classes, and 3 features. At the root, all the classes are grouped into a single set. An arc is present for every computable feature which can reduce the set size, so there are three arcs at the root. The darkened path in the tree is the minimum cost path.

There are cases for which is not possible, given the set of features, to distinguish two classes. Indistinguishable classes are referred to as *congruent classes*, and the corresponding nodes as *congruent nodes*. Since the classes in an aspect classification tree represent aspects of some object, the existence of congruent classes means that the corresponding aspects cannot be distinguished with the available features; such aspects are referred to as *congruent aspects*. Congruent aspects do not represent a failure of the search procedure, but rather indicate a fundamental limitation of the feature set. In many cases, the linear shape change determination step corrects for ambiguous aspect classification and determines the correct object pose.

3.3.2. Linear Shape Change Determination

The aspect classification step results in the classification of an input image as an instance of an aspect. Since an aspect consists of a contiguous collection of views of an object, aspect classification is equivalent to rough localization; the possible collection of object poses are limited to those consistent with the observed aspect.

The next step in the localization process determines the exact pose, given the initial estimate obtained from aspect classification. The same set of features is visible throughout an aspect, so no non-linear events such as the appearance or disappearance of a feature occur. Therefore, the second step consists of determining the exact pose within an aspect, subject only to linear changes (rotation and translation) of features; this step in the localization process is known as *linear shape change determination* (LSCD).

One way to perform LSCD is to utilize a model-based approach in which image features are matched to model features, a pose is hypothesized and used to predict locations of features in the image, and the pose is refined by computing the error in predictions and observations and updating the pose appropriately. In most model-based systems, the matching stage is very difficult, since all possible matches between image and model features must be investigated. Since aspect classification has been performed, however, the correspondence between some set of image and model features has already been established. In particular, the assumptions underlying the LSCD method presented here are:

- the correspondences between model and image faces are known;

- the correspondences between model and image edges are unknown.

The aspect classification strategy presented in the previous subsection was encoded in the form of a tree, the classification tree. Each leaf node of the tree represents an aspect or collection of congruent aspects, and for each leaf node a different LSCD strategy may be appropriate. The VAC presented here computes a separate LSCD strategy for each leaf node of the aspect classification tree. The steps in each LSCD strategy are encoded as nodes that are attached to the leaf nodes of the aspect classification tree. Although the particular computational procedures vary between aspects, the same steps are followed in the same order for each aspect:

1. determine the coordinate system of the primal face (the visible face with the largest 3D area):

(a) determine the origin of the primal face;

(b) determine the z-axis orientation of the primal face;

(c) determine the x-axis orientation of the primal face;

2. estimate the body coordinate system;

3. establish correspondences between image and model edges;

4. recover exact body coordinates by numerical minimization.

The LSCD strategy is determined off-line through the following steps:

1. For each aspect, the visible face with the largest 3D area is selected as the primal face;

2. Each primal face is analyzed, and a method for defining the face coordinate system is determined. Separate nodes are attached to the classification tree which define the exact procedures used in each individual step of determining the origin, z-axis, and x-axis of the primal face.

3. Given the estimated coordinate system of the primal face and the transformation between primal face coordinates and model coordinates, the body coordinate system can be estimated with respect to the sensor coordinates. This is encoded as a separate node of the tree.

4. Knowing the object aspect and a rough estimate of the body coordinate system enables the prediction of the location of model edges in the image. The predicted edges can then be matched to observed edges. This process is encoded as a separate node of the tree.

5. A fine-tuning procedure is used to determine the exact body coordinates by adjusting the estimated body coordinates so that image edges exactly match predicted model edges. An exact match is not possible, so the procedure finds the body coordinates that minimizes the error between predicted and observed edge locations. This procedure is encoded as a separate node of the tree.

3.3.3. The Interpretation Tree

The overall strategy for object localization is encoded in the form of a tree, the interpretation tree. The top part of the interpretation tree consists of the aspect classification tree, and consists of directions for a series of feature value computations and tests that result in the classification of an input image into an instance of an aspect. The bottom part of the interpretation tree is six nodes deep, and consists of the steps in the LSCD strategy that are appropriate for each aspect. Figure 5 illustrates the interpretation trees for a sample object: a toy car.

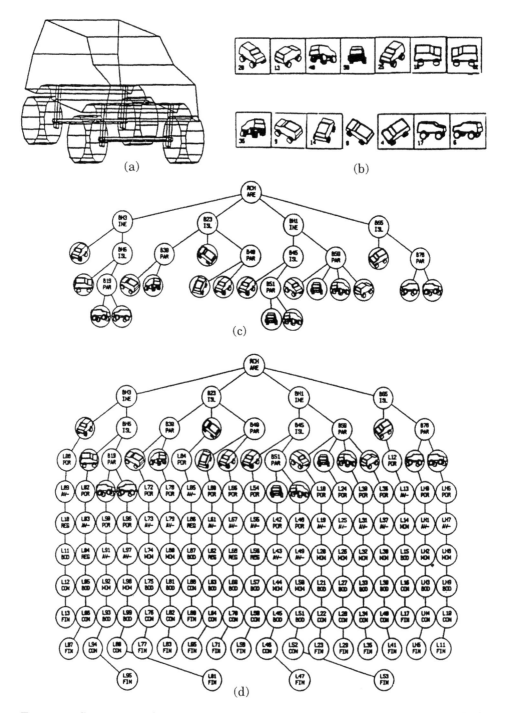

Figure 5. Generation of a Recognition Strategy for a Toy Car. (a) Object model. (b) Object aspects. (c) Aspect classification tree. (d) Complete interpretation tree.

3.4. Run-time Execution

Each of the procedures represented by a node of an interpretation tree corresponds to an executable object stored in a program library. During the off-line phase, the creation of a node of the interpretation tree is accompanied by the instantiation of an executable object from the program library, and the insertion of the object at the node. During the on-line phase, these instantiated objects are executed in order to perform object recognition. Message-passing is used for communication between objects.

For each execution of the object recognition strategy during the on-line phase, a preprocessed image is passed to the root of the interpretation tree and the object stored at the root is invoked. The execution of the root object results in the computation of some feature value and the application of a test. The outcome of the test results in the preprocessed image being passed to the appropriate node at the next level of the tree, and the corresponding stored object is invoked. The sequence of message-passing and object invocation proceeds until a leaf node is reached, indicating that processing is complete. The results of processing are then passed back up the tree and returned from the root.

At congruent nodes of the classification part of the tree, no unique aspect is identified. Instead, several aspects are determined to be possible. For each congruent aspect, the LSCD part of the tree is executed, and the results returned are compared by the congruent node; the aspect yielding the minimum error is selected as the correct interpretation and this result is passed back up to the root of the tree.

3.5. Application

In this section, we illustrate the execution of the compiled strategy. As stated previously, the VAC presented here was constructed to utilize sensors that produce dense range images; that is, the features used are those that can be determined given an input range image. In order to show the sensor independence of the VAC, the results have been demonstrated using two different range sensors: dual photometric stereo [14], and an ERIM laser range finder [12]. The results for dual photometric stereo are reported below.

The complete interpretation tree for the toy car is presented in Figure 5. The toy car was placed in a scene on top of a pile of other objects. The input car scene is shown in Figure 6. Preprocessing stages results in the computation of:

- A needle map containing the gradient space values at each pixel.

- An edge map.

- A label map indicating the region to which each pixel belongs.

The aspect classification steps performed are illustrated by the black nodes shown in the interpretation tree of Figure 7 . Starting below the node at which the aspect is identified, the LSCD processing begins. The first node determines the mass center of the target region and declares that position to be the origin of the face coordinate system. The next node determines the average surface orientation of

Figure 6. Input Scene for Dual Photometric Stereo Experiment.

the target region and declares that to be the orientation of the z-axis of the primal face coordinate system.

The next two nodes complete the determination of the face coordinate system and estimate the body coordinate system. Figure 8 illustrates the estimated body coordinates overlaid on the input image.

The next step in the process is the determination of the edge pairs and fine-tuning of the object pose. Figure 9 illustrates the results of these processes.

4. A VAC Utilizing Image Synthesis for Prediction

The appearance-based system of Sato et al [21] was built for the purpose of recognizing specular objects. Specular objects pose a special problem for computer vision. Specularities are often the most prominent image features, and yet contain no brightness variations which can be used for edge detection or 3D shape analysis. Moreover, specular features may appear, disappear, or change shape abruptly with small variations in viewpoint. The presence of a specularity requires a precise configuration of illuminant, surface normal, and sensor, and therefore provides a powerful constraint on the underlying surface geometry. However, the constraint is purely local, and does little to constrain the object pose.

Specular reflections are found in nearly every imaging scenario. Metal, glass, plastic, and many other materials are highly specular. In addition to optical images, there are other imaging systems that are based on specular reflection. For example, radar is based on specular reflection, as are ultrasonic underwater imaging and

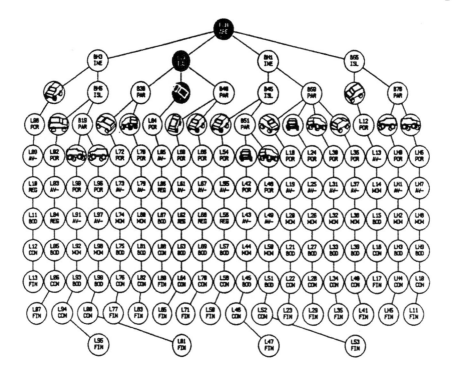

Figure 7. Execution of Aspect Classification Tree. Black nodes indicate the path followed during execution.

Figure 8. Estimated Body Coordinates Overlaid on Input Scene.

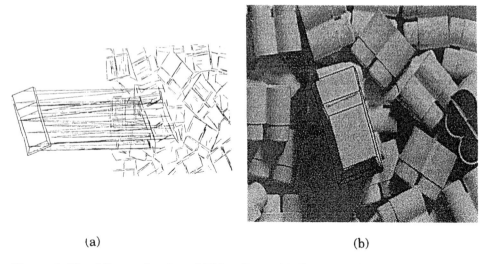

(a)   (b)

Figure 9. Final Determination of Object Pose. (a) Correspondences between model and image edges. (b) Final pose overlaid on input scene.

medical sonography. Therefore, it is important to establish techniques to recognize objects using specular images.

4.1. Explicit Object and Sensor Models

An analytic approach to appearance-based vision is impractical in the domain of specular objects. Because of interreflections between shiny surfaces, analytic prediction of specularities is extremely difficult. An alternative is the use of a sensor simulator, which generates object appearances based on both object and sensor models.

Sensor simulators are very similar to ray tracers of computer graphics. A 3D scene is described by a geometric modeler. A sensor simulator traces the path of light rays from pixels into the scene. Every time a ray hits an object surface, the ratios of reflected and transmitted energy are computed on the basis of the surface reflectance function and coefficient of refraction, respectively. The sensor simulator then traces both the reflected and refracted rays. When a ray reaches a light source, the energy emitted by the source is specified by the sensor model. The incident energies of all the rays towards a pixel are summed to determine the brightness value at the pixel and therefore predict the object appearance at the pixel.

One critical difference between a sensor simulator and a ray tracer is that sensor simulators maintain the symbolic correspondences between regions of the image and the object surfaces underlying the regions. In the case of specular objects, this correspondence permits the analysis of which object surfaces produce strong, stable specularities, and from which directions specularities are visible on each surface.

Specular features are characterized by strong, distinct, and saturated brightness. In many cases, specularities can be extracted by the simple procedure of binary thresholding. Some specularities are easier to detect than others, however. In general, the size of a feature determines the ease with which it can be detected. For example, elongated specularities on a cylindrical surface are easy to detect, while a specular spot on a small sphere is difficult to detect. Thus, the detectability of a specular feature is related to the 3D shape of the surface underlying the feature.

While a specular feature might be easily detected, it could be quite unstable; that is, the specularity might be visible only over a small range of viewpoints, and a slight change in viewpoint could cause it to disappear. Such specularities are termed unstable, and are poor choices for use in recognition. The stability of a specular feature is related to the size of the collection of viewpoints from which the feature can be detected.

To make the concept of stability clearer, consider a co-located camera and light source. Specular reflections are detectable when the object surfaces are nearly perpendicular to the camera line-of-sight. Consider the motion of the camera/light system around the surface of a sphere centered on an object, with the line of sight always toward the center of the sphere. If a small specular sphere is being imaged, a specular spot will be observed, and will continue to be observed for all viewpoints; hence the specularity arising from a spherical surface is extremely stable. Now consider a cube being imaged. Each planar surface only yields a specularity when the line of sight is perpendicular to the surface, and the specularity disappears for small changes in viewpoint; hence, specularities from planar surfaces are unstable. The area on the viewing sphere corresponding to detectable viewpoints is a measure of stability of a specular feature.

Figure 10 illustrates the detectability and stability for specular features over four different surface types: planar, cylindrical, conical, and elliptical. As can be seen in the figure, planar surfaces have easily detectable specularities that are low in stability, while spherical surfaces have low detectability but high stability. Cylindrical and conical surfaces fall somewhere in between.

4.2. Predict and Analyze Appearances

The system employs an exhaustive method for appearance analysis. Assuming that range to the object is constant, all possible viewing directions can be represented as points on the unit sphere. A geodesic partition of the viewing sphere uniformly tessellates the sphere into small triangles. The center of each triangle is chosen to represent a viewing direction. Triangles can be further subdivided into smaller triangles to make the sampling as fine as desired. The sensor simulator is then used to generate synthetic images at each representative viewpoint.

Each image is processed to extract specularities, and the data structure created by the sensor simulator is used to determine the primitive component underlying each specularity. For N primitive components $P_1, P_2, \ldots, P_N$, a cell label can be defined as an N-tuple $(X_1, X_2, \ldots, X_N)$ such that $X_i = 1$ or 0 according to whether or not

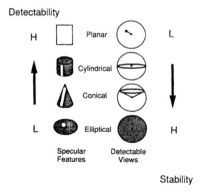

Figure 10. Detectability and Stability of Specular Features

component P_i gives rise to a detectable specularity at the viewpoint represented by the cell. Cells with identical cell labels are grouped together to form aspects. Thus, in the case of specular objects, aspects are defined with respect to detectable specularities.

4.3. Generate Recognition Strategy

As discussed above, specular features can vary in their characteristics of detectability and stability. For the purposes of object recognition, the system sorts specular features on the basis of detectability and stability in order to select the most effective features for aspect classification.

Detectability was defined above as the measure of the ease with which a specularity can be detected, and was related to the area of the specularity. The system uses as a measure of detectability the number of pixels of the largest appearance of a specularity, normalized by dividing by the area of the largest detected specularity.

Stability was defined as the area of the viewing sphere over which a given specularity is detected. In the case of a tessellated viewing sphere, this measure can be approximated by counting the number of cells within which the specularity is detected, normalized by the total number of cells.

An evaluation function is required to combine the measures of detectability and stability into a measure of overall feature utility. For each aspect, the features are ordered by decreasing utility. At run-time, matches are made in order of decreasing utility.

Aspect classification is equivalent to rough localization. Finer localization is difficult in the case of specular images because specular features change their shape drastically with small changes in viewpoint. Moreover, the exhaustive approach used in aspect determination may miss an unstable specularity that is only visible between two cells. Consequently, deformable template matching was selected as the procedure for fine localization.

Deformable template matching permits the template to deform according to certain constraints. An appearance is described as a combination of templates, each of which describes a specularity. The templates are interconnected conceptually by springs. The quality of match is measured by the sum of the internal deformation energy of the springs, and the external energy needed to fit each template to a real specularity. Thus, a deformable template can deform to find a match, even when a specularity changes shape or position. Moreover, matches can still be made even in the presence of accidental appearances or missing features.

A deformable template is prepared for each aspect using the appearance which is located at the center of the aspect. Specularities appear as spots or line segments, so each template consists of spots and line segments. Specular features are extracted from the central appearance. For an elongated feature, a line is fit to the feature and used to represent it in the template. For a spot feature, a point located at the center of the feature, is used to represent the feature in the template. A conceptual spring is located at each endpoint of a line feature, and at the point representing a spot feature. The spring energy is calculated from the displacement between the original and current location of the spring. Thus, the energy of a spot feature is a function of the displacement between the current and original position. For a line feature, the energy is a function of the displacement energy of the two endpoints.

4.4. Run-time Execution

Run-time execution is broken into two distinct stages: aspect classification and verification. In contrast to the system discussed in section 3, aspect classification does not uniquely classify an input image as an instance of an aspect. Rather, aspect classification is used to eliminate impossible aspects. Remaining aspects are input into the next stage, in which deformable template matching is used for verification.

4.4.1. Aspect Classification

Specular features can be very unstable. Small changes in viewpoint may cause a given specularity to appear, disappear, or change shape. Consequently, it is difficult to identify a single specularity, or the set of specularities that define an aspect, with complete confidence. Therefore, rather than employ a binary classification of an input image as an instance of an aspect, the run-time aspect classification system employs a continuous classification method based on the Dempster-Shafer methodology [22]. Figure 11 illustrates the classification method. For simplicity, the illustration is limited to aspects lying on a single great circle of the viewing sphere. Each match with a template from a single feature generates a likelihood distribution, in which a value close to 1 means that the corresponding aspect is very likely. Likelihood distributions from separate features are merged using Dempster-Shafer theory. As shown in the figure, each additional feature reduces the number of likely aspects and sharpens the peaks of the remaining ones. The likelihood values for impossible aspects decrease with each additional feature, while the likelihood values of possible aspects increase. After the evidence from every available feature has been applied, the overall likelihood distribution may still contain several peaks, each of which represents a possible aspect classification for the input image.

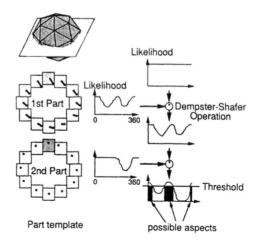

Figure 11. Aspect Classification Based on Evidential Reasoning.

4.4.2. Verification

The verification process determines the correct aspect by matching the input image to the complete template for each of the possible aspects. Each template can move over the entire image to minimize the total energy. The total energy is comprised of a weighted sum of constraint energy, and potential energy:

$$E_{total} = W_{constraints} E_{constraints} + W_{potential} E_{potential}$$

Potential energy represents the energy of the position of the template, while constraint energy represents the energy of the relations between template components. Potential energy is readily visualized as the height of the template in a potential field defined by the detected specularities in the image. Constraint energy is modeled by springs connecting the template points to image feature points; as the template deforms, the springs stretch and the constraint energy increases. Figure 12 illustrates template matching.

An optimization procedure is used to find the energy minimum. To avoid getting trapped in local minima, some noise is added to the total energy. The global minimum energy for each template specifies the quality of fit of the input image to the template. The best match is chosen by comparing the minimum energies for each of the candidate aspects.

4.5. Application

The VAC for specular object recognition has been applied to two different kinds of specular images: real optical specular images, and synthesized synthetic aperture radar (SAR) images. In this section, the results for real optical specular images are reported. Results for SAR images are reported in Sato et al [21].

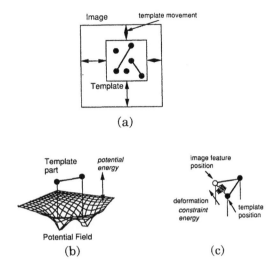

Figure 12. Deformable Template Matching. (a) Template. (b) Potential energy. (c) Constraint energy.

For this experiment, a real toy airplane was constructed. The sensor used was a tv camera with a co-located light source. The sensor parameters were obtained by calibration. Figure 13 shows a real specular image of the toy airplane.

The object was modeled using the Vantage geometric modeler [2]. Object aspects were determined using the exhaustive method. Since man-made objects such as the airplane have only a few stable poses, an aspect map over the entire viewing sphere was not generated. Instead, only the appearances of the airplane from the equator of the viewing sphere were considered. Sample images were generated on the equator at 5 deg increments for a total of 72 samples. Appearances were generated using a sensor simulator [9]. Figure 14 illustrates the object model and some sample appearances.

The concentric arcs in Figure 15 correspond to the visibility maps for each primitive surface. The outermost arc corresponds to the detectable directions of the rear fuselage. The arc is unbroken - the part can be observed from all viewing directions. The top-left image in shows the set of possible appearances of specular features arising from the rear fuselage as a function of viewing direction. The other images correspond to other arcs of the visibility map. Some of the arcs in the map are broken; the missing arc regions correspond to viewpoints from which the primitive surface is not visible. The figure shows the features in their computed order of significance.

To test the resulting recognition program, a real specular image of the toy airplane was obtained and input to the system. The first step in the recognition process is aspect classification, in which possible aspects are searched by matching with partial templates. Figure 16 shows the input image and the results of matching to the first

Figure 13. Real Specular Image of Toy Airplane.

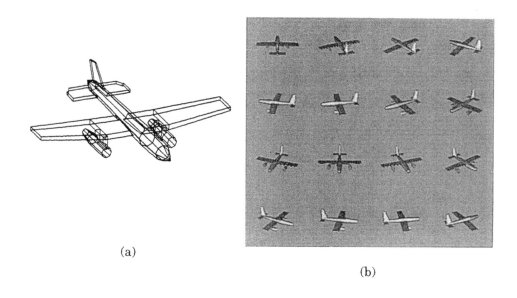

(a)

(b)

Figure 14. Model Airplane and Predicted Specular Appearances. (a) Airplane model. (b) Sample appearances.

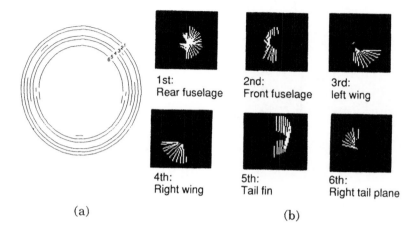

1st: Rear fuselage	2nd: Front fuselage	3rd: left wing
4th: Right wing	5th: Tail fin	6th: Right tail plane

(a)   (b)

Figure 15. Visibility and Significance of Features in Optical Experiment. (a) Visibility map. (b) Specular features in order of significance.

three partial templates. The figure clearly shows the narrowing of the likelihood distribution as additional matching is performed. The result of the aspect classification step was the selection of aspects at 45 deg, 115 deg, 125 deg, 165 deg, and 170 deg.

Following aspect classification, verification was performed using deformable template matching. Figure 17 illustrates the verification step. One template was used for each of the aspects selected. In each case the template changed its shape to match the specularities in the input image and converged to the shapes shown by the white lines superimposed on the copies of the real image in the figure. The minimum energy, and hence the best match, was obtained for the aspect at 170 deg.

5. Summary

In this paper, we presented the paradigm of appearance-based vision, which is a paradigm for building object recognition systems. The paradigm is called appearance-based, since an integral step is the prediction and analysis of object appearances. An appearance-based system is called a vision algorithm compiler, or VAC. The input to a VAC is a set of object and sensor models, and the output is an executable object recognition program.

Appearance-based systems share four principle defining characteristics:

- *Two-stage Process* A VAC operates in two distinct stages. The first stage is performed off-line, and consists of the analysis of appearances and the generation of an object recognition program. The second stage is performed on-line, and consists of the execution of the previously generated program.

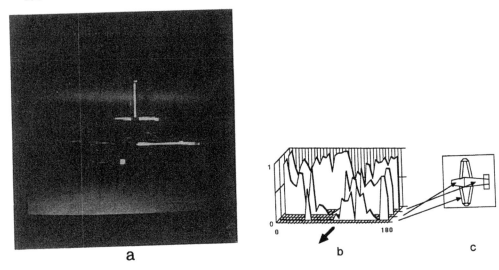

a

b

c

Figure 16. Aspect Classification Stage for Optical Experiment. (a) Input optical image. (b) Accumulation of evidence. (c) Airplane parts.

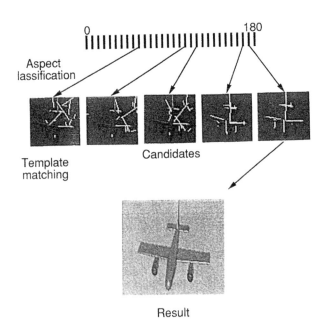

Figure 17. Verification Stage for Optical Image.

- *Explicit Object and Sensor Models* A VAC embodies an overall approach to object recognition that can be applied to a variety of objects and sensors. Therefore, explicit and exchangeable models are utilized. Sensor models specifies the features detectable by the sensor, along with procedures to compute the detectability and reliability of each feature. Object models include geometric and photometric properties of the objects.

- *Appearance Prediction and Analysis* Objects are recognized based on their appearances in images. Therefore, the prediction and analysis of appearance is fundamental to generating competent object recognition programs. A VAC predicts appearances based on the information in object and sensor models. The range of appearances may be determined analytically or exhaustively, and the appearances may be predicted analytically or through image synthesis.

Acknowledgements

The authors thank Takeo Kanade and the members of VASC (Vision and Autonomous System Center) of Carnegie Mellon University for their valuable comments and discussions.

REFERENCES

1   F. Arman and J. K. Aggarwal. Automatic generation of recognition strategies using cad models. In *Proc. of the IEEE Workshop on Directions in Automated CAD-Based Vision*, pages 124–133. IEEE, 1991.

2   P. Balakumar, J.C. Robert, R. Hoffman, K. Ikeuchi, and T. Kanade. Vantage: A frame-based geometric/sensor modeling system – programmer/user's manual v1.0. Technical Report CMU-RI-TR-91-31, Carnegie Mellon University, Robotics Institute, 1991.

3   P.J. Besl and R.C. Jain. Three-dimensional object recognition. *ACM Computing Surveys*, 17(1):75–145, March 1985.

4   R. C. Bolles and P. Horaud. 3DPO: A three-dimensional part orientation system. In T. Kanade, editor, *Three-Dimensional Machine Vision*, pages 399–450. Kluwer, Boston, MA, 1987.

5   O. I. Camps, Shapiro. L. G., and R. M. Haralick. PREMIO: an overview. In *Proc. of the IEEE Workshop on Directions in Automated CAD-Based Vision*, pages 11–21. IEEE, 1991.

6   S. Chen and H. Freeman. On the characteristic views of quadric-surfaced solids. In *Proc. of the IEEE Workshop on Directions in Automated CAD-Based Vision*, pages 34–43. IEEE, 1991.

7   R.T. Chin and C.R. Dyer. Model-based recognition in robot vision. *ACM Computing Surveys*, 18(1):67–108, March 1986.

8   P. J. Flynn and A. K. Jain. BONSAI: 3D object recognition using constrained search. *IEEE Trans. on Pattern Analysis and Machine Intelligence*, 13(10):1066–1075, 1991.

9   Y. Fujiwara, S. Nayar, and K. Ikeuchi. Appearance simulator for computer vision research. Technical Report CMU-RI-TR-91-16, Carnegie Mellon University, The Robotics Institute, 1991.

10 C. Goad. Special purpose automatic programming for 3D model-based vision. In *Proc. of DARPA Image Understanding Workshop*, pages 94–104. DARPA, 1983.

11 C. Hansen and T. C. Henderson. CAGD-based computer vision. *IEEE Trans. on Pattern Analysis and Machine Intelligence*, 11(11):1181–1193, 1989.

12 M. Hebert and T. Kanade. Outdoor scene analysis using range data. In *Proc. of Int. Conf. on Robotics and Automation*, pages 1426–1432, San Francisco, April 1986. IEEE Computer Society.

13 K. S. Hong, K. Ikeuchi, and K. D. Gremban. Minimum cost aspect classification: a module of a vision algorithm compiler. In *Proc. 10th Int. Conf. on Pattern Recognition*, pages 65–69, 1990. A longer version is available as CMU-CS-90-124, School of Computer Science, Carnegie Mellon University (1990).

14 K. Ikeuchi. Determining a depth map using a dual photometric stereo. *The International Journal of Robotics Research*, 6(1):15–31, 1987.

15 K. Ikeuchi. Generating an interpretation tree from a CAD model for 3-D object recognition in bin-picking tasks. *International Journal of Computer Vision*, 1(2):145–165, 1987.

16 K. Ikeuchi and K. S. Hong. Determining linear shape change: toward automatic generation of object recognition programs. *CVGIP: Image Understanding*, 53(2):154–170, 1991.

17 K. Ikeuchi and T. Kanade. Towards automatic generation of object recognition programs. *Proc. of IEEE*, 76(8):1016–1035, August 1988. A slightly longer version is avaiable as CMU-CS-88-138, School of Computer Science, Carnegie Mellon University (1988).

18 J. J. Koenderink and A. J. Van Doorn. Internal representation of solid shape with respect to vision. *Biological Cybernetics*, 32(4):211–216, 1979.

19 D. J. Kriegman and J. Ponce. Computing exact aspect graphs of curved objects: solids of revolution. *Int. Journal of Computer Vision*, 5(2):119–135, 1990.

20 H. Platinga and C. R. Dyer. Visibility, occlusion, and the aspect graph. *Int. Journal of Computer Vision*, 5(2):137–160, 1990.

21 K. Sato, K. Ikeuchi, and T. Kanade. Model based recognition of specular objects using sensor models. *CVGIP: Image Understanding*, 55(2):155–169, 1992.

22 G. A. Shafer. *A Mathematical Theory of Evidence*. Princeton Univ. Press, Princeton, NJ, 1976.

23 P. Suetens, P. Fua, and A. J. Hanson. Computational strategies for object recognition. *ACM Computing Surveys*, 24(1):5–61, 1992.

Three-Dimensional Object Recognition Systems
A.K. Jain and P.J. Flynn (Editors)
© 1993 Elsevier Science Publishers B.V. All rights reserved.
259

Recognizing 3D Objects Using Constrained Search[1]

W. Eric. L. Grimson, Tomás Lozano-Pérez, Steven J. White[a] and Norman Noble[b]

[a]Artificial Intelligence Laboratory
Massachusetts Institute of Technology
Cambridge, Massachusetts 02139 USA

[b]SOCS Research
Los Gatos, California 95032 USA

1. Introduction

In the model-based approach to object recognition, a set of stored geometric models are compared with features extracted from an image of a scene (cf. [7, 17]). The comparison generally involves finding a valid correspondence between a subset of the model features and a subset of the image features, where valid means there exists some transformation of a given type mapping each model feature (roughly) onto its corresponding image feature. This transformation specifies the object's *pose* (its position and orientation with respect to the image coordinate system). Thus, the goal is to find any instances of the transformed model in the scene, and to determine the extent of the model in the data.

More formally, let $\{F_i | 1 \leq i \leq m\}$ be a set of model features measured in coordinate frame $\mathcal{M}$, $\{f_i | 1 \leq i \leq s\}$ be a set of sensory features measured in coordinate frame $\mathcal{S}$, and $\mathcal{T} : \mathcal{M} \to \mathcal{S}$ be a legal transformation from model to sensor coordinates. The goal is to identify a correspondence, $I \subseteq 2^{m \times s}$, that pairs model and sensor features. Each correspondence I specifies some transformation $\mathcal{T}_I$ which maps each model feature close to its corresponding image feature:[2]

$$I = \{(m_i, s_j) | \rho(\mathcal{T}_I m_i, s_j) \leq \epsilon\},$$

where ρ is some distance measure (e.g. Euclidean distance for point features, or maximum Euclidean separation for line features). In general the quality of an interpretation is measured by the size of the correspondence, $|I|$, and the goal is

---

[1]This report describes research done in part at the Artificial Intelligence Laboratory of the Massachusetts Institute of Technology. Support for the laboratory's artificial intelligence research is provided in part by the Advanced Research Projects Agency of the Department of Defense under Army contract number DACA76-85-C-0010 and under Office of Naval Research contract N00014-85-K-0124. WELG is supported in part by NSF contract number IRI-8900267.

[2]A given interpretation I will in fact generally define a range of 'equivalent' transformations in the sense that there are a number of transformations that generate the same set I.

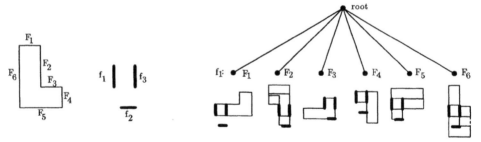

Figure 1. We can build a tree of possible interpretations, by first considering all the ways of matching the first data feature, f_1, to each of the model features, $F_j, j = 1, \ldots, m$. In the right part of the figure, we show an example of these pairings for the model and data shown at the left. Below each pairing is shown the range of legal associated poses.

either to find the interpretation maximizing $|I|$, or all interpretations where $|I| > t$ for some threshold t.

Approaches to recognition can be categorized by how they search for solutions. One class of methods focuses on finding the correspondence I, typically by searching a potentially exponential space of pairings of model and data features (e.g. [9, 11, 15, 17]). A second class focuses on finding the pose T, typically by searching a potentially infinite resolution space of possible transformations, (e.g. [2, 12, 13, 25–27]). A third class is a hybrid of the other two, in that correspondences of a small number of features are used to explicitly transform a model into image coordinates (e.g. [1, 5, 10, 22, 23]). Here, we focus on the first class of methods, specifically on the Interpretation Tree approach to correspondence space search. We illustrate this method on the particular application of recognizing and locating 3D cylindrical tubing from laser range data.

2. Searching the interpretation tree

Our goal is to find correct correspondences between image and model features, while avoiding the exploration of the entire space of possible correspondences. We do this by using an interpretation tree (e.g.[18, 19, 3, 14, 30–32, 24]).

Suppose we arbitrarily order the data features. We match the first data feature with each model feature, represented as nodes at the first level of the tree (Figure 1). Note the range of possible poses associated with each pairing. For each hypothesized assignment to data feature f_1, we consider possible assignments of the second data feature f_2 to model features (Figure 2). The example on the left is consistent with a single rigid transformation of the object, while the example on the right is not, as indicated by the fact that the two ranges of possible poses do not intersect.

We can continue adding new levels to the tree, one for each data feature. A node at level n describes an n-interpretation, as the nodes on the path from the current

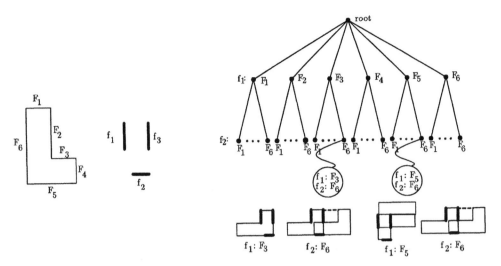

Figure 2. Each node in the second level of the tree defines a pairing for the first two data features, found by tracing up the tree to the root. Two examples are shown. The example on the left is consistent with a single rigid transformation, as the two ranges of poses specified by each of the data-model feature pairings have a common pose. The example on the right is not consistent with a single transformation, as its two ranges of poses do not have a common pose.

node to the tree root identify an assignment of model features to the first n data features.

We want to find consistent s-interpretations, where s is the number of sensor features and where consistent means there exists a legal transformation mapping each model feature into rough agreement with its corresponding data feature. A simple method would test each leaf of the tree for such a legal transformation. This is overly expensive. The key is to find constraints that will allow us to exclude entire subtrees, and hence entire subspaces of the search space, without explicitly exploring them. The interpretation tree enables us to focus our search in a coherent manner, by starting at the tree root, and testing interpretations as we move downward in the tree. As soon as we find a node that is not consistent, i.e. for which no rigid transformation will correctly align model and data feature, we can terminate any further downward search below that node. In testing for consistency at a node, we seek simple constraints that will correctly preserve consistent interpretations, while removing some, if not all, of the inconsistent ones, e.g. in Figure 2 we need a simple method of deducing that the second example is not consistent, without directly solving for the best transformation mapping model features to data features.

2.1. Constraints reduce the search

To search the tree, we use depth-first backtracking. Starting at the root of the tree, we move down the first available branch at each node. At each node, we check the consistency of that node by examining the constraints associated with that node. Notationally (where i is the node's level in the tree, p is the branch chosen from the previous node, j identifies any level above i and q is the branch associated with j) we use:

- **unary**(i, p) = True iff pairing i^{th} data feature to p^{th} model feature is consistent.

- **binary**(i, j, p, q) = True iff pairing i^{th} data feature to p^{th} model feature and pairing of jth data feature to q^{th} model feature is consistent.

If all the constraints are satisfied, we continue downward. For a level n node, there is one new unary constraint and n-1 new binary constraints. Thus, the lower we go in the tree, the more constraints that must hold true.

If we reach a node at which a constraint does not hold, we abandon the remaining subtree below that node, and backtrack to the previous node. We then explore the next branch of that node. If there are no more branches, we backtrack another level, and so on. If we reach a leaf of the tree, we accumulate that possible interpretation, backtrack and continue, until the entire tree has been explored, and all the possible interpretations have been found. We will then subject these interpretations to additional testing.

While we have chosen to attack our search problem using backtracking search, there are other methods available for finding solutions to a consistent labeling problem. These include best first search, beam search, full and partial look-ahead, forward checking, backchecking, and back marking (see, for example, [20] for a discussion of these methods).

2.2. Model tests to verify hypotheses

Once the search reaches a leaf of the tree, we have accounted for all of the data features, and it would seem that we thus have a solution to the recognition problem. The interpretation defined by this node need not be a globally consistent one, however, and we must therefore check each leaf of the tree reached by the constrained search process to verify that the interpretation at that leaf is globally valid.

To do this, we solve for a rigid transformation mapping points $\vec{V}$ in model coordinates into points $\vec{v}$ in sensor coordinates, given by $\vec{v} = sR\vec{V} + \vec{v}_0$ where R is a rotation matrix, $\vec{v}_0$ is a translation vector, and s is a scale factor. Typically, we use a least-squares method to find the transformation minimizing the error between the transformed model features and the data features. We then verify the interpretation by checking that the transformation maps each model feature to a position that agrees (within error bounds) with the corresponding data feature.

2.3. Summary of the tree search approach

In short, we execute a depth-first search of an interpretation tree, applying constraints to pairings of data and model features to cut off fruitless paths in the tree. Any leaf reached by the process hypothesizes a feasible interpretation, which we check by solving for the associated pose, and verifying that it is consistent with the interpretation.

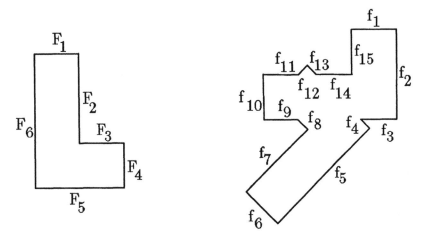

Figure 3. A simple example of a scene with occlusion and spurious data. Since there is no rigid rotation of the object model on the left that will align it with all of the data features, the basic constrained tree search method will not find any interpretations of the data. Note that the example data on the right does not include the effects of error, but does include occlusion.

3. Dealing with spurious data

This tree search method only works when all of the data comes from one object. For images such as the simple example in Figure 3, the method will find no interpretations, as no rigid transformation of the object will align it with all of the data features, even though there are clearly two subgroups of features consistent with the model.

The assumption of data from a single object either requires that the observed scene contain only one object, or that some mechanism has partitioned the data from a cluttered scene into subsets known to correspond to single objects. The first solution, while restrictive, is possible in situations in which one has some control over the sensing environment. The second solution is more general, and in this section we extend our search method to separate out subsets of the data that have come from a single object, and determine the correspondence between that subset and the model features.

3.1. Extending the tree search method

An easy method for selecting relevant sensory data is to introduce a new model feature, called a *null character*. At each node of the interpretation tree, we add as a last resort an extra branch corresponding to this feature, which indicates that the data feature to which it is matched is to be excluded and treated as spurious data. Since the data feature is excluded, it cannot affect the current interpretation, and hence any constraint involving a data feature matched to the null character is consistent. With this addition, it is possible for the constrained search method to deal with spurious data.

Figure 4 summarizes the search method. We use it to hypothesize interpretations which are verified by applying a model test. We use our model test method to determine the coordinate frame transformation, and to test that the transformed model features agree with the data features, provided we only apply it to those data-model pairings in the interpretation that actually involve a true model feature, and

```
(0)  d←0  ; initialize at root of tree, d denotes depth in tree
     Mode←search   ; start in search mode
     I₀ ← {}    ; initial interpretation is empty
     Status←consistent ; initial interpretation is consistent

(1)  If mode=search & status=consistent ; still okay
     Then d←d+1   ; increment depth
          jd ←1   ; start with first model feature
     (1.1) Id ←  Id−1 ∪ {(fd, Fjd)}    ;add new pairing
     (1.2) If jd =m+1 ; null character match, always consistent
              Then if d=s, save Id & set mode←backtrack ; save possible solution
                          Go to (3) ; backtrack
                  Else go to (1) ; continue downward
             If jd ≠m+1  ; real model feature
                 Then if      unary(d, jd)       =True & ∀ i        =        1,...,d-1
        binary(i, d, ji, jd) =True
                          ; all new constraints satisfied, still consistent
                          Then if d=s, save Id & set mode←backtrack ; save possible solu-
tion
                              Go to (3) ; backtrack
                          Else go to (1) ;more features, so continue downward search
                      Else status←inconsistent, go to (2)

(2)  If mode = search & status = inconsistent
         Then if jd=m  ; no more choices at this d
                  set mode←backtrack, go to (3)
              else jd ← jd + 1, go to (1.1) ; try next model feature

(3)  If mode=backtrack  ; backtracking to find next alternative
         then if jd ≠m+1 ; still have choices at this depth
                  set jd ← jd+1 & mode←search,
     go to (1.1) ; continue search
              else if d=0  ; at end of choices
                  then halt, return saved Id's from (1.2)
                  else d←d-1 ; decrement depth
                      Id ← Id+1/{(fd+1, Fjd+1)} ; remove most recent pairing
                      go to (3)    ; continue backtracking
```

Figure 4. Pseudo-code description of tree search method, with null character.

not the null character.

3.2. Additional search reduction methods

While the basic method of Figure 4 is adequate, it is also computationally slow. There are several additions to the method that can considerably speed up its performance. These include keeping track of the best interpretation seen so far, and terminating further downward search when it cannot possibly lead to a better interpretation, extending this idea to terminate all search as soon as an interpretation that is "good enough" is found, and preselecting "good" subspaces of the search space on which to initially concentrate our efforts. Each of these methods is discussed in detail in [17].

4. Three dimensional lines.

The basic search method applies to [17]: 2D edges, 2D vertices, 2D circular arcs, 3D lines, and 3D surface patches. Here, we focus on matching 3D linear features, defined as

$$\texttt{linear}_i = (\hat{t}_i, (\vec{b}_i, \vec{e}_i)) \qquad \texttt{LINEAR}_p = (\hat{T}_p, (\vec{B}_p, \vec{E}_p)).$$

where $\vec{b}_i, \vec{e}_i$ are the endpoints of a data edge, and $\hat{t}_i$ is a unit vector tangent to the edge, and where $\vec{B}_p, \vec{E}_p, \hat{T}_p$ are the corresponding vectors for a model edge.

4.1. Constraints for 3D edges.

To determine the consistency of a node in the interpretation tree, we use a set of constraints on the relative geometry of data and model features.

As a unary constraint, we use the length of an edge. If $\ell_i = \|\vec{e}_i - \vec{b}_i\|$ and $L_p = \|\vec{E}_p - \vec{B}_p\|$, then

length-constraint(i, p) = True iff $\ell_i \leq L_p + \epsilon_L$

i.e., if the length of the ith image edge is less than the length of the pth model edge, modulo some predefined error in the measurement, ϵ_L, then it is possible to match these two edges. Note that the constraint explicitly allows for occlusion and uncertainty, and hence may still be satisfied in cases that are not, in fact, correct.

For binary constraints, we have the following. Given two data edges and two model edges, the angle between each pair must be roughly the same:

angle-constraint(i, j, p, q) = True iff $\theta'_{ij} \in [\,'_{pq} - 2\epsilon_a, \,'_{pq} + 2\epsilon_a]$

where θ'_{ij} denotes the angle between $\hat{t}_i$ and $\hat{t}_j$, and $'_{pq}$ denotes the angle between $\hat{T}_p$ and $\hat{T}_q$. Similarly, the range of distances between pairs of edges must be in agreement:

distance-constraint(i, j, p, q) = True iff $[d_{\ell,ij}, d_{h,ij}] \subseteq [D_{\ell,pq} - 2\epsilon_p, D_{h,pq} + 2\epsilon_p]$.

Here $[D_{\ell,pq} - 2\epsilon_p, D_{h,pq} + 2\epsilon_p]$ denotes the range of distances for model edges, and $[d_{\ell,ij}, d_{h,ij}]$ denotes the range of distances for corresponding data edges.

The maximum distance is given by checking vectors between endpoints of the edges:

$$d_{h,ij} = \max\{\rho(\vec{b}_i, \vec{b}_j), \rho(\vec{b}_i, \vec{e}_j), \rho(\vec{e}_i, \vec{b}_j), \rho(\vec{e}_i, \vec{e}_j)\}.$$

For the minimum distance, we must consider the smallest distance between endpoints of the edges, the possibility that the projection from an endpoint of one edge in the direction of the normal of the second edge intersects that edge, and the possibility that the two edges have a minimal separation occuring strictly within their interiors. Hence, the minimum distance is given by:

$$
\begin{aligned}
\min\{ \quad & \rho\left(\vec{b}_i, \vec{b}_j\right), \quad \rho\left(\vec{b}_i, \vec{e}_j\right), \quad \rho\left(\vec{e}_i, \vec{b}_j\right), \quad \rho\left(\vec{e}_i, \vec{e}_j\right) \\
& \rho\left(\vec{b}_j, \vec{b}_i + \left\langle \vec{b}_j - \vec{b}_i, \hat{t}_i \right\rangle \hat{t}_i\right) && \text{if } \left\langle \vec{b}_j - \vec{b}_i, \hat{t}_i \right\rangle \in [0, \ell_i] \\
& \rho\left(\vec{e}_j, \vec{b}_i + \left\langle \vec{e}_j - \vec{b}_i, \hat{t}_i \right\rangle \hat{t}_i\right) && \text{if } \left\langle \vec{e}_j - \vec{b}_i, \hat{t}_i \right\rangle \in [0, \ell_i] \\
& \rho\left(\vec{b}_i, \vec{b}_j + \left\langle \vec{b}_i - \vec{b}_j, \hat{t}_j \right\rangle \hat{t}_j\right) && \text{if } \left\langle \vec{b}_i - \vec{b}_j, \hat{t}_j \right\rangle \in [0, \ell_j] \\
& \rho\left(\vec{e}_i, \vec{b}_j + \left\langle \vec{e}_i - \vec{b}_j, \hat{t}_j \right\rangle \hat{t}_j\right) && \text{if } \left\langle \vec{e}_i - \vec{b}_j, \hat{t}_j \right\rangle \in [0, \ell_j] \\
& \rho\left(\vec{b}_i + \alpha\hat{t}_i, \vec{b}_j + \beta\hat{t}_j\right) && \text{if } \alpha \in [0, \ell_i], \beta \in [0, \ell_j]\}
\end{aligned}
$$

where

$$\alpha = \frac{\left\langle \vec{b}_i - \vec{b}_j, \left\langle \hat{t}_i, \hat{t}_j \right\rangle \hat{t}_j - \hat{t}_i \right\rangle}{1 - \left\langle \hat{t}_i, \hat{t}_j \right\rangle^2}, \qquad \beta = \frac{\left\langle \vec{b}_i - \vec{b}_j, \hat{t}_j - \left\langle \hat{t}_i, \hat{t}_j \right\rangle \hat{t}_i \right\rangle}{1 - \left\langle \hat{t}_i, \hat{t}_j \right\rangle^2}$$

where the terms followed by if clauses are included in the set only if the clause is true, and where $\langle .,. \rangle$ denotes inner product.

Additional constraints can be obtained by representing each edge in terms of a local coordinate frame, and requiring that the components of each data edge as decomposed in that frame agree with the corresponding components of the model edge. For example, we can use the measurements

$$\left\langle \vec{b}_i + \alpha\hat{t}_i - \vec{b}_j - \beta\hat{t}_j, \hat{t}_i \right\rangle, \left\langle \vec{b}_i + \alpha\hat{t}_i - \vec{b}_j - \beta\hat{t}_j, \hat{t}_j \right\rangle, \left\langle \vec{b}_i + \alpha\hat{t}_i - \vec{b}_j - \beta\hat{t}_j, \hat{t}_i \times \hat{t}_j \right\rangle$$

where the ranges are derived as α ranges from 0 to ℓ_i and β ranges from 0 to ℓ_j. We let

$$d^{\perp}_{\ell,ij}(\vec{u}) = \min_{\alpha,\beta} \left\langle \vec{b}_i + \alpha\hat{t}_i - \vec{b}_j - \beta\hat{t}_j, \vec{u} \right\rangle \quad \text{and} \quad d^{\perp}_{h,ij}(\vec{u}) = \max_{\alpha,\beta} \left\langle \vec{b}_i + \alpha\hat{t}_i - \vec{b}_j - \beta\hat{t}_j, \vec{u} \right\rangle$$

denote the extremal values of these ranges, where $\vec{u}$ is one of $\hat{t}_i, \hat{t}_j, \hat{t}_i \times \hat{t}_j$.

We could measure that same ranges for the model edges, but we still need to account for error, however. If ϵ_p is the maximum deviation of a position measurement, and ϵ_t is the maximum perpendicular deviation of a tangent vector, then the range of values for the component measurement, modulo error, can be bounded by[17]:

$$
\begin{aligned}
& \frac{1}{\sqrt{1 + \epsilon_t^2}} \left(\left\langle \vec{B}_i - \vec{B}_j, \vec{u} \right\rangle + \epsilon_t \left\| \vec{B}_i - \vec{B}_j \right\| + 2\epsilon_p(1 + \epsilon_t) \right) + \\
& + \frac{\alpha}{\left(1 + \epsilon_t^2\right)} \left(\left\langle \vec{U}, \hat{T}_i \right\rangle + 2\epsilon_t + \epsilon_t^2 \right) - \frac{\beta}{\left(1 + \epsilon_t^2\right)} \left(\left\langle \vec{U}, \hat{T}_j \right\rangle + 2\epsilon_t + \epsilon_t^2 \right)
\end{aligned}
$$

```
unary = length-constraint(k + 1, j_{k+1})
binary = angle-constraint(i, k + 1, j_i, j_{k+1})
        & distance-constraint(i, k + 1, j_i, j_{k+1})
        & component-constraint(i, k + 1, j_i, j_{k+1}, t̂_i)
        & component-constraint(i, k + 1, j_i, j_{k+1}, t̂_j)
        & component-constraint(i, k + 1, j_i, j_{k+1}, t̂_i × t̂_j)
```

Figure 5. Additions to search method for 3d edges.

where α ranges from 0 to L_p, β ranges from 0 to L_q and where $\vec{U}$ is taken as $\hat{T}_p, \hat{T}_q$ and $\hat{T}_p \times \hat{T}_q$ respectively for the three different component constraints. We can use this expression to find the minimum and maximum values for each component measurement, which we denote by the range $[M^{\perp}_{\ell,pq}(\vec{u}), M^{\perp}_{h,pq}(\vec{u})]$. With these new derivations for the component measurements, we can then use:

component-constraint$(i, j, p, q, \vec{u})$ = True iff

$$[d^{\perp}_{\ell,ij}(\vec{u}), d^{\perp}_{h,ij}(\vec{u})] \subseteq [M^{\perp}_{\ell,pq}(\vec{u}) M^{\perp}_{h,pq}(\vec{u})].$$

Figure 5 shows the modifications needed for recognition from linear features.

4.2. Getting the edges from real data.

We must ensure that we can extract approximations of 3D edges from real scenes. There are several methods for doing this.

The first approach essentially applies edge detection to 3D data, looking for sharp changes in depth or orientation, rather than sharp changes in brightness. The 3D data can be extracted using active ranging systems. For example, by using a laser striping system, we can compute a dense depth map for a scene under the sensor. Edges can be extracted using 3D edge detectors based on smoothing and differentiation, directly analyzing and categorizing discontinuities in the range data, or analyzing the differential geometry of the range surface. Given such 3D edges, and object models based on the same features, we can then apply our constrained search method, using the constraints derived above. One such system is used in the HANDEY hand-eye system [29].

A second method for getting 3D edges is to extract them directly from visual data, without first constructing a dense depth map. One common way to do this is through stereo vision. Some stereo methods extract dense estimates of depth, in which case we can either apply 3D edge detectors, or use surface patches as our matching features. Most recent stereo methods, however, only produce explicit 3D estimates along image contours with sharp intensity changes. Given such depth estimates, split-and-merge techniques can be used to obtain estimates of linear 3D edge fragments in the scene. An example of object recognition using linear features from stereo processing is given by the TINA system developed by [35]. They describe a variation on the constrained search approach developed here, using stereo data as the sensory input, and matching both linear and simply curved segments, via a set of constraints that varies slightly from the ones described here. A related approach is described in [24].

A third method for obtaining 3D edges is to use motion analysis. By processing optical flow information, one can obtain either dense estimates of the motion and the shape of points in the scene, or estimates of the motion and position of edges in an image. In the first case, additional processing will obtain estimates of edges in the data. In both cases, one can only extract relative distance information, since there is an inherent speed–distance ambiguity in motion data that cannot be resolved. One way to deal with this is to modify the constraints. In particular, the distance constraint may be dropped and the component constraints may be replaced with simpler constraints involving unit vectors in the direction of the separation vectors, rather than the actual separation vectors [30]. With these variations, the constrained search method still applies to this case.

5. A Practical Cylinder Recognition System.

Other 3D linear features, such as the axes of cylinders, can be used for recognition. As a practical demonstration of these methods, we now focus on recognition of objects composed of segments of cylinders, connected by toroidal bends, e.g. tubing.

5.1. The Problem

The motivation for this system is given by the following scenario. Tubing for an aircraft are formed by inserting a straight piece of pipe stock, of some constant radius, into an automatic tube bending machine. This machine then executes a numerical control algorithm, in which a portion of the pipe is passed through a bending device, then the pipe is bent by some predetermined amount, creating a toroidal section of pipe. The pipe is then rotated through some angle, advanced, and bent again, and so on, according to some control algorithm. The specification of the feed length, bend rotation and bend angle is called a LRA (for Length, Rotate, Angle) model. It can be used to specify any tube form which can be manufactured using the bending machines.

The completed tube consists of a series of straight legs and toroidal bends. Once the complete pipe is created, it is compared against a part specification. Prior to the development of the system described here, this comparison was done by hand. In particular, in one common measurement method, a human operator would place the pipe on a flat table over a master blueprint, align the pipe by eye with several key points in the blueprint, then visually check that the rest of the pipe agreed with the outlines of the blueprint. Since the tubes can bend in 3D, this method may need several alignments and may still be inaccurate on portions of the tube that cannot be placed flat on the support surface under any orientation of the tube.

The goal of the system described here was to automate the inspection process. The desired specifications of such an inspection system were:

- Accuracy of .004 inches.

- Speed of 1 minute per part.

- Arbitrary part placement and pose in the inspection station.

- No part "programming" allowed — only the part specification or bending program is available to the inspection system.

The input to the system is:

- A known object model, i.e. a blueprint or physical model of the ideal tube, or

- a set of specifications on the components of the ideal object. This includes specifications on the tube radius, length of the straight tube segments, the angle and radius of bend of the toroidal connections and the relative orientation of the straight segments.

The output of the system is simply another LRA model which reflects the actual geometry of the measured tube. This, when compared to the specification LRA model, can be used for quality control. It can also be used to adjust the bending program to bring the process closer to specification.

5.2. The Approach

Our approach to this problem can be described in a series of stages.

- We use an active laser scanning device as our sensory input. In the first stage, we use the scanner to obtain a sparse set of measurements from the tube.

- These measurements are then matched against our known model, to obtain an estimate of the position and orientation of the tube.

- Given this knowledge, we then use a path planner to derive a scanning strategy. This strategy plans a path for the scanner that will trace along the tube, obtaining samples at regular intervals, while keeping the scanner oriented with the current orientation of the tube section, and while avoiding collisions with the tube and occlusions of the tube with respect to the scanner.

- By tracking the scanner at close range along the tube, we obtain high accuracy information about the location of the tube axis. This can be used to generate a tube model.

- This model can be matched to the specification LRA model to insure no deviation exceeds specified limits. The deviations can also be used to adjust the bending program such that subsequent parts manufactured come closer to matching the specified LRA model.

We now describe each of the components in turn.

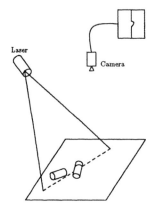

Figure 6. Schematic of laser sensing system. For each position of the plane of light, a scan of points in the camera can be processed to determine three dimensional position.

5.3. The Laser Scanner

Our input device is a structured-light laser scanner manufactured by Technical Arts Corporation in Redmond, Washington. It operates by scanning a laser beam using an oscillating mirror such that a plane of light is produced. A video camera is placed at an angle to this plane such that a portion of the plane is in the camera field of view (FOV). When an object is placed in this visible region such that it intersects the laser plane, points in the camera image illuminated by the laser unambiguously correspond to fixed 3D scene points. The correspondences between the scene points and image points are calculable by using a nonlinear projective transform, which can be determined by scanning an object of known form. Due to the fact that a multitude of such points can be viewed in any one scan, the device actually produces 240 3D measurements for any single scan.

When the scanner is transported, any point in a 3D volume can be measured by adding the scanner data to the position information of the transport. When the scanner is not moving, we call the scan data **static** (See Figure 6). In this case, the scan data corresponds to a planar cross-section of the scene. When the scanning takes place while the transport is moving, the scan is called **dynamic** and the cross-section information is more complex. In dynamic scans the sweep profile of the oscillating laser mirror determines the scan profile, as shown in Figure 7. This can be quite complex and difficult to predict at times.

5.4. The Coarse Scan

Because the tubes can bend in three dimensions, and because we allow the tubes to be placed in arbitrary position on the support table, within a 2.5 by 4 by 14 foot volume, our scanning system faces a quandary. We would like to obtain accurate information about the geometry of the tube, but to do this we need the scanner to be

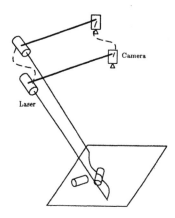

Figure 7. Schematic of laser sensing system. If the laser system is dynamically controlled, a surface of illumination is intersected with objects in the scene, to obtain three dimensional position.

close to the tube in order to get accurate and dense scan data from the tube surface. It would also be impractical from the standpoint of scanning and processing speed to scan the entire work volume. Since we do not know the position and orientation of the tube, we cannot safely do this without the potential for colliding with the tube. Hence, in a first pass, we fly the laser and camera system along a gantry, at a high enough standoff from the support table that we are guaranteed not to hit the tube, no matter what its orientation. For efficiency, since we need to scan the entire work volume, we do not attempt to densely measure the surface shape of the tube. Rather, we scan the scene at regular intervals, obtaining a sparse sampling of the objects under the scanner.

More specifically, the first step of the two pass locate/inspect process involves locating and determining the pose of a formed tube in the work volume. Four structured-light sensors are transported over the work surface. These scanners sample any object surfaces above the table height using a plane of laser light and solid-state cameras. These planar cross-sections of the work surface are sampled at approximately a 0.08 inch spacing.

Each scan cross-section is segmented by searching for discontinuities larger than 0.5 inches in height or 0.25 inches in width. Data within these scan segments are assumed to be from a single cylindrical cross-section of the tube (see Figure 8). Although we could execute any path for the sensing system (see Figure 7), the sparse scanners are arrayed with their scan planes aligned along a line perpendicular to the direction of travel. The fields of view of adjacent scanners are partially overlapped, but the scans are acquired sequentially over time. The output from this stage is a set of data segments from the surface of the tube. An example is shown in Figure 9.

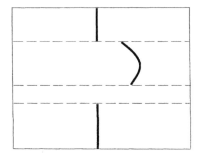

Figure 8. Example of a laser slice. A representative image for a cylindrical object under the scanner is shown in dark. A jump in distance greater than 0.5 inches in height (the top dotted line) or a gap in the data greater than 0.25 inches (the bottom two dotted lines) signifies an end to a data segment.

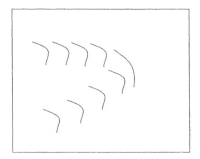

Figure 9. Example of segments of data from the surface of the tube.

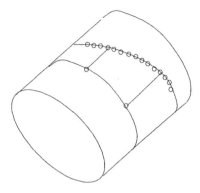

Figure 10. Illustration of a linear time algorithm for finding the cylinder axis.

5.5. Finding the axes

Our first task is to process the data to determine the location of the tube. To do this, we want to find the positions and orientations of the axes of each of the straight legs. By segmenting the individual scans, we obtain connected sets of points that come from the surface of the tubing. The trick is to separate out data due to the straight legs from data due to the toroidal bends, and then determine the parameters of the cylindrical sections.

We focus on the issue of locating the cylindrical sections from the range data. There are many methods for segmenting range data into coherent components: classifying surface patches, (e.g. [36, 16, 15, 6]); extracting cylinders (including generalized cylinders) cones and spheres, (e.g. [33, 10, 21]); extracting superquadrics (e.g. [34, 4]). Here, we find cylinders by hypothesis accumulation, i.e. by testing one or more scans to construct a hypothesis for a cylinder axis and then identifying large consistent sets of hypotheses. Therefore, the crucial step is constructing a cylinder axis from one or more scans. A common method fits an ellipse to the scan, and uses the parameters of the ellipse to solve for the cylinder axis [8]. While this only requires a single scan and is independent of sensor location, the technique is limited to static scans and is subject to numerical accuracy problems, especially when only small segments of the ellipse are available. In this section, we develop a very simple technique using two or three scans on a cylinder to find the orientation of its axis. The technique is very stable numerically. Additionally, it works both with static scans and with a broad class of dynamic scans.

5.5.1. Characterizing the cylinder axis

Basically, we look for surface rulings on the cylinder, since such rulings are parallel to the axis of the cylinder.

Step 1: Consider the situation shown in Figure 10. Given two scans (A and B), pick two points on one scan, $a_i (i = 1, 2)$; the points should be widely separated on the scan, but they must lie on the common overlap of the two scans (where by common overlap, we mean the intersection of the projection of each of the scans onto the base of the cylinder). We must choose $b_i (i = 1, 2)$ on the B scan such that the lines $\overline{a_1 b_1}$ and $\overline{a_2 b_2}$ are parallel. We will write $b_{i,k} (k \leq n)$ to indicate the choice of the kth point on the B scan as b_i. Then, for each point $b_{i,k} (k \leq n)$ on the B scan, construct the unit vector $v_{i,k}$ from a_i pointing at the point $b_{i,k}$. The dot product $v_{1,k} \cdot v_{2,l}$ (for any choice of points $b_{1,k}$ and $b_{2,l}$) measures the cosine of the angle between the two line segments. A brute force algorithm simply finds all the dot products $v_{1,k} \cdot v_{2,l}$ and picks the one closest to 1.0. If there are n points on scan B this operation requires computing $2n$ unit vectors and performing n^2 dot products. But, we can do better.

A more efficient algorithm exploits what we know of the geometry. We start with $k = 0, l = 0$, that is, the initial B points are $b_{1,0}$ and $b_{2,0}$; note that these are the *same* endpoint of the B scan. The basic loop of the algorithm increments k and then executes a loop that increments l and evaluates $v_{1,k} \cdot v_{2,l}$ until it finds a maximum. This loop is repeated until the value of the best dot product starts dropping.

Finding the parallel surface lines

```
l_best  =  0
dot_best  =  v_1,0 · v_2,0
Loop for k from 0 to n
    dot  =  dot_best
    Loop for l from l_best to n
        If v_1,k · v_2,l  ≥  dot
            Then dot  =  v_1,k · v_2,l;  l_best  =  l
            Else Exit Loop
    End Loop
    If dot  ≥  dot_best
        Then dot_best  =  dot
        Else Return (v_1,k−1 and v_2,l_best)
End Loop
```

The value of dot_{best} should approach 1.0 monotonically and then start to drop off. At the peak, both the $v_{i,k}$ are estimates of the cylinder axis direction. This estimate can be refined to sub-sample resolution by interpolating between the points $b_{i,k}$. In the worst case this algorithm computes $2n$ unit vectors and n dot products. In fact, one expects significantly less than these bounds. If there is relatively little rotation of the surface of illumination between the A and B scans, then the initial choice of k and l to be greater than 0 can significantly reduce the computation.

One important caveat in using the above algorithm is that the presence of noise can lead to false local extrema in the value of the dot product. Our implementation actually looks ahead on the scan a bit to make sure that the extremum is the global one.

Noise can also affect the accuracy of the computed rulings. Since we are mainly interested in the direction of the axis, which should be parallel to the rulings, we can reduce the effects of error by averaging the results obtained from a variety of choices of the points a_i. For example, given points a_1, a_2, a_3, the operation described above can be repeated for the three different combinations (a_1, a_2), (a_2, a_3) and (a_1, a_3), and

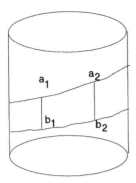 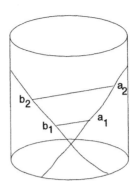

Figure 11. Two solutions to the problem of parallel lines, whose endpoints lie on the surface of the cylinder. In the case on the left, the lines are also parallel to the cylinder axis. In the case on the right, they are not.

the computed axis directions for each pairing can be averaged. Generally speaking, the wider apart the a_i points are, the more accurate the results.

Step 2: Once we find parallel lines, we can hypothesize that these are actually rulings on the surface of the cylinder. Next we must verify that this is actually true, i.e. that we are in the case shown in the left part of Figure 11, and not the right part of the figure, i.e. that we have a true ruling. In [28] we show that this can be done by intersecting the infinite line through either of the hypothesized rules with a third scan. If either extension does intersect the scan, then we assume that we have a pair of rulings that are parallel to the axis of the cylinder. If neither extension intersects the third scan, we may still be able to use the hypothesized rulings, but we cannot guarantee that they are in fact rulings. Furthermore, if the scans are nearly parallel (e.g. with a straightforward translation of the sensor configuration), there is no possibility that the scans can cross on a single leg. In this case, two scans are sufficient to uniquely determine the axis orientation.

Step 3: Given parallel rulings, we know that the axis direction is the same as the rulings, (see Figure 12). Now we can cluster axes hypotheses from successive pairs of scans. Each cluster of consecutive scans giving rise to a similar axis constitutes a different hypothesized cylinder segment. Note that if a sequence of scans pairwise give rise to roughly the same axis direction, then the likelihood of pairs of line segments being parallel to one another, but not parallel to the axis cylinder becomes vanishingly small. As a consequence, it is possible to omit step 2 above.

Step 4: We have found that the axis direction is sufficient to do the grouping of cylindrical segments, but once the axis is known it is straightforward to compute numerically an estimate of the axis displacement. In other words, points on the

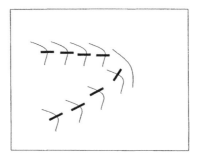

Figure 12. Example of segments of data from the surface of the tube.

actual axis are given in sensor coordinates as

$$\vec{a}(\alpha) = \vec{p} + \alpha\hat{t}$$

where $\hat{t}$ is a unit vector in the direction of the axis and $\vec{p}$ is a displacement vector from the origin of the coordinate system. The method in the previous steps allows us to determine $\hat{t}$. Any of several methods will serve to determine $\vec{p}$. In our testing, we have simply used the position of the average point of the scan as a (very) rough estimate of the center of the cylinder. Alternatively, we could use a straightforward Hough transform technique to determine the offset of the cylinder axis.

5.6. Surface Rulings

Note that this technique for finding cylinder axes from data essentially reduces to finding rulings on the cylinder surface from sparse scans. If we can find such rulings, we can use the simple relationship between the surface rulings and the axis of the cylinder to deduce the position and orientation of that axis. This observation suggests that more complex surfaces could also be analyzed in this manner. For example, cones could be deduced by finding rulings on the surface that intersect at a common point. It may also be possible to analyze generalized cylinders in this manner.

5.7. Getting the bends

We have used the described method to segment a tube made up of cylindrical and toroidal sections. The desired result is a list of the scans on the straight (or cylindrical) sections. We detect the onset of a bend by examining the angles that the computed axes make with the global coordinate axes. The axes on the straight sections will have nearly constant values of these angles while the axes computed along the toroidal bends will have rapidly changing values. Figure 14 shows the axis angles computed from the examples in Figure 13 and the segmentation into linear segments. The angles are relatively noisy due to the inaccuracies in the scan data, but during the straight segments the deviation from the average angle is within 2 or 3 degrees. We use a simple split and merge algorithm to find segments of nearly constant angle. We have used this method in the results in Figure 13 with quite satisfactory performance. The output from this stage is a set of cylinder axes, including an axis direction, an offset from the origin, and the positions of the endpoints of the axis.

5.8. Matching the Model

Given a set of measured cylinder axes, we can now match these against a model of the part to determine its location and orientation.

For the particular case of cylinder axes, we use a series of constraints. First, we use a unary length constraint, which requires that a candidate matching of a sensory and model leg have the lengths match within some error bounds. The sensory feature leg length is generally allowed to be significantly shorter than the model leg (10%), and up to twice the model leg length.

The binary constraints used are an angle constraint and distance constraints based on pairs of sensory and model features. A pair of model axes can match a pair of data axes if the angle between the model axes is roughly the same as that of the data axes. We implement this by computing the dot product of the unit vectors in the direction of the axes, and requiring that the computed values for the model and data axes agree, within a fixed angular error tolerance.

We can also use the distance between legs as a constraint, which we incorporate based on the shortest chord joining the two leg axes.

Figure 15(a) shows this chord XY, referred to as the "base" of the leg pair. It is orthogonal to both leg axes. The points X and Y do not necessarily lie between the endpoints of the feature legs (AB or CD). There are four distances saved as feature distance constraints. These are the minimum and maximum distances between the intersection of the base and the legs (X and Y) and the endpoints. These are shown in the figure as the distances a, b, c, and d. Another binary constraint used in the match pruning is the base length. In the figure, this is the distance between X and Y. All the sensory binary length constraints are compared to the candidate model pair values with error bounds built into the model data set.

Legs pairs in the model or sensory sets which are nearly parallel are pruned from the search tree due to the instability of any transform between model and sensory features that would result by using them. When the leg lengths are nearly equal, there are two possible transforms to align the model and sensory leg pairs. When the legs are also nearly orthogonal, an additional matching ambiguity exists, raising the number of possible transforms to four. When such conditions arise, each possible transform is tested.

Of course any data leg matched to the $*$ character is ignored in this process.

5.9. Model Transformation and Testing

Combinations of legs which survive the constraint pruning are used to create the transforms of the model to the sensory data. These transformations carry the "base point" — the midpoint of the base — of the model legs to the base point of the sensory legs. This is shown in Figure 15(b) as point O. The transforms also rotate the base of the model legs into alignment with the base of sensory legs (indicted as the X axis in the figure), and they carry a bisector of the model legs to one of the

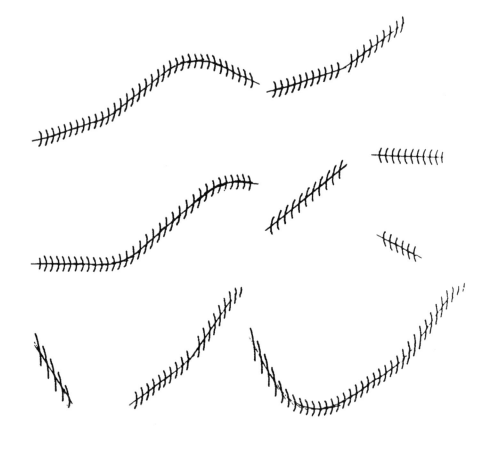

Figure 13. Examples of tube axes computed by the described method.

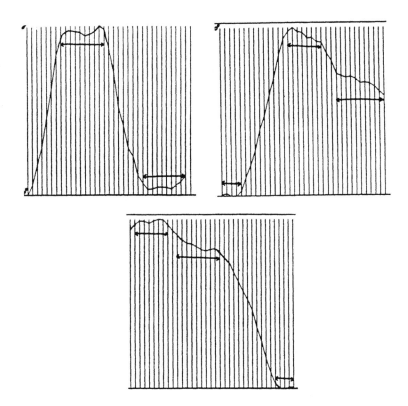

Figure 14. Axis angles computed for the examples of Figure 13. The straight segments are indicated in each graph, and correspond to the cylinder segments found in Figure 13.

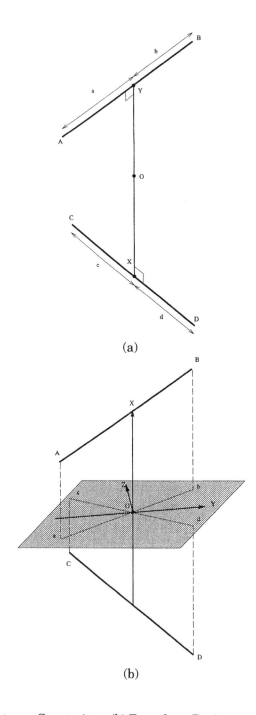

Figure 15. (a) Distance Constraints, (b) Transform Basis

four bisectors of the sensory legs. A bisector, shown as the Y axis in the figure, is a ray emanating from the base point which bisects one of the angles formed by the projections, along the base and onto a plane normal to the base, of the axes of the legs.

If such a transformation brings any leg substantially below the table surface of the or otherwise outside the work volume, the transform is rejected. Once the entire model data set is transformed, each corresponding model and sensory leg is tested for good alignment. Each model/sensory pair is tested and assigned a numeric measure of fit. This measure is based on the angle between the two leg axes as well as the distance between the legs, determined in much the same way as the base distance was calculated in the constraint steps described above. If any pair mismatch exceeds a threshold, the transform is rejected. Of the accepted transforms, the one with the best match measure is chosen as the solution. This measure is based on the base distance matching and a measure of the total model leg length matched.

Least squares methods were also tested. Perhaps surprisingly, no measurable improvement in the computed alignments was found over the base leg methods.

The purpose of matching is to determine the alignment of the tube on the inspection table. Although the transformed model would normally suit this purpose, the physical tube often is incorrectly formed, or distorted by gravity or friction, such that the best match is still measurably different than the model. To correct for this, a final step transforms each model leg to bring it into as close a fit as possible to the measured data. This complex heuristic, which includes unmatched and partially matched model legs, is called 'warping' the model. This completes the model location processing, the output of which is a complete transformed model which is very close to the form and orientation of the tube placed arbitrarily on the table. The next pass — inspection — involves producing an inspection plan and executing that plan to produce a precise sensory model.

5.10. Planning a Path

In order to measure the tube form, a large amount of data is needed from the cylindrical legs. The toroidal bend sections are not used in the measurement, since the bending process tends to deform the tube cross-section. It is important to get as much data as possible on the cylindrical sections to insure accurate results. Since the accuracy of the scan data is inversely proportional to the measurement range of the detail sensor and its standoff from the part, the scanner must be transported over each leg of the part in close proximity to the leg. The laser plane orientation must also be roughly aligned to the tube axis. This requirement means that an inspection plan must be generated from the localized model information which involves 4-axis (three Cartesian and z-axis rotation) gantry moves which transport the detail scanner over the tube sections. While the scanner is being transported, scan data is acquired dynamically.

The first pass produces a model of the location and orientations of all of the tube sections. From this information it isn't difficult to generate a path plan for the detail scanner. For instance, the scanner standoff was 18". The scanner path is therefore the same as the localized tube model with the Z-axis information translated by 18". The scanner orientation is simply the projection of the tube model leg axis projected on the X-Y plane.

Once the path plan is generated, it is tested against the localized model and the known size of the scan head to insure collision-free scanning.

5.11. The Second Scan

Once the inspection plan is complete, the detail scan data is acquired. Since the accuracy of the axis measurements depends, in part, on getting a large amount of data covering as much of the straight cylindrical tube sections as possible, about 200 scan points are acquired for each scan cross-section. A scan is taken about every 0.1" along the length of the tube leg. Therefore, on a 6" leg, 120,000 individual 3D data points are acquired.

Cylinders are fit to the data, using least squares methods, and a detail scanner sensory model produced from these data. This model is the end-product of the inspection process. It is used both to determine the acceptability of the part and to produce a "corrected" bending program for subsequent part runs.

5.12. Testing the method

We have tested this method as follows:

- The scanner was tested on approximately 400 different designs. Although the original specifications were to be able to inspect 95% of the different designs, all of the designs tested were inspected correctly.

- Speed. The ATS inspected a typical 6' long 5 bend tube in just under 2 minutes (total spase/detail scan and data processing time).

- Repeatablity. When a tube is inspected multiple times, the data vary by less than 0.003" (3 sigma).

- Accuracy. The system was accurate on tube axis measurements to +/- 0.008" (3 sigma) (Raw data was accurate to 0.003").

- Sizes: Tube diameters varied between 3/16" and 3".

- Bends: Models with up to 13 bends were inspected.

- Overall Tube length: up to 13'

6. Summary

We have described a general framework for recognizing 3D objects from 3D sensory data, using constrained search of an interpretation tree. Such methods have been applied to a variety of data, including laser range data, stereo data and motion data. We have demonstrated the method by describing a working system that inspects tubing composed of cylindrical and toroidal shapes.

REFERENCES

1 Ayache, N. & O.D. Faugeras, 1986, "HYPER: A new approach for the recognition and positioning of two-dimensional objects," *IEEE Trans. Patt. Anal. & Mach. Intell.*, Vol. 8, no. 1, pp. 44–54.

2 Ballard, D.H., 1981, "Generalizing the Hough transform to detect arbitrary patterns," *Pattern Recognition* 13(2): 111–122.

3 Baird, H.S., 1985, *Model-Based Image Matching using Location*, MIT Press, Cambridge, MA.

4 Bajcsy, R. & F. Solina, 1987, "Three dimensional object representation revisited," First ICCV, London, UK, pp. 231–240.

5 Basri, R. & S. Ullman, 1988, "The Alignment of Objects with Smooth Surfaces," *Second Int. Conf. Comp. Vision*, 482–488.

6 Besl, P.J. & R.C. Jain, 1988, "Segmentation through variable-order surface fitting, *IEEE Trans. Patt. Anal. & Mach. Intell.* Vol. 10, no. 2, pp. 167–192.

7 Besl, P.J. & R.C. Jain, 1985, "Three-dimensional object recognition," *ACM Computing Surveys*, Vol. 17, no. 1, pp. 75–154.

8 Bolles, R.C., 1986, "Three-dimensional locating of industrial parts," in *Robot Sensors: Vol 1, Vision*, ed. Alan Pugh, IFS Publications Ltd. and Springer Verlag

9 Bolles, R.C. & R.A. Cain, 1982, "Recognizing and locating partially visible objects: The Local Feature Focus Method," *Int. J. Robotics Res.*, Vol. 1, no. 3, pp. 57–82.

10 Bolles, R.C. & M. A. Fischler, 1981, "A RANSAC-based approach to model fitting and its application to finding cylinders in range data," *7th IJCAI*, Vancouver, B.C., Canada, pp. 637–643.

11 Bolles, R.C. & P. Horaud, 1986, "3DPO: A Three-dimensional Part Orientation System," *Int. J. Robotics Res.*, Vol. 5, no. 3, pp. 3–26.

12 Cass, T.A., 1988, "A robust parallel implementation of 2D model-based recognition," *IEEE Conf. Comp. Vision, Patt. Recog.*, Ann Arbor, MI, pp. 879–884.

13 Cass, T.A., 1990, "Feature matching for object localization in the presence of uncertainty," MIT AI Lab Memo 1133.

14 Ettinger, G.J., 1988, "Large hierarchical object recognition using libraries of parameterized model sub-parts", *IEEE Conf. Comp. Vision, Patt. Recog.*, pp. 32–41

15 Faugeras, O.D. & M. Hebert, 1986, "The representation, recognition and locating of 3-D objects," *Int. J. Robotics Res.* Vol. 5, no. 3, pp. 27–52.

16 Faugeras, O.D., M. Hebert & E. Pauchon, 1983, "Segmentation of range data into planar and quadric patches," *Third CVPR*, Arlington, VA, pp. 8–13.

17 Grimson, W.E.L., 1990, *Object Recognition by Computer: The role of geometric constraints*, MIT Press, Cambridge.

18 Grimson, W.E.L. & T. Lozano-Pérez, 1984, "Model-based recognition and localization from sparse range or tactile data", *Int. Journ. Rob. Res.*, 3(3):3–35.

19 Grimson, W.E.L. & T. Lozano-Pérez, 1987, "Localizing overlapping parts by searching the interpretation tree", *IEEE Trans. PAMI* 9(4):469–482.

20 Haralick, R.M. & G.L. Elliot, 1980, "Increasing tree search efficiency for constraint satisfaction problems", *Artificial Intelligence* **14**:263–313.

21 Hebert, M. & J. Ponce, 1982, "A new method for segmenting 3-D scenes into primitives," *Sixth ICPR*, Munich, West Germany, pp. 836–838.

22 Huttenlocher, D.P. and S. Ullman, 1987, "Object Recognition Using Alignment", *Proceedings of the First International Conference on Computer Vision*, pp. 102-111.

23 Huttenlocher, D.P. & S. Ullman, 1990, "Recognizing Solid Objects by Alignment with an Image," *Inter. Journ. Comp. Vision* **5**(2):195–212.

24 Knapman, J., 1987, "3D model identification from stereo data", *Proceedings of the First International Conference on Computer Vision*, London, pp. 547–551.

25 Lamdan, Y., J.T. Schwartz & H.J. Wolfson, 1988, "Object Recognition by Affine Invariant Matching," *IEEE Conf. on Comp. Vis. and Patt. Recog.* pp. 335–344.

26 Lamdan, Y., J.T. Schwartz & H.J. Wolfson, 1990, "Affine Invariant Model-Based Object Recognition," *IEEE Trans. Robotics and Automation*, vol. 6, pp. 578–589.

27 Lamdan, Y. & H.J. Wolfson, 1988, "Geometric Hashing: A General and Efficient Model-Based Recognition Scheme," *Second Int. Conf. on Comp. Vis.* pp. 238–249.

28 Lozano-Pérez, T., W.E.L. Grimson, & S.J. White, 1987, "Finding cylinders in range data," *Int. Conf. Rob. Autom.*, Raleigh, NC, pp. 202–207.

29 Lozano-Pérez, T., J.L. Jones, E. Mazer, P.A. O'Donnell, W.E.L. Grimson, P. Tournassoud, & A. Lanusse, 1987, "HANDEY: A robot system that recognizes, plans, and manipulates", *Proc. Int. Conf. Robot. Autom.*, Raleigh, NC, pp. 843–849.

30 Murray, D.W., 1987, "Model-based recognition using 3D structure from motion", *Image and Vision Computing*, 85–90.

31 Murray, D.W., D.A. Castelow & B.F. Buxton, 1989, "From images sequences to recognized moving polyhedral objects", *Inter. Journ. Comp. Vision* **3**(3):181–209.

32 Murray, D.W. & D.B. Cook 1988, "Using the orientation of fragmentary 3D edge segments for polyhedral object recognition", *Inter. Journ. Comp. Vision* **2**(2):153–169.

33 Nevatia, R. & T. O. Binford, 1973, "Structured descriptions of complex objects," *Third IJCAI*, Stanford, CA, pp. 641–647.

34 Pentland, A., 1986, "Perceptual organization and the representation of natural form," *Art. Intel.* **28**(3):293–331.

35 Pollard, S.B., J. Porrill, J.E.W. Mayhew, & J.P. Frisby, 1987, "Matching geometrical descriptions in three-space", *Image and Vision Computing*, **5**(2):73–78.

36 Shirai, Y. & M. Suwa, 1971, "Recognition of polyhedra with a range finder," *Second IJCAI*, London, pp. 80–87.

Three-Dimensional Object Recognition Systems
A.K. Jain and P.J. Flynn (Editors)
© 1993 Elsevier Science Publishers B.V. All rights reserved.

Recognition of Superquadric Models in Dense Range Data

Alok Gupta[a] and Ruzena Bajcsy[b]

[a]Siemens Corporate Research, Inc.
755 College Road East
Princeton, New Jersey 08540 USA

[b]GRASP Laboratory
Department of Computer and Information Sciences
University of Pennsylvania
Philadelphia, Pennsylvania 19104 USA

Abstract

Recognition and representation of three-dimensional structure of complex objects is essential for a vision system. This chapter deals with the recovery of superquadric part-structure of objects in dense range data by a residual-driven recursive algorithm. Only the geometric shape information inherent in superquadric and bi-quadric models is used to perform the segmentation. The superquadric segmentation (SUPERSEG) system integrates the piecewise bi-quadric segmentation obtained by a local-to-global surface segmentation method with the global-to-local residual driven superquadric fitting method. A set of acceptance criteria provide the objective evaluation of intermediate descriptions, and decide whether to terminate the procedure, or selectively refine the segmentation. The control module generates hypotheses about superquadric models at clusters of underestimated data and performs controlled extrapolation of part-models by shrinking the global model. Results are presented for real range images of scenes of varying complexity, including objects with occluding parts, and scenes where surface segmentation is not sufficient to guide the volumetric segmentation.

1. Introduction

Recognition and representation of three-dimensional structure of complex objects is essential for a vision system. In this chapter, we present an integrated system for recovering superquadric models in dense range data by a residual-driven recursive algorithm. Only the geometric shape information inherent in the superquadric and bi-quadric models is used to perform decomposition of a complex scene into parts corresponding to superquadric models. Thus, instead of matching stored models, the shape vocabulary includes a continuum of shapes that can be recovered from the data. The resulting decomposition into parts is very useful for the high-level symbolic reasoning object-recognition processes, which can attach domain specific labels to the parts, and reason at a level where the visual input is structured in terms of primitives, rather than cope with the difficulties of low-level vision and huge pile of unstructured data.

The SUPERSEG (SUPERquadric SEGmentation) system (shown in Figure 1) has five major components. The bi-quadric *surface segmentation* and the recovery of the global *superquadric model* are performed independently. The segmented surfaces are refined and analyzed by the *surface description* module to extract region adjacency information, surface discontinuities, and global shape properties, which are used to guide the volumetric segmentation. After surface analysis is complete and the global superquadric model is recovered, the *control module* begins the global-to-local residual-driven procedure to recursively derive the part-structure. A set of acceptance criteria provide the objective evaluation of intermediate descriptions, and decide whether to terminate the procedure, or selectively refine the segmentation. Both qualitative (local distribution of residuals) and quantitative measures (normalized deviation of data from the model) are used for the complete evaluation of the volumetric models. The control module generates hypotheses about part-models at selected clusters of underestimated data. The global model discards the underestimated data, which is used for controlled extrapolation of part-models. This in turn shrinks the global model and makes it converge on a part of the object.

For the sake of completeness, we will summarize all the components of the SUPERSEG system in this chapter. The control module for the volumetric segmentation will be presented in some detail. For a complete description of the SUPERSEG system the reader is referred to Gupta [3, 4]. The next section introduces the bi-quadric model and surface analysis, and the formulation for the recovery of one superquadric model. The model evaluation criteria is summarized in section 3. Section 4 presents various issues in volumetric segmentation, and the general strategy for the residual-driven recursive procedure. Results are presented in section 5, with conclusions and future directions in section 6.

2. Shape Primitives and Segmentation

To obtain a global shape description from single-viewpoint 3-D data, we need to address shape at the volumetric and surface levels. The volumetric level describes the global shape of parts in terms of 3-D primitives, while the surface primitives account for the internal surface boundaries and surface patches which are difficult to model with volumetric primitives. With computability, simplicity, and the utility of the shape representation as our major concerns, we decided to use bi-variate polynomials (up to second-order) and superquadrics as our surface and volumetric models respectively. Bi-quadrics achieve $2\frac{1}{2}$-D clustering, while superquadrics achieve 3-D clustering of the $2\frac{1}{2}$-D data.

2.1. The Surface Model: Bi-quadrics
The variable-order bi-variate model is given by :

$$\hat{f}(r, \mathbf{a}, \mathbf{x}) = \sum_{0 \le i+j \le r} \mathbf{a}_{ij} x^i y^j \tag{1}$$

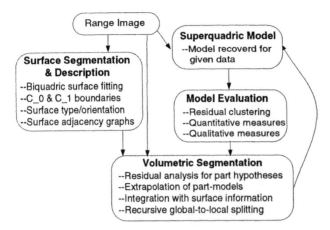

Figure 1. The control flow of the SUPERSEG system. In this chapter we summarize the surface segmentation module, and focus on model evaluation and superquadric volumetric segmentation.

where the vector a is defined in the parameter space $\mathcal{A}$. We restrict the order r of the model to $0 \leq r \leq 2$, thereby admitting planar and bi-quadric surfaces. Surfaces of higher-order can introduce oscillations, are computationally expensive, and are difficult to interpret qualitatively. Bi-quadric patches can be qualitatively described as convex or concave, and they can be used to compute discontinuities and orientation of the major axis of a curved surface.

2.1.1 Surface Segmentation

The literature review and the surface segmentation procedure are described in detail in Leonardis, Gupta and Bajcsy [8]. Data aggregation is performed via model recovery in terms of variable-order bi-variate polynomials using iterative regression. The process starts by placing a grid pattern of seed surface patches across the image. For each seed patch a plane is fit and the fit error is compared to an error threshold T_s. Those seed patches whose fit error is better than the threshold are used to start the iterative processing. The simplest form of an iteration involves 1) growing current surface patches beyond their present borders, 2) updating the parameters of the surface patch, and 3) pruning the number of patches based on an objective function.

If a surface patch can not be grown as it is, then the system tests a higher order polynomial model to see if it would provide a better description of the image. The segmentation algorithm attempts to find the best description of the image it can using the lowest order models possible. The system grows surface patches by testing the image points neighboring the border of the surface patch for compatibility by comparing them to the compatibility constraint C. The square of the difference

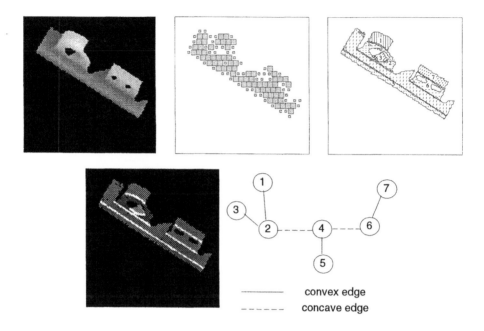

Figure 2. Surface analysis of the NIST object. Top: The range image; seed regions; and the bi-quadric surface segmentation. Bottom: The C_1 (surface normal) edges marked at the overlapping parts of the surfaces; and the surface adjacency graph (SAG) for the object.

between the image value and the extrapolated value of the surface at that point determines the compatibility. At each iteration, each surface patch is grown at most a fixed distance from its boundaries at the start of the iteration.

If a surface patch grew during the current iteration then the coefficients of the polynomial need to be updated to reflect the influence of the new data points found during the growing stage. Bi-variate polynomial surface fitting is accomplished through general linear least squares regression. Growing can produce overlapping surface patches. If two surface patches completely overlap and one provides a significantly better description of the data as determined by the global least squared error, then there is no need to keep the surface patch with the larger error. This is part of the motivation for doing surface patch selection interleaved with surface patch recovery (*Recover-and-Select* paradigm). Selecting the surface patches on which to continue processing is based on optimizing an objective function. The objective function for each particular surface patch model is a weighted linear combination of: number of points in the surface patch (benefit), global least squared error (cost), and the order of the polynomial (cost). The function is in the form of a Boolean quadratic problem which is optimized by a greedy algorithm that selects an optimal combination of patches to derive the current best description. The complexity of the procedure is proportional to the number of currently valid models multiplied by the time needed to evaluate the optimization matrix. The Recover-and-Select procedure terminates when all the valid models are completely grown.

2.1.2. Surface Description and Refinement

The surface segmentation module gives a piecewise, and possibly overlapping segmentation of the range data into bi-quadric patches. This description is used by the SUPERSEG system after refinement of surface boundaries and extraction of surface properties by the surface description module. The reader is referred to Gupta [3, 4] for detailed algorithms. The surface description module refines the segmentation by analysis of the intersection curves of each pair of intersecting regions. This analysis removes the artifacts of geometric region growing, and also provides an elegant algorithm for 3-D edge localization and labeling into convex and concave types. A surface adjacency graph (SAG) is constructed to encode spatial relationships among surfaces and the edges between them. The general forms of the biquadrics is reduced to their standard forms to reveal the type of curved surface, and label it as convex and concave. Additionally, information about the major axis (for curved objects) is extracted by analyzing the bi-quadric orientation with respect to the object-centered-system determined using eigenvector analysis. This allows correct placement of the major axis (Z) for cylindrical superquadrics where diameter is greater than the length. This analysis prepares the surface segmentation for use by volumetric segmentation. The complete surface analysis for a machined object (from NIST) is shown in Figure 2.

2.2. Superquadric Part-Models

Parametric models like generalized cylinders and their derivatives have been popular choice of volumetric primitives. Generalized cylinders have a rich vocabulary of shapes, but in practice it is limited to simple linear-straight-homogeneous-cylinders. Deformable models based on generalized cylinders [15] or superquadrics [14] have the disadvantage that they are too complex and have so far been shown to work only on pre-segmented data. The descriptions generated by our method can be used as starting approximations for the complex deformable models. Superquadrics have been used in vision [11, 13] to represent natural part-structure. Existing methods for partitioning data into superquadric primitives include that by Solina [13], Pentland [10, 11], Gupta et al [6], and Ferrie et al [2]. One of the major drawbacks of all the previous methods (except Solina's) is that they assume a one-to-one correspondence between superquadric models and surface models. Solina attempted to achieve segmentation during model recovery, which is unpredictable and not general in application. We tackle these problems by *using* the superquadrics to drive the segmentation process, and drawing support from surface description where possible.

2.2.1. Formulation for Recovering one model

Solina [13] has developed a model recovery procedure to fit tapered and bent models to given data. However, the procedure does not have segmentation capabilities. The SUPERSEG system uses his formulation for the recovery of a single superquadric model for a given set of 3-D points. The superquadric implicit equation is given by:

$$F(x, y, z) = \left[\left[\left(\frac{x}{a_1} \right)^{\frac{2}{\varepsilon_2}} + \left(\frac{y}{a_2} \right)^{\frac{2}{\varepsilon_2}} \right]^{\frac{\varepsilon_2}{\varepsilon_1}} + \left[\frac{z}{a_3} \right]^{\frac{2}{\varepsilon_1}} \right]^{\varepsilon_1} \tag{2}$$

a_1, a_2, and a_3 define the superquadric size, and ε_1 and ε_2 describe the shape. If the inside-outside function (IO function for short) $F(x, y, z) = 1$, the point $P(x, y, z)$ lies on the surface of the superquadric. If $F(x, y, z) < 1$, the point lies inside and if $F(x, y, z) > 1$, the point lies outside the superquadric. In Gupta et al [5] we have shown that $F(x, y, z)$ is square of a factor β by which a model has to scaled to make it pass through the point.

Solina [13] has formulated the superquadric model recovery problem in general position and orientation by using Euler angles ϕ, θ, ψ to define the orientation, and p_x, p_y, p_z to define position of the superquadric in a world coordinate system. Global deformations like linear tapering along X and Y axes is specified by K_x and K_y respectively, and symmetric bending by curvature k and angle α. Thus, bending and tapering introduce two parameters each in the final superquadric equation, bringing the total parameter count to 15. Solina used the Levenberg-Marquardt method [12] to minimize the goodness-of-fit function of deformed superquadrics in general position given by :

$$GOF = \frac{1}{N} \sum_{i=1}^{N} R_i^2 \tag{3}$$

where

$$R_i = \sqrt{a_1 a_2 a_3} (F(x_i, y_i, z_i; a_1, a_2, a_3, \varepsilon_1, \varepsilon_2, \phi, \theta, \psi, p_x, p_y, p_z, K_x, K_y, k, \alpha) - 1) \tag{4}$$

The model recovery starts with fitting an ellipsoidal shape on 3-D points and converges on a shape that minimizes the least-squares error. We have found that the procedure needs correct estimation of Z-axis for cylindrical objects, and converges faster if the initial orientation is close to the final one.

3. Model evaluation: Residual analysis

We now have a fine-to-coarse surface segmentation procedure and a procedure to recover the global superquadric model for the given data. We want to develop a control structure that will segment the given data set by constant evaluation of the intermediate superquadric approximations of the data and by using the information from the biquadric segmentation and other geometric constraints. In this section we summarize a set of criteria for the complete evaluation of a superquadric model.

The superquadric recovery formulation imposes two constraints on the recovering model:

1. **Surface Constraint:** A point should satisfy the inside-outside function.

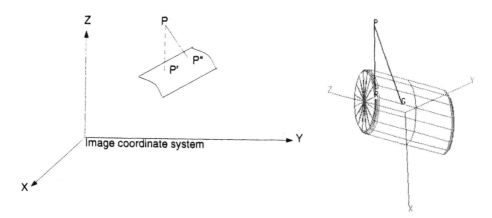

Figure 3. Computation of deviation of a point from the superquadric model. Left: The Z-residual is computed along the viewing direction in the image coordinate system. PP' is the distance along Z, while PP'' is the minimum distance. Right: The IO residual is based on the inside-outside function, measuring the distance corresponding to PG and not PR, the minimum distance.

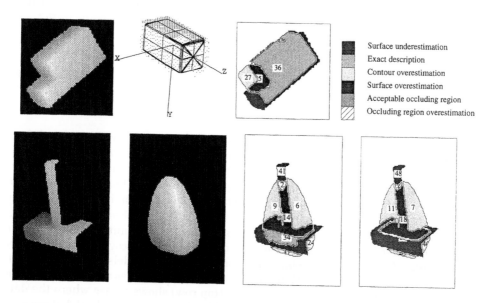

Figure 4. **Top:** Block_1. Left to right: Range image; its global model; and the Z-residual of the global model. The missing volume results in local residuals but the global model has acceptable error-of-fit. Top right: Legend for the interpretation of the residual maps in this chapter. **Bottom:** The composite object: The range image; the projection of the global model; the Z-residual map; and the IO-residual map.

2. **Volume Constraint:** The $\sqrt{a_1 a_2 a_3}$ factor provides for the smallest volume satisfying the surface constraint.

Volumetric segmentation imposes acceptance criteria for the recovered models, reflecting the scale considerations, which dictate that all the data points must correspond to the model within the given error tolerance. When an arbitrary collection of points (non-convex) is presented to the single-model formulation, and if there exists no model that will satisfy the surface constraint, the model averages out the IO function value to leave certain points outside the model (F > 1, underestimated) and some inside the model (F < 1, overestimated). If the concavities (or convex deficiencies) are significant, then cluster of points have values away from the ideal value of 1. In such cases, the recovered model is not a satisfactory description of the underlying data, and the presence of such clusters signals the need for decomposition of data into smaller pieces to satisfy the modeling constraints. Therefore, the model is fully analyzed, both qualitatively (using local distribution of residual types) and quantitatively (using normalized global residuals), to aid in determining a further course of action. The deviation of data points from the model surface can be measured by the following three methods (illustrated in Figure 3):

1. Goodness-of-fit ($\mathcal{G}$) measure based on the IO function, where $\mathcal{G} = F - 1$, without the volume factor. In Figure 3, the IO function value corresponds to PG.

2. Deviation ($\mathcal{D}$) along Z direction: Corresponds to PP'.

3. Minimum distance ($\mathcal{M}$) of a point from the model. (PP'' & PR). Both PP' and PG overestimate $\mathcal{M}$.

3.1. Quantitative Measures
The normalized values of $\mathcal{G}, \mathcal{D}$ and $\mathcal{M}$ for all the points give the global deviation of the model from data. $\mathcal{G}$ is computed during model recovery.$\mathcal{D}$ and $\mathcal{M}$ require additional computations for each point. $\mathcal{M}$ is computed by an iterative method using β to provide a good initial approximation [5]. The pointwise comparison between the projected model and the image gives $\mathcal{D}$. Empirically determined thresholds for a "good" fit are 0.32 value of $\mathcal{G}$ indicating a 15% scaling of the model ($0.32 = F - 1 = \beta^2 - 1 \Rightarrow \beta = 1.15$), and a value of 2 to 3 units for $\mathcal{D}$. However, relying on these global thresholds for the evaluation of a recovered model can be misleading. An acceptable global error can result from models with local details that are averaged out in the global consideration. This requires analysis of the type and the distribution of residuals. Figure 4 (top row) shows a case where the data points are overestimated by the global model having an acceptable global error-of-fit. Therefore, to evaluate a model completely we need the following qualitative measures.

3.2. Qualitative Measures
The deviation of points from the fitted model can be used to cluster points to obtain residual maps:

1. IO-residual map: IO function-based clustering, and assign $\mathcal{G}_i$ to each P_i.

2. Z-residual map: Clustering along viewing direction, and assign $\mathcal{D}_i$ to each P_i.

3. $\mathcal{M}$-residual map: IO function-based clustering, and assign $\mathcal{M}_i$ to each P_i.

Depending on data-model relationships, six types of clusters can arise in a residual map (Z- and IO-residual maps are shown for the composite object in Figure 4, bottom row) :

1. **Surface underestimation** (s_under): The model surface underestimates the point (leaves it outside) when viewed along Z direction or in the IO sense. Regions 41 and 48 in Z-residual and IO-residual maps respectively, represent s_under regions of the composite object (Figure 4, bottom row).

2. **Surface overestimation** (s_over): The model surface overestimates the point (leaves it inside) when viewed along Z direction or in the IO sense. Points that appear to be underestimated in IO sense can actually appear overestimated along Z because of the directionality constraint of the Z residuals. For example, the underestimated region for the composite object in Figure 4 has a number of points hidden behind the model that appear to be overestimated along Z (region 14) but are underestimated in the inside-outside sense.

3. **Acceptable description** (s_exact): The model estimates the data points within the specified tolerance. Again, due to the non-directionality of the IO function, concave surfaces that are modeled by the hidden side of the model (the side that is not visible from the viewing direction) will be labeled as acceptable, whereas the Z-residual map will show them as overestimated. Due to the presence of parts, the composite object (Figure 4, bottom row) has small s_exact regions, while the global model for block_1 (Figure 4, top row) is a good approximation for the majority of the surface points (region 36).

4. **Contour overestimation** (c_over): Due to the symmetry and shape constraints of the rigid model, the projection of the model on the image coordinate system can result in overestimation of the silhouette of the data. These regions predict extra data which does not exist in the image. c_over is similar in both Z-residuals and IO-residuals (regions 6 and 9, and 7 and 11 respectively in Figure 4).

5. **Acceptable Occluding regions** (occ_ok): When data is decomposed to arrive at a piecewise description, it is desirable to allow the volumes of the models to occlude each other such that the occluded model *underestimates* or *exactly describes* the data points that do not belong to it. For example, the base in the NIST object in Figure 5 underestimates the points belonging to the other parts. These residuals should not adversely affect the model for the base.

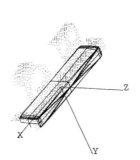
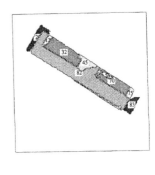

Figure 5. The NIST object: The range image; the partmodel for the base; and its Z-residual map. The occ_ok regions (32 and 70) are acceptable occlusions since they underestimate the data not belonging to the base.

6. **Occluding region overestimated** (occ_over): Regions where the model overestimates the data points not belonging to it are undesirable and therefore count against the model during evaluation.

The residuals generated by comparing the recovered model on the given data form the basis of the coarse-to-fine volumetric segmentation approach. The next step labels the connected pixels in clusters of one of the six types. A residual adjacency graph (RAG) is constructed to illustrate the spatial relationships among the clusters. Figure 6 shows the residual adjacency graph for the global model of Scene_1 containing 4 clusters of data. The RAG encodes connectivity information of the residual regions, and therefore can be used to isolate disconnected data clusters by the connected-component analysis of the RAG. RAGs dynamically change as the models evolve and data clusters become disconnected during the segmentation process. For a detailed description of residual analysis, and its role in model evaluation, the reader is referred to Gupta [3].

3.3. Using the Residuals for Superquadric Evaluation

For a data set of cardinality n, we know the number of points that are exactly described, underestimated, and overestimated. To enforce scaleability and size invariance, we use the relative measures stored as percentage of data that is exactly described, overestimated or underestimated. Various thresholds can be put on these fractions to define the acceptance criteria. Similarly, the c_over can be studied relative to the original data. Usually there is a 10 to 20% c_over regions due to noisy data near the edges and due to the fact that superquadric only approximately follows the boundaries of real data. A 20% value means that in order to describe 100 points in the data, an overestimation of 20 points occurred in volumetric sense. Similarly, an acceptance condition for occ_over regions can be enforced. A combination of these conditions gives the acceptance criteria to the control module, and defines the termination conditions for the model recovery.

In addition to the residuals for the entire data, residuals for individual surfaces

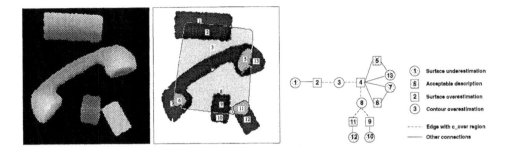

Figure 6. Scene_1: The range image; the Z-residual map of the global model; and its Residual adjacency graph (RAG). The connected-component analysis of the RAG after removing the dashed edges gives individual data clusters.

can also be obtained to further refine the acceptance criteria. The residuals are computed only for the domain of the biquadric surface included in the superquadric model. Thus, an acceptable model will describe all its constituent surfaces with high confidence. It is also possible to selectively enforce the thresholds, for example, some of the surfaces (e.g. smaller than a fixed size) can be ignored during the evaluation, while others can be given more weight. Together with the globally-relative acceptance criteria outlined above, the surface-relative criteria form a comprehensive criteria for superquadric model evaluation.

4. Volumetric Segmentation

The description available at this stage is in the form of the piecewise bi-quadric patches along with the information about the surface-type, orientation and edges. Also available is the global superquadric model and a set of criteria to exhaustively evaluate it given the original dataset. The main task of the control structure for volumetric segmentation can be defined as systematically integrating surface descriptions with the global-to-local superquadric recovery approach, evaluating the intermediate descriptions, and deciding on the strategies for segmentation.

4.1 Issues in Volumetric Segmentation

Surfaces have more local support and hence provide reliable intermediate-level clustering in terms of bi-quadrics. Unfortunately, the analytical correspondence between biquadrics and superquadrics is minimal since biquadrics belong to the class of non-central quadrics while superquadrics are more appropriately non-linear deformations of ellipsoids which belong to the class of central quadrics. Earlier, we established some correspondence between the two for the purpose of superquadric axis alignment for curved objects.

4.1.1. Superquadric Recovery Formulation for Segmentation

Regardless of how the model recovery is formulated, the most difficult aspect of using the recovery-based formulation is that the domain of the model (data points for which the model has to be recovered) has to be defined before the model is recovered. This requires that segmentation process be interleaved with model recovery. Skeletonization is a popular approach [9, 11, 15] for segmentation, but it is sensitive to occlusion and requires the knowledge of internal boundaries for complex objects. Besides, skeletonization works best for curved surfaces, but is ambiguous and distracting when the volume consists of planar surfaces or has surface discontinuities.

4.1.2. Coping with the Missing Information

Data can be missing due to shadows cast due to the scanner geometry, or due to occlusion with other parts, or due to self-occlusion. The first case is handled by allowing the model to predict the missing data. As mentioned before, the second case can be conveniently handled by considering the residuals for those points to be acceptable. Self-occlusion is not a problem if the view is not degenerate. In case of degenerate views, additional data or reasoning is required to get volume estimates.

4.1.3. Orienting the Initial Superquadric Model

As a rule of thumb, the initial ellipsoidal model is oriented along the eigenvectors of the moment matrix, and the Z-axis is aligned along the eigenvector with maximum eigenvalue (corresponding to least axis of inertia). The model is sensitive to the selection of the Z-axis (axis along a_3 dimension) for cylindrical models. This problem is resolved by using axis information from bi-quadric surfaces. In addition, we make the orientation of the part model the same as one of the constituent surfaces. This heuristic was found to give consistently better solution than if the orientation is not corrected.

4.1.4. Surface Support for the Superquadric Data

The volumetric segmentation procedure considers only those data points that are included in one or more surface patches. The logic behind this requirement is that if a data point cannot gather surface support then it can be excluded from the volumetric consideration as well. It also has the desirable effect of leaving out the outliers (filtered by the iterative regression approach of biquadric recovery), thereby providing clean data for the least-squares procedure for the superquadric recovery.

4.2. The Strategy for Volumetric Segmentation

A schematic diagram of our approach for volumetric segmentation is shown in Figure 1. The objective of the control module is to evaluate the global superquadric model and devise the appropriate strategy to either segment the data by hypothesizing part-models as indicated by the residuals, or terminate the procedure. The segmentation at surface level can be used to guide the volumetric segmentation, but when to rely on surface information, and what surface information to use is not clear at a first glance. Depending on whether or not to invoke surface information, there are three basic strategies:

1. For every surface model one superquadric model is recovered. This is the most popular but the weakest strategy to follow, but it can be used as a planarity check for a patch as also to determine its orientation.

2. Segmenting the object at concave discontinuities found during the segmentation, and recovering a superquadric model for each convex component. This is not a general strategy, but if applicable, it should be recognized and used by the control procedure.

3. The residual-driven global-to-local approach, that generates part hypotheses at the s_under (underestimated) regions, and extrapolates the part-models, as the global model shrinks by discarding the points that were underestimated by the earlier fit.

The third strategy is the most general one. However, it is too slow and tends to generate more false hypotheses than if surface information was also taken into account. Crucial information that surfaces provide is the existence of step edges and the concave surface normal discontinuities, as also the orientation of individual surface patches, which is of crucial importance in orienting the part (seed) superquadric models placed on s_under regions. We will first explain the three strategies as independent methods and then describe the integration of their best features to obtain a general control structure. We will illustrate the results of following these strategies for the composite object in Figure 7.

4.2.1. Strategy 1: One Superquadric Model for Every Bi-quadric Surface

If there is a 1-to-1 correspondence between the biquadric and superquadric surfaces, then a superquadric model can then be recovered for each surface patch. However, this is not a general strategy, as it does not allow combination of surfaces across convex discontinuities, and hence works only for convex blob-like parts [11, 1, 14]. The result of this strategy for the composite object gives an unrealistic looking description, as shown in Figure 7. The box is better represented as one volume rather than two "flat" volumes. This strategy is not acceptable in general because it does not maximize the positive volume of the data. However, it can be used by the general segmentation procedure for the following purposes:

1. As an unambiguous planarity check for the planar patches that are recovered as curved surfaces due to noise and distortion in data. This approach is similar to the planarity check based on computing eigenvectors of the moment matrix [7], but it also gives the size and shape of the surface. The Z-axis is aligned along the shortest axis for this test.

2. A superquadric model for each surface gives an estimate of the global orientation of the patch. Biquadrics don't have this information since their Z-axis is fixed. It is crucial in orienting the part-models hypothesized by the control module.

We want the control structure to enforce the minimum volume constraint for a given set of points, but maximize the positive volume by extending the model as much as possible without creating contour overestimations (c_over regions).

4.2.2. Strategy 2: Grouping Convex Surfaces

A more sophisticated strategy will allow the models to combine surfaces along convex discontinuities, thereby allowing individual surfaces to be grouped to get a volumetric fit. Ferrie et al [2] have used this strategy to segment objects at concave edges derived from differential geometric analysis. If the grouping of surfaces is sufficient (after analyzing the SAG) to form the volumetric model, then further segmentation of surfaces is not necessary. But this is not true in general, because the surface patches may need further segmentation for a volume description. For example, an L-shaped object needs the surface with the L-shape to be broken into at least two parts. Therefore, a more elaborate strategy is needed that will recognize the possibility of further surface segmentation, and at the same time use the convexity information if possible. The composite object can be completely segmented by grouping the convex connected components together, as shown in Figure 7, (bottom row, center).

4.2.3. Strategy 3: Global-to-Local Superquadric Fitting

By following a global-to-local approach, driven solely by residual analysis, it is possible to generate part hypotheses at the s_under regions, place local superquadrics there, and let them grow (extrapolate) as the global model shrinks by discarding the points that were underestimated by the earlier fit. No surface information is assumed to be available.

The strategy is illustrated for the composite object in Figure 8 where the global model (untapered), begins to discard the underestimated points on the cylinder (corresponding to OU-region 41). We will later describe a procedure for selecting the starting OU-regions. The global model shrinks iteratively as the local model on the cylinder accepts the points rejected by the global model. When the global model converges to the box shape, it has no s_under residuals to discard in the direction of the local model, so it stops shrinking and the local model stops growing. Note that residuals change drastically as the model begins to approach the the cluster of points that it can model without significant s_over or s_under residuals. The final models look similar to those obtained using the convex-combination of the surface patches. The qualitative residuals of the global and part models are shown in Figure 9. The residuals of the part-model (corresponding to the cylinder) remain acceptable throughout the iterative process.

This strategy is general in application and fits in our global to local approach to volumetric segmentation. In addition, it can use the surface information effectively by rejecting false hypotheses early in the segmentation process. We will now present our integrated approach, combining the elements of strategies 1, 2 and 3 to give an efficient control structure that systematically generates volumetric descriptions.

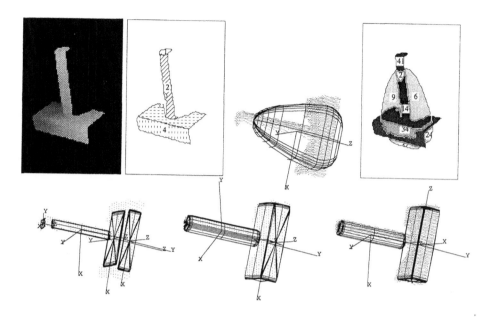

Figure 7. The composite object. Top: Range image; bi-quadric segmentation; The global model; and its Z-residuals. Bottom: Left: strategy 1: One superquadric model for every surface; Center: strategy 2: Convex combination of surfaces; Right: strategy3: Result of shrinking the global model.

300

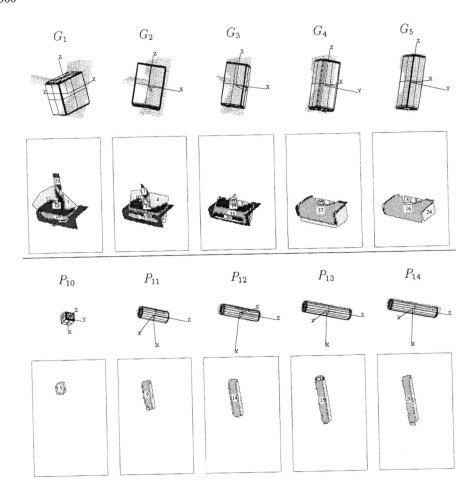

Figure 8. The composite object: Global to local model growing and shrinking. Top: Shrinking the global model and the corresponding residuals; Bottom: Growing the local model and the corresponding residuals

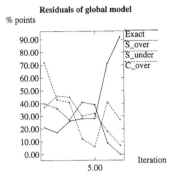
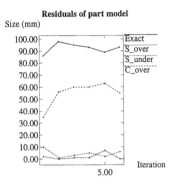

Figure 9. The composite object: Residuals of the global and part models as they shrink and grow respectively.

4.3. General Strategy: Integrating the Three Strategies

As noted earlier, some of the surfaces require further segmentation to conform to any superquadric model in a combination of surface patches. This is the most difficult scenario and the one we focus most of our attention on. But at the same time, we don't want to ignore the possibility of cases 1 and 2, which simplify the volumetric description for simpler objects. Therefore, our strategy is to control the flow of the segmentation in such a way that the easier strategies can be identified early on, and at least their results can be used later if the object turns out to be more complex.

The global strategy beginning with the biquadric segmentation is outlined in the algorithm outlined in Figure 10. Figure 11 shows the hierarchical structure of the superquadric model representation. The global model for the complete data set is G_0, and $[G_1...G_n]$ represent the evolution of the global model as it is refined by the global-to-local process. The first refinement gives G_1 with q part-models, $P_{10}..P_{q0}$, describing clusters of data taken away from G_0. Thus, P_{i0} models are stored as children of G_1 and are inherited by all the subsequent refinements of G_1. The global model either terminates as one model, or can break into more global models $[G_{n1}...G_{np}]$ as dictated by the dynamic connected-component analysis of the residual adjacency graph. Each global model behaves just like G_0, but inherits all the children of its previous iterations. The children are likewise represented as a chain of evolution from P_{10} to P_{1m}. A child model can also become a global model if further segmentation becomes necessary for the data in its domain. Thus, the representation is recursive by definition, reflecting the control flow of the procedure which is recursive in its global-to-local approach.

The control structure branches out to the appropriate strategy depending on the relative amount of each type of residual. Figure 10 shows all the cases that can exist after the residual analysis. Each of these cases indicate the type of data being modeled, and the appropriate strategies can be invoked. For example,

1. Recover bi-quadric models for the range data.

2. Recover global superquadric model (S_G) for the data.

3. Recover superquadric models for each bi-quadric surface.

4. Analyze (S_G) for the Z-residuals and perform connected component analysis of the OU-regions.

5. **if**(multiple clusters)
   then foreach (cluster i)

   Determine orientation.

   Recover global superquadric S_G^i

   goto step 4.

   else /* single cluster */

   If (Fit == OKAY)
       output current S_G
   else if (fit == SU) /* s_under regions exist */
       Provide for flat object during the general analysis.
   else if (fit == CO) /* c_over regions exist */
       Invoke contour constraint.
   else if (fit == SO) /* s_over regions exist */
       Analyze for negative volume.
   else if (fit == CO_SO) /* c_over and s_over regions exist */
       Invisible side of superquadric modeling data.
   else if (fit == SO_SU) /* s_over and s_under regions exist */
       Object with surface details. Do general analysis.
   else if (fit == SU_CO) /* s_under and c_over regions exist */
       Do general analysis of underestimated regions.
   else if (fit == SU_SO_CO) /* s_under, s_over, and c_over regions exist */
       Do general analysis of underestimated regions.

6. Done with volumetric segmentation.

Figure 10. The control flow for volumetric segmentation.

if only surface overestimations (s_over) are the significant residuals in the Z-residual map, then the underlying data is concave and requires negative volume descriptions. Flat objects result in significant surface underestimations. However, if a flat object has to broken into parts, it will also have contour overestimations along the Z direction.

4.3.1. Selection of Part-Models

s_under regions corresponding to underestimated points are the indication of convexities or part-structure in the global cluster of data. They *protrude* from the global model, suggesting the existence of a separate part, part of which is under-estimated due to global averaging by the minimization procedure. Unfortunately, since the global model tends to average out residual errors, these regions can be odd-shaped and can be elongated in one direction, or surround the object completely (as with flat objects). We want to start new partmodels at regions that constrain the model at least in two dimensions and allow a good approximation of the orientation of the model by extracting it from the orientation of the constituent surfaces.

For this purpose, the s_under regions are positioned in the object-coordinate system of the parent model in such a way that the region extremities are known in terms of the parent model. Thus, if a regions extends beyond an axis completely, like region 41 for the composite object in Figure 7, and its size is large enough to place a superquadric model, then it is selected as a partmodel. Additionally, the direction (which is negative Z-axis for the composite object) in which the residual lies is also stored, such that all the residuals in that direction are considered as part-models. This has the desirable effect of removing data from the parent model only in one direction, thereby letting the model shrink in a stable manner. Later, when the residuals in a particular direction do not exist any more (due to the current global shape), the other direction of shrinking is chosen following the same method.

This strategy can be easily refined to include surface information in deciding which s_under regions are the best starting points. For the composite object in Figure 7 (top row, left), surface 1 is completely underestimated by the global model and the curved surface is partly included. Therefore the s_under region due to the cylinder is a better choice for new part than the region formed due to the box. Other considerations include the size and location of the regions, and the types of surfaces they underestimate.

4.3.2. Growing or Extrapolating the Part-Models

The partmodels placed at the s_under regions are constrained only along the direction they have data. The directions (with respect to the part-model's coordinate system) along which the model is not constrained are the prime candidates for extrapolation of the model. Thus a model can grow only if it is not constrained in the direction of the available data. The extrapolation has the effect of changing the cross-section of the model (the a_1 and a_2 dimensions) and the length (a_3) of the model. Either of the shape parameters for a box shaped object will not change, while both or one of them will change for a curved object. But if a curved surface is present, then the surface segmentation will result in a curved patch and the local

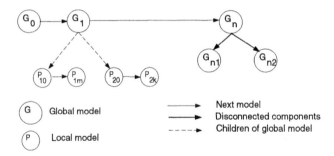

Figure 11. The hierarchical superquadric representation.

model can be placed on the entire curved patch rather than a part of it. Thus, the real issue is in growing of part-models having planar patches. The question is, how to extrapolate a model given an initial set of data, initial orientation and the neighboring unclaimed data?

One approach is to check the neighborhood connectivity, and slowly increment the model domain by constantly evaluating the refined model. In general, all the parameters (including translation and orientation) are allowed to change, though at times it may be necessary to fix some of the parameters. If changes result in an increased error or in a model that is not acceptable, then the previous model is considered final and a new model is started at the underestimated points. However, this approach has a problem that it extends the model equally in all directions, which may not work in general.

A more controlled method would extrapolate the model only in one dimension and ignore data altogether in other directions. By direction, we mean the six axis directions of the object centered coordinate system. We can divide the problem into extrapolation along the length of the model (Z-axis) or along the cross-section of the model. For box shaped objects ($\varepsilon_2 < 0.5$), the growth along a particular direction is possible by simply observing the coordinates of the potential data point and comparing it with the dimensions of the part-model. The model is thus grown along only one dimension at a time and constantly evaluated. The model grows only if there are unclaimed data points along that dimension *and* including those points does not conflict with the points already accepted by the model.

Following the above considerations, the control structure recursively shrinks and grows global and part models respectively, and terminates individual global models when their fit is acceptable. The global models shrink till they converge on a part or disappear after the data is completely accounted for. Part-models grow for as long as they satisfy the acceptance criteria. In the next section we present detailed results of applying the general control structure on complex objects, and discuss the salient aspects of our segmentation schema.

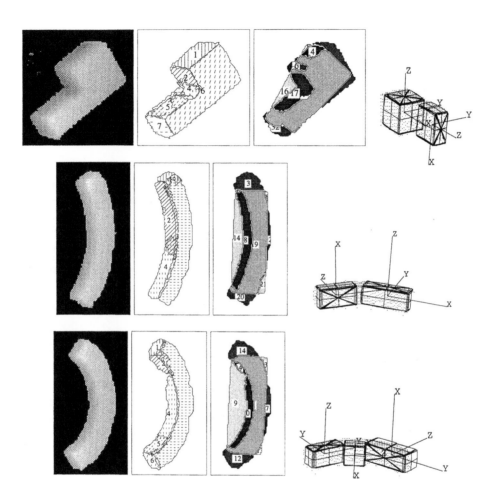

Figure 12. Surface and volumetric segmentation of L_1, Bent_1, Bent_2 shapes: The range image; bi-quadric segmentation; the Z-residuals of the global model; and the final volumetric description obtained by the residual-driven recursive procedure.

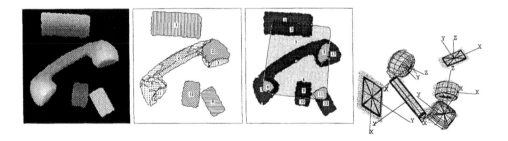

Figure 13. Scene_1: The range image; its bi-quadric segmentation, the Z-residual map for its global model; and the final description of the scene.

5. Experimental Results

The SUPERSEG system was tested on a number of objects of varying complexity. We will present results on six scenes:

1. L and bent objects: Strategy 3 is required for segmentation, however strategy 1 is used to orient part models.

2. Scene_1: contains multiple objects.

3. Wrench: has a part structure.

4. NIST object: a machined object.

The SUPERSEG system is composed of more than 20,000 lines of documented C code, with an X display interface for real-time visualization, and a PostScript interface for output generation. The range data was obtained from a structured lighting laser-scanner.

System Parameters:

All the values were determined empirically: Maximum number of iterations for model recovery: 15; tolerance for deviation from model: 0.32 for the I/O function ($\mathcal{G}$), and 3mm for the Z-residuals; qualitative acceptance criteria: at least 80% points exactly modeled and, $< 10\%$ each of other residuals.

The L_1, Bent_1, and Bent_2 shapes:

These simple shapes (Figure 12) are actually quite complex to segment at a volumetric level because the surface segmentation does not directly result into convex combination of surfaces. The residual-driven global-to-local method begins by introducing part models at underestimated regions. Two parts are obtained for L_1 and Bent_1 objects, while three parts are obtained for Bent_2 since it curves more than the Bent_1 object. These examples demonstrate the capability of the control structure in generating piece-wise descriptions of the objects which do not have a model in the superquadric vocabulary.

307

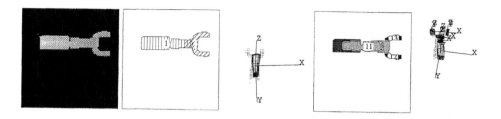

Figure 14. The Wrench: Range image; its bi-quadric segmentation; the global model; Z-residuals of the global model; and the final description.

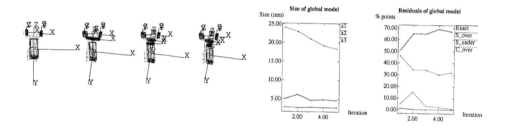

Figure 15. The Wrench. left to right: the four iterations needed to extract the part-structure; Graphs of change in the size of the global model during shrinking; and qualitative residuals.

Scene_1 (Multiple Clusters):

As shown in Figure 13, the scene consists of four clusters that are segmented by connected component analysis of the residual adjacency graph. The description for the two boxes and the cylindrical object is final after first component separation, while the phone requires further segmentation. The axis of the cylindrical object is correctly oriented using the bi-quadric information, even though the diameter of the cylinder is greater than its height.

The Wrench:

Shown in Figure 14, the surface segmentation of wrench is inadequate to capture its part-structure. Starting with the global model, the SUPERSEG procedure places the part-models at s_under regions labeled 12 and 13. (Figure 14), and begins to shrink the global model. The four iterations and the corresponding qualitative residuals are shown in Figure 15, where the head of the wrench is segmented into 3 parts, and neck is separated from the handle of the wrench.

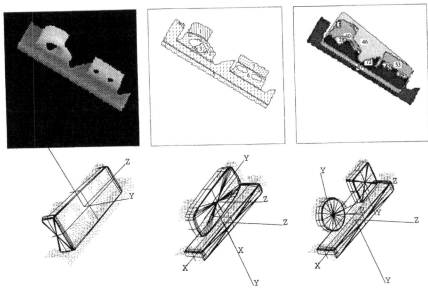

Figure 16. The NIST Object: Top: The range image, its bi-quadric segmentation, and the Z-residuals of the global model. Bottom: the global model; Description after the first component separation; and the final description.

The NIST Object:

Shown in Figure 16, the NIST object has holes and a definite part structure. The bi-quadric surface description correctly segments the surface into planar patches (2, 4, 5, 6, and 7) and second-order patches (1 and 3). The process (Figure 16, bottom row) starts by hypothesizing the base of the object in convex sense, which is accepted. The two remaining parts are separated by RAG analysis during the second iteration. Thus the control structure integrates surface information with the global-to-local surface approach and dynamically segments the data clusters. The residuals for the base show that occlusion are accepted by the acceptance criteria, thereby allowing a single model for the base.

6. Conclusions and Future Directions

We presented a fine-to-coarse surface segmentation strategy and a coarse-to-fine volume segmentation strategy, and a control structure that integrates these descriptions in a systematic manner. The control procedure is able to adapt to the complexity of the input and generate suitable descriptions. It can be adapted to prefer certain descriptions to other, as also to give coarse or fine descriptions based on scale considerations, which determine the thresholds in the residual analysis. Since relative error measures are used for evaluation, the method is not sensitive to the absolute differences in size. The superquadric models are scale, position, and orientation invariant as long as the viewpoint for the parts is not degenerate. They are not sensitive to occlusion due to other parts if the available information can constrain all the model parameters. Curved bi-quadric models, on the other hand, are sensitive to orientation since the Z-axis is fixed along the viewing direction. Since volumetric segmentation is completely data-driven and does not have a priori knowledge about the objects or object domain, segmented descriptions of an object viewed from different angles may differ in part-structure. However, the number of parts, and their relationships will still be qualitatively similar. While this is undesirable if precise object-recognition is the goal, such descriptions are adequate for a qualitative description of the object; for designing grasping strategies for object manipulation; and for navigation or path planning applications. The invariance demanded by object recognition tasks can be imposed by using the high-level knowledge (in the form object models) to guide the segmentation of the object. At the very least, these descriptions can reduce the complexity of the search by narrowing down the list of potential candidates during the database search.

Due to the Z-depth ($2\frac{1}{2}$-D) nature of the range data, the volumetric models may not be fully constrained for curved objects. This will result in a perceptually acceptable box-shaped description for two perpendicular planes, while an elliptical cylinder may be obtained for a circular cylinder. Whaite and Ferrie [1990] define an "ellipsoid of confidence", derived from the covariance matrix of the fitting procedure, within which a number of superquadric models fit the data equally well. Model recovery can also be constrained by using the physical constraints of support and stability, as investigated in Gupta et al[6].

Extension of the paradigm includes: incorporation of domain specific knowledge from stored models or in the form of geometric constraints; use of sparse depth data; and use of multiple views to supplementing and integrating information. The surface and volumetric descriptions form the basis of 3-D object models. It is possible to use the surface segmentation as a clustering mechanism for an automatic model generation system or for further refinement by an operator-assisted model generation system in the spirit of reverse engineering, thereby reducing the complexity of model building process significantly. The description generated by the SUPERSEG system can also be used as the first approximation for more complex deformable models, which can then refine the segmentation by modeling the data more closely than that is possible by rigid models.

Acknowledgements

This research was performed at the GRASP laboratory of the University of Pennsylvania. It was supported in part by: AFOSR Grants 88-0244, 88-0296; Army/DAAL 03-89-C-0031PRI; NSF Grants CISE/CDA 88-22719, IRI 89-06770; ARPA Grant N0014-88-K-0630, and United States Postal Service.

REFERENCES

1 Trevor Darrell, Stan Sclaroff, and Alex Pentland. Segmentation by minimal

310

description. In *Proceedings of the Third International Conference on Computer Vision*, pages 112–116, Osaka, Japan, December 1990. IEEE.

2   F. P. Ferrie, J. Lagarde, and P. Whaite. Darboux frames, snakes, and superquadrics: Geometry from the bottom-up. In *Proceedings of the IEEE Workshop on Interpretation of 3D Scenes*, pages 170–176, Austin, TX, November 1989.

3   Alok Gupta. Surface and volumetric segmentation of complex 3-d objects using parametric shape models. Technical Report MS-CIS-91-45, Department of Computer and Information Science, University of Pennsylvania, 1991.

4   Alok Gupta and Ruzena Bajcsy. Volumetric segmentation of range images of 3-d objects using superquadrics. *CVGIP: Image Understanding*, To Appear.

5   Alok Gupta, Luca Bogoni, and Ruzena Bajcsy. Quantitative and qualitative measures for the evaluation of the superquadric models. In *Proceedings of the IEEE Workshop on Interpretation of 3D Scenes*, pages 162–169, Austin, TX, November 1989.

6   Alok Gupta, Gareth Funka-Lea, and Kwangyoen Wohn. Segmentation, modeling and classification of the compact objects in a pile. In Hatem Nasr, editor, *Selected Papers on Automatic Object Recognition*, pages 333–343. SPIE, Volume MS-41, 1991.

7   R. Hoffman and A. K. Jain. Segmentation and classification of range images. *IEEE Trans. Pattern Analysis and Machine Intelligence*, 9, 1987.

8   Ales Leonardis, Alok Gupta, and Ruzena Bajcsy. Segmentation as the search for the best description of the image in terms of primitives. In *Proceedings of the Third International Conference on Computer Vision*, pages 121–125, Osaka, Japan, December 1990. IEEE.

9   R. Nevatia and T.O. Binford. Description and recognition of complex-curved objects. *Artificial Intelligence*, 8:77–98, 1977.

10 A. P. Pentland. Automatic extraction of deformable part models. *International Journal of Computer Vision*, 4:107–126, 1990.

11 Alex P. Pentland. Recognition by parts. In *International Conference on Computer Vision*, pages 612–620, London, 1987.

12 William H. Press, Brian P. Flannery, Saul A. Teukolsky, and William T. Vetterling. *Numerical Recipes in C*. Cambridge University Press, 1988.

13 F. Solina and R. Bajcsy. Recovery of parametric models from range images: The case for superquadrics with global deformations. *IEEE Trans. on Pattern Analysis and Machine Intelligence*, 12(2):131–147, February 1990.

14 D. Terzopoulos and D. Metaxas. Dynamic 3d models with local and global deformations: Deformable superquadrics. *IEEE Trans. Pattern Analysis and Machine Intelligence*, 13(7):703–711, 1991.

15 D. Terzopoulos, A. Witkin, and M. Kass. Constraints on deformable models: Recovering 3d shape and nonrigid motion. *Artificial Intelligence*, pages 91–123, August 1988.

16 P. Whaite and F.P. Ferrie. From uncertainty to visual exploration. In *Proceedings of the Third International Conference on Computer Vision*, pages 690–697, Osaka, Japan, December 1990. IEEE.

Recognition by Alignment[1]

Daniel P. Huttenlocher [a]

[a]Computer Science Department
Cornell University
Ithaca, New York 14853 USA

Abstract

We consider the problem of recognizing solid objects from a single two-dimensional image of a three-dimensional scene. We present a closed-form method for computing a transformation from a three-dimensional model coordinate frame to the two-dimensional image coordinate frame, using three pairs of model and image points. Then we describe a recognition system that uses this transformation method to determine possible *alignments* of a model with an image. Each of these hypothesized matches is verified by comparing the entire edge contours of the aligned object with the image edges. Using the entire edge contours for verification, rather than a few local feature points, reduces the chance of finding false matches. The system has been tested on partly occluded objects in highly cluttered scenes.

1. Introduction

A key problem in computer vision is the recognition and localization of objects from a single two-dimensional image of a three-dimensional scene. The model-based recognition paradigm has emerged as a promising approach to this problem (e.g., [1, 11, 14, 16, 19, 20]). In the model-based approach, stored geometric models are matched against features extracted from an image, such as vertices and edges. An interpretation of an image consists of a set of model and image feature pairs, such that there exists a particular type of transformation that maps each model feature onto its corresponding image feature.

In this chapter we present a model-based method for recognizing solid objects with unknown three-dimensional position and orientation from a single two-dimensional image. The method consists of two stages: (i) computing possible *alignments* — transformations from model to image coordinate frames — using a minimum number of corresponding model and image features, and (ii) verifying each of these alignments by transforming the model to image coordinates and comparing it with the image. In the first stage, local features derived from corners and inflection

---

[1]This work was done at the MIT Artificial Intelligence Laboratory, and was supported in part by DARPA under Army contract DACA76-85-C-0010, in part by the ONR University Research Initiative Program under contract N00014-86-K-0685, and in part by DARPA under ONR contract N00014-85-K-0124.

points along edge contours are used to compute the possible alignments. In the second stage, complete edge contours are then used to verify the hypothesized matches.

Central to the method is a means of computing a transformation from a three-dimensional model coordinate frame to the two-dimensional image coordinate frame. This method determines a possible *alignment* of a model with an image on the basis of just three corresponding model and image points (or two corresponding points and unit orientation vectors). The method is based on an affine approximation to perspective projection, which has been used by a number of other researchers (e.g., [4, 13, 19, 20]). Unlike earlier approaches to this problem, however, we have derived a simple closed-form solution for computing the position and orientation of an object under this imaging model.

There are two key observations underlying the transformation method. First, the position and orientation of a rigid, solid object is determined up to a reflective ambiguity by the position and orientation of some plane of the object (under the affine imaging model). This plane need not be a surface of the object, but rather can be any "virtual" plane defined by features of the object. Second, the three-dimensional position and orientation of an object can be recovered from the affine transformation that relates such a plane of the model to the image plane.

1.1. Model-Based Recognition by Alignment

Model-based recognition systems vary along a number of major dimensions, including: (i) the types of features extracted from an image, (ii) the class of allowable transformations from a model to an image, (iii) the method of computing or estimating a transformation, (iv) the technique used to hypothesize possible transformations, and (v) the criteria for verifying the correctness of hypothesized matches. In this section we briefly discuss the design of the alignment recognition system in terms of these five dimensions.

In the alignment recognition system, local features are extracted from intensity edge contours. These features are based on inflections and corners in the edge contours. The location of a corner or inflection defines the *position* of the feature, and the orientation of the tangent to the contour at that point defines the *orientation* of the feature. While extended features such as symmetry axes can also be used to align a model with an image, they are relatively sensitive to partial occlusion. Therefore, the current system relies only on local features.

A valid transformation from a model coordinate frame to the image coordinate frame consists of a rigid three-dimensional motion and a projection. We use a "weak perspective" imaging model in which true perspective projection is approximated by orthographic projection plus a scale factor. The underlying idea is that under most viewing conditions, a single scale factor suffices to capture the size change that occurs with increasing distance (such a model has also been used by [14, 20]). Under this affine model, a good approximation to perspective projection is obtained except when an object is deep (i.e., large in the viewing direction) with respect to its distance from the viewer (e.g., railroad tracks going off to the horizon).

The recognition system that we have implemented operates by considering the possible alignments of a model with an image. The maximum possible number of such alignments is polynomial in the number of model and image features. These features are based on extracting corners and inflection points in two-dimensional edge images. Each feature specifies both a location (the (x, y) position of the corner or inflection) and an orientation (the tangent to the curve at that point). From features of this type, it is possible to compute the alignment using just two corresponding model and image features. Thus for m model features and n image features there are $\binom{m}{2}\binom{n}{2}2!$, or $O(m^2n^2)$, possible different alignments, resulting from each corresponding pair of two model and image features. On the other hand for point features (where there is no orientation) the worst case number of hypotheses is $O(m^3n^3)$. No labelling or grouping of features is performed by the current implementation, thus reliable perceptual grouping methods could be used to improve the expected-time performance of the system.

Each hypothesized alignment can be verified in time proportional to the size the model, by comparing the transformed model with the image. Thus the worst-case running time of the method is $O(m^3n^2\log n)$ for features such as inflections and corners that specify both location and orientation — $O(m\log n)$ time to verify each of $O(m^2n^2)$ hypotheses. If on the other hand the features specify only location information, then the number of hypotheses is $O(m^3n^3)$ and the overall running time is $O(m^4n^3\log n)$. This relatively low-order polynomial running time contrasts with a number of other major methods of hypothesizing transformations: search for sets of corresponding features [8, 16], and search for clusters of transformations [19, 20]. The former type of method is exponential in the number of features in the worst case. The latter type of method involves multi-dimensional clustering, for which the solution methods are either approximate (e.g., the generalized Hough transform) or iterative (e.g., k-means). Methods of this latter type are often not well-suited to recognition in cluttered scenes (see [9]).

In the final stage of the recognition process, each hypothesized alignment of a model with an image is verified by comparing the transformed edge contours of the model with nearby edge contours in the image. This reduces the likelihood of false matches by using a more complete representation of an object than just the local features from which an alignment is hypothesized.

1.2. The Recognition Task

The recognition problem addressed in this chapter is that of identifying a solid object at an unknown position and orientation, from a single two-dimensional view of a three-dimensional scene (3D from 2D recognition). Thus an object has three translational and three rotational degrees of freedom, which must be recovered from two-dimensional sensory data. The input to the recognizer is a grey-level image and a set of three-dimensional models. The models are represented as a combination of a wire frame and local point features (vertices and inflection points because these points are relatively stable under projection). The output of the recognition process is a set of model and transformation pairs, where each transformation maps the specified model onto an instance in the image.

The imaging process is assumed to be well approximated by orthographic projection plus a scale factor, as discussed in the following section. It is also assumed that the transformation from an object to an image is rigid, or is composed of a set of locally rigid parts. Under these assumptions there are six (unknown) parameters for the image of an object under a rigid-body motion: a three-dimensional rotation, a two-dimensional translation in the image plane, and a scale factor.

Edge contour features are used for hypothesizing and verifying alignments of a model with an image, so it is assumed that objects are identifiable based on the shape of their edge contours (surface discontinuities). Thus an object must have edge contours that are relatively stable over small changes in viewpoint, which is true of most objects that do not have smoothly changing surfaces. Work by Basri [2] addresses the problem of aligning smooth solid objects with a two-dimensional image.

2. The Transformation Method

A central problem in model-based recognition is the efficient and accurate determination of possible transformations from a model to an image. We have shown [12, 21] that three pairs of corresponding model and image points specify a unique (up to a reflection) transformation from a three-dimensional object coordinate frame to a two-dimensional image coordinate frame. This transformation specifies a rotation and scaling in three-space (a similarity transformation), and a translation in the plane. Based on this result, we present here a closed form solution for computing the transformation that aligns a model with an image using three pairs of corresponding model and image points. The method only involves solving a second order system of two equations.

2.1. The Imaging Model

3D from 2D recognition involves inverting the projection that occurs from the three-dimensional world into a two-dimensional image, so the imaging process must be modeled in order to recover the position and orientation of an object with respect to an image. We assume a coordinate system with the origin in the image plane, I, and with the z-axis (the optic axis) normal to I.

Under the perspective projection model of the imaging process (with an uncalibrated camera) a transformation from a solid model to an image can be computed from six pairs of model and image points. The equations for solving this problem are relatively unstable, and the most successful methods use more than six points and an error minimization procedure such as least squares [16]. When the parameters of the camera (the focal length and the location of the viewing center in image coordinates) are known, then three corresponding model and image points can be used to solve for up to four possible transformations from a model plane to the image plane [5, 7]. One restriction of this model is that a calibration operation is required to determine the camera parameters, so if the camera has a variable focal length it is necessary to recalibrate when the focal length is changed.

While perspective projection is an accurate model of the imaging process, the magnitude of the perspective effects in most images is relatively small compared with the magnitude of the sensor noise (such as the uncertainty in edge location). This suggests that a more robust transformation may be obtained by ignoring perspective effects in computing a transformation between a model and an image. Furthermore, even when a perspective transformation is computed, other properties of a recognition method may prevent perspective effects from being recovered. For instance, the SCERPO system [16] uses approximately parallel line segments as features, but parallelism is not preserved under perspective transformation. Thus the choice of features limits the recognition process to cases of low perspective distortion, even though a perspective transformation is being computed.

Under orthographic projection, the direction of projection is perpendicular to the image plane, and thus an object does not change size as it moves further from the camera. If a linear scale factor is added to orthographic projection, then a relatively good approximation to perspective is obtained. The approximation becomes poor when an object is deep with respect to its distance from the viewer, because a single scale factor cannot be used for the entire object.

Under the weak perspective imaging model used here, the transformation from a three-dimensional model point $x = (x, y, z)$ to a two-dimensional image point $x' = (x', y')$ is given by $x' = \pi(s\mathbf{R}x + \mathbf{b})$, where π is orthographic projection (i.e., $\pi : \Re^3 \rightarrow \Re^2$), s is a scale factor, $\mathbf{R}$ is a three-dimensional rotation, and $\mathbf{b}$ is a two-dimensional translation in x and y. We denote the transformation parameters by $Q = (s, \mathbf{R}, \mathbf{b})$. Several recognition systems [4, 14, 20] have used this imaging model. Unlike the current method, however, they have not solved the problem of how to compute a three-dimensional transformation directly from corresponding model and image points. Instead they use a heuristic approach such as storing multiple views of an object in order to approximate the transformation [20], or restrict recognition to planar objects [4, 14].

We have shown [12, 21] that the correspondence of three non-collinear points is sufficient to determine (up to a reflection) the transformation $Q = (s, \mathbf{R}, \mathbf{b})$ mapping the three-dimensional model coordinate frame to the two-dimensional image coordinate frame. The result is based on the well known fact that an affine transformation of the plane is uniquely defined by three pairs of non-collinear points. What we have shown is that an affine transformation of the plane defines a unique transformation of space. In effect this transformation gives the orientation of one plane with respect to another, up to a reflection (see [12] for a proof).

2.2. Computing the Transformation

Given that there exists a unique (up to a reflection) alignment transformation, Q, specified by three corresponding model and image points, we now show how to compute this transformation (and the corresponding two-dimensional affine transformation A). Consider three pairs of points (a_m, a_i), (b_m, b_i) and (c_m, c_i), where the image points (subscripted with i) are in two-dimensional sensor coordinates and the model points (subscripted with m) are in three-dimensional object coordinates.

The three pairs of model and image points specify an affine transformation from the plane defined by the three model points to the image plane. This affine transformation maps each point x in the plane defined by a_m, b_m and c_m to an image point according to $x' = Lx + b$, where L is a nonsingular 2×2 matrix and b is a 2-vector. We now give a method for computing Q, and in the process also computing $A = (L, b)$.

Step 0. Rotate and translate the model so that a_m is at the origin, $(0, 0, 0)$, and the b_m and c_m are in the x–y plane. This operation can be performed offline for each triple of model points.

Step 1. Using the transformed model from step 0, define the translation vector $b = -a_i$, and translate the image points so that a_i is at the origin, b_i is at $b_i + b$ and c_i is at $c_i + b$.

Step 2. With the model and image points transformed as described in the previous two steps, now solve for the linear transformation

$$L = \begin{pmatrix} l_{11} & l_{12} \\ l_{21} & l_{22} \end{pmatrix}.$$

given by the two pairs of equations in two unknowns

$$\begin{aligned} Lb_m &= b_i \\ Lc_m &= c_i \end{aligned}$$

This yields the linear transformation component, L of a unique affine transformation $A = (L, b)$ mapping the plane specified by the three model points into the image plane. The vector b was computed in Step 1. This transformation exists as long as the points are not collinear.

Step 3. The scale factor s and the rotation matrix R of Q are specified by the entries of L, and two additional parameters c_1 and c_2. These two parameters can be thought of as specifying the relative orientation (slant and tilt) of the model plane with respect to the image plane. We solve for the two parameters c_1 and c_2, using

$$c_1 = \pm\sqrt{\frac{1}{2}(w + \sqrt{w^2 + 4q^2})},$$

and

$$c_2 = \frac{-q}{c_1},$$

where

$$w = l_{12}^2 + l_{22}^2 - (l_{11}^2 + l_{21}^2),$$

and

$$q = l_{11}l_{12} + l_{21}l_{22}.$$

Specifically these two values give the height (z-distance) of the points b_m and c_m above the image plane, once they are aligned such that they project to b_i and c_i respectively.

Step 4. There are two possible rotations of space that (corresponding to the reflective ambiguity in the solution) such that the model is mapped onto the image. Call the 3×3 matrices representing these transformations R^+ and R^-. These two matrices each have the 2×2 matrix L embedded in the upper left, because they effect the same two-dimensional transformation as L. We can use the results of [12] to solve for the remaining entries of sR^+ and sR^- given L. This yields

$$sR^+ = \begin{pmatrix} l_{11} & l_{12} & (c_2 l_{21} - c_1 l_{22})/s \\ l_{21} & l_{22} & (c_1 l_{12} - c_2 l_{11})/s \\ c_1 & c_2 & (l_{11} l_{22} - l_{21} l_{12})/s \end{pmatrix},$$

where

$$s = \sqrt{l_{11}^2 + l_{21}^2 + c_1^2}.$$

This specifies the complete alignment transformation, Q, with two-dimensional translation b, scale s and three-dimensional rotation R^+. The image coordinates of a transformed model point x are then given by the x and y coordinates of $x' = \pi(sR^+x + b)$.

The reflected solution sR^- is identical to sR^+ except the entries r_{13}, r_{23}, r_{31}, and r_{32} are negated.

This method of computing a transformation is relatively fast, because it involves a small number of terms, none of which are more than quadratic. Our implementation using floating point numbers on a Symbolics 3650 takes about 3 milliseconds.

3. Feature Extraction

The previous section describes a method that uses three corresponding model and image points to compute a transformation from a model coordinate frame to the image coordinate frame. In this section we present a method for extracting simple, local point information from the intensity edges of an unknown image. The method relies on connected intensity edge contours, so a grey-level image is first processed using a standard edge detector [3] and the edge pixels are then chained together into singly-connected contours. For each resulting contour, local orientation and curvature of the contour are computed. The curvature is used to identify two types of local events along a contour: an inflection or a corner. The location of each inflection or corner and the tangent orientation at that point are then used as features for computing possible alignments of a model with an image. As noted above, two pairs of such model and image features are sufficient to compute an alignment (by either implicitly or explicitly using the orientations to define a third point).

3.1. Chaining Edge Contours

The output of an edge detector is a binary array indicating the presence or absence of an edge at each pixel. This section describes a simple method of chaining neighboring eight-connected pixels together into contours. When a given pixel has only one or two neighbors the chaining process in unambiguous. Otherwise, a decision must be made as to which pixels belong together as part of an edge. The chaining algorithm works in two stages. First, all the unambiguous chains are formed; any pixel that has more than two neighbors is ambiguous and thus defines the start of a new chain. The second stage uses a process that we call *mutual favorite pairing* in order to merge neighboring chains into a single chain. This merging process continues until no two neighboring chains can be joined together.

The mutual favorite pairing procedure works as follows. Given a set of edge contours $\{x_i\}_{i=1,...,n}$, each contour x_i has an ordered sequence of possible matching contours. This ordering is computed using a cost function that measures how bad the resulting contour would be if the two contours were joined. Each contour x_i is considered in turn. A contour x_i is only merged with its best matching contour x_j if x_j in turn has x_i ranked as its best match (hence the name "mutual favorite"). This process continues until no two contours (including the resulting merged contours) can be joined together. One slight complication over the above description is that each contour has two endpoints that are considered separately, because merging two contours actually occurs at a specific endpoint.

The cost function for sorting the candidate merges is based on the distances and tangent orientations between endpoints of a contour. For a given endpoint e_i of some contour with local tangent orientation v_i at that endpoint and another endpoint e_j of another contour with tangent orientation v_j, the cost of matching those endpoints is a function of: (i) how far apart the endpoints are, and (ii) what the angle is between the each of the tangents and the segment connecting the two endpoints. That is, the cost of matching e_i with e_j is defined as $D = \|e_i - e_j\|(\angle v_i v + \angle v_j \overline{v}))$, where $\|e_i - e_j\|$ is the distance between the endpoints, v is the unit vector in the direction $e_i - e_j$, $\overline{v}$ is the unit vector in the opposite direction of v, and $\angle$ measures the angle between two unit vectors.

An iteration of the mutual favorite pairing procedure considers the two endpoints of each contour separately. If two contours each rank the other as the best match (based on D) then they are merged together, and the merged ends of the contours are removed from further consideration. The iteration stops when no more endpoints to be merged are found. Clearly the process terminates, because there are a finite number of chains, and at each step either two contours are merged into one, or there are no merges and the iteration stops. Each iteration takes $O(n \log n)$ time for n edges, and at least one contour is removed at each step, so the worst case running time is $O(n^2 \log n)$. This worst case only occurs if all the contours have an endpoint near the same place in the image. In general, there are a small number of contours with endpoints near any given point, so the expected running time is effectively linear in the number of contours rather than quadratic.

3.2. Edge Orientation and Curvature

For a curve g, parameterized by its arclength s, the local orientation at a point $g(s)$ along the curve is defined by the unit vector t_s that is tangent to the curve at that point. There are several ways of estimating the local orientation of a curve. The simplest method is to define a local neighborhood d, and estimate the orientation as $t_s = g(s - d) - g(s + d)$. The major shortcoming of this method is sensitivity to noise, because each tangent is estimated using only two points. A method that is less sensitive to noise is to estimate the orientation at $g(s)$ by fitting a line to the points in the neighborhood, $g(s - d), \ldots, g(s + d)$. A standard line fitting technique is the least squares method. In this case, we are concerned with minimizing the

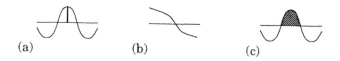

(a)   (b)   (c)

Figure 1. Methods of filtering zero crossings: (a) the height of the largest peak between two zeroes, (b) the slope of the curve at the zero crossing, and (c) the area under the peak.

squared distance in the direction normal to the best fitting line [6].

Given the local tangent orientations, the curvature, κ, of a path in space $x = g(s)$, parameterized by arc length, s, is given by $\kappa n = dt/ds$, where t is the unit tangent vector and n is the unit normal vector in the direction in which the path is turning. This formula has a special case for plane curves, $\kappa = d\psi/ds = 1/\rho$, where ψ is the angle of the slope of the tangent line, and ρ is the radius of the osculating circle, tangent to the curve. We use this method to compute the curvature from the tangent vectors. In contrast many computational vision algorithms rather parameterize $g(s)$ in terms of separate functions $x(s)$ and $y(s)$, and then compute curvature locally on a pixel-by-pixel basis. This is quite sensitive to noise, however, due to the computation of second derivative quantities from pointwise image measurements.

A number of methods have been proposed for smoothing curvature [10, 18, 17]. These methods are concerned with reconstructing a smoothed version of a curve. In contrast, we are interested only in extracting certain local interest points along a curve, such as corners and inflections. Thus the following rather simple smoothing procedure is sufficient. The curvature of a path parameterized by arclength, $\kappa(s)$, is smoothed. Events such as zeroes of curvature and high curvature points are then identified in the smoothed curvature. Each of these events occurs at some particular location along the arc, s. For each event, the corresponding location s along the original contour is marked as the point where the event occurs.

In areas of low curvature, there may be a number of small inflections that are relatively unimportant. To filter out these insignificant zeroes of curvature we use a method based on the magnitude of the area of $\kappa(s)$ between two successive zero crossings. When the area is smaller than a threshold value, the neighboring zero crossings are discarded as insignificant (see Figure 1c). This method is less sensitive to noise than using a threshold on the height of the peak between two zero crossings (shown in part a of the Figure), because it does not rely on a value at a single point. Also, unlike using a threshold on the slope at the zero crossing (shown in part b of the Figure) the method does not exclude low-slope zero crossings for which there is still a large peak between the two zero crossings. It should be noted that the computation of the area of $\kappa(s)$ between two successive zero crossings is trivial. It is just the total turning angle of the contour over that region of arclength (i.e., the angle between the tangent orientations to the original curve at the those two corresponding locations).

Using the edge curvature information, inflection points and corners are identified

Figure 2. Local corners, marked as dots, are found by identifying local maxima in the smoothed curvature function.

along the edge contours. A *feature* for alignment consists of the location of a corner or inflection, and the tangent orientation at that point. Each feature provides both location and orientation information, and thus a pair of model and image features defines an alignment transformation. Inflections are located where the curvature changes sign, and corners are located at high curvature points. Figure 2 shows a contour and the corners identified by marking the peaks in the smoothed curvature function.

4. Verification

Once a potential alignment transformation has been computed, it must be determined whether or not the transformation actually brings the object model into correspondence with an instance in the image. This is done by transforming the model to image coordinates, and checking that the transformed model edges correspond to edges in the image. The verification process is of critical importance because it must filter out incorrect alignments without missing correct ones.

The verification method is hierarchical, starting with a relatively simple and rapid check to eliminate many false matches, and then continuing with a more accurate and slower check. The initial verification procedure compares the segment endpoints of the model with the segment endpoints of the image. Recall that a segment endpoint is defined at inflections and at the ends of straight segments. Each point thus has an associated orientation, up to a reflective ambiguity, defined by the orientation of the edge contour at that point. A model point correctly matches an image point if both the position and orientation of the transformed model point are within allowable error ranges of a corresponding image feature (currently 10 pixels and $\frac{\pi}{10}$ radians). Those alignments where a certain percentage (currently half) of the model points are correctly matched are then verified in more detail.

The detailed matching procedure compares the entire edge contours of a transformed model with the image. Each segment of the transformed model contour is matched against the image segments that are nearby. The comparison of contours is relatively time consuming, but reduces the likelihood that an incorrect match will accidentally be accepted. The initial verification generally eliminates many hypotheses, however, so this more time consuming procedure is only applied to a small number of hypotheses.

The comparison of a transformed model segment with an image is intended to distinguish between an accidental and a non-accidental match. For instance, an accidental match is highly unlikely if a transformed model segment exactly overlaps an image segment such that they coterminate. If a transformed model segment lies near some image segments that do not coterminate with it, then there is a higher likelihood that the match is accidental. We define three types of evidence for a match: positive evidence, neutral evidence, and negative evidence. A model edge that lies near and approximately coterminates with one or more image edges is positive evidence of a match, as shown in the first part of the figure. A model edge that lies near an image edge which is much longer or much shorter than the model edge is neutral evidence of a match, and is treated as if there were no nearby edge. A model edge that is crossed by one or more image edges, at a sufficiently different orientation than the model edge, is negative evidence of a match.

The verification process uses a table of image edge segments stored according to x position, y position, and orientation. A given image segment, i, is entered into the table by taking each point, $g(s)$, along its edge contour and the corresponding tangent orientation, t_s, and storing the triple $(i, g(s), t_s)$ in the table using quantized values of $g(s)$ and t_s. A positional error of 10 pixels and an orientation error of $\frac{\pi}{10}$ radians are allowed, so a given point and orientation will generally specify a set of table entries in which to store a given triple.

In order to match a model segment to an image, each location and orientation along the transformed model contour is considered in turn. A given location and orientation are quantized and used to retrieve possible matching image segments from the table.

5. The Matching Method

We have implemented a 3D from 2D recognition system based on the alignment method. The system uses simple, local features as described in the preceding sections: corners and inflections extracted from intensity edges. Each feature specifies location (and perhaps orientation) information. For features such as inflections and corners, specifying both location and orientation information, the orientations can be used to define a third pair of points for computing the alignment (or the alignment computation can be modified somewhat). For features specifying location information only, each triple of model and image features defines a possible alignment of the model with the image. The alignments computed from all pairs (or triples) of features are verified by transforming the edge contours of the model into image coordinates, and comparing the transformed model edges with nearby image edges. Alignments that account for a high percentage of a model's contour are kept as correct matches.

A model consists of a three-dimensional wire-frame, together with the locations of the corner and inflection features that are used for computing possible alignments. In the current implementation we use several different models of an object from different viewpoints. (It would also be possible to use a single three-dimensional model and a hidden line elimination algorithm.) For example, in the current system a cube is represented by 9 line segments in 3D coordinates and by 12 corner features at the vertices (one feature for each corner of each of the three faces). The edge contours of each model are represented as chains of three-dimensional edge-pixel locations. Properties of a surface itself are not used (e.g., the curvature of the surface), only the bounding contours.

Once the features are extracted from an image, all pairs of two model and image features are used to solve for potential alignments, and each alignment is verified using the complete model edge contour. If an alignment maps a sufficient percentage of the model edges onto image edges, then it is accepted as a correct match of an object model to an instance in the image. In order to speed up the matching process on a serial machine the image feature pairs are ordered, and once an image feature is accounted for by a verified match it is removed from further consideration. The ordering of features makes use of two principles. First, those pairs of image features that lie on the same contour are considered before pairs that do not. Second, feature pairs near the center of the image are tried before those near the periphery.

An alignment specifies two possible transformations, that differ by a reflection of the model about the three model points. A further ambiguity is introduced by not knowing which of the two model features in a pair corresponds to which of the two image features. Thus each feature pair defines four possible alignment transformations.

To verify an alignment, each model edge segment is transformed into image coordinates in order to determine if there are corresponding image edges. The image edges are stored in a table according to position and orientation, so only a small number of image edges are considered for each model edge, as discussed in Section 4. When an alignment matches a model to an image, each alignment feature in the image that is part of an accounted for edge is removed from further consideration by the matcher. This can greatly reduce the set of matches considered, at the cost of missing a match if image features are incorrectly incorporated into verified alignments of other objects. This is unlikely, however, because the verifier has a low chance of accepting false matches.

The recognizer has been tested on images of simple polyhedral objects in relatively complex scenes, and on piles of laminar parts. The method is not restricted to the recognition of polyhedral objects, only to objects with well-defined local features (see [2] for extensions to matching smooth solids).

Figure 3 illustrates the performance of the recognizer on an image of a complex scene. Part (a) of the figure shows the grey-level image, part (b) shows the Canny edges and the features (detected corners are marked by dots), and part (c) shows the verified instances of the models in the image. The model is projected into the image, and the matched image edges are shown in bold. All of the hypotheses that survived verification are shown in part (c) of the figures. The test image contains about 100 corners, yielding approximately 150,000 matches of the model to the image. The matching time (after feature extraction) is about 6 minutes on a Symbolics 3650. The initial edge detection and feature extraction operations take approximately 4 minutes.

It can be seen from this example that the method can identify a correct match of a model to an image under a variety of viewing positions and angles. Furthermore, a correct match can be found in highly cluttered images, with no false matches. There are, however, slight alignment errors due to noise in the locations of the features. An alignment is computed from only three corresponding points, so there are not many sample points over which to average out the sensor noise. Thus it may be desirable to first compute an estimate of the transformation from three corresponding points, as is done here, and if the initial match is good enough then perform some sort of least squares fit to improve the final position estimate. One possibility is to compute a least squares matching under perspective projection (e.g., [16]). However the magnitude of the perspective effects is generally smaller than the magnitude of the sensor noise. Thus another possibility is to compute a least squares matching under a degenerate three-dimensional affine transformation (e.g., an axonometric transformation). Finding corresponding features using our transformation method minimizes the number of transformations that must be considered, and simplifies the computation of each transformation. Then solving for a least-squares transformation provides a more accurate match.

6. Summary

We have shown that when the imaging process is approximated by orthographic projection plus a scale factor, three corresponding model and image points are necessary and sufficient to align a rigid solid object with an image (up to a reflective ambiguity). We then used this result to develop a simple method for computing the alignment from any triple of non-collinear model and image points. Most other 3D from 2D recognition systems either use approximate solution methods [20], are restricted to recognizing planar objects [4, 14], or solve the perspective viewing equations [7, 16], which are relatively sensitive to sensor noise because the perspective effects in most images are small relative to the errors in feature localization.

The recognition system described in this chapter uses two or three pairs of corresponding model and image features to align a model with an image, and then verifies the alignment by transforming the model edge contours into image coordinates. The alignment features are local and are obtained by identifying corners and inflections in edge contours. Thus the features are relatively insensitive to partial occlusion and stable over changes in viewpoint. By identifying the minimum amount of information necessary to compute a transformation, the system is able to use simple local features, and only considers only a low-order polynomial number of possible matches. We have shown some examples of recognizing partially occluded rigid objects in relatively complex indoor scenes, under normal lighting conditions.

REFERENCES

1   N. Ayache and O.D. Faugeras "HYPER: A New Approach for the Recognition and

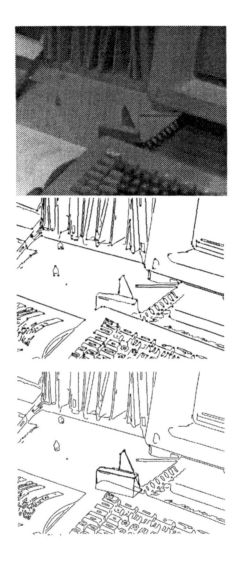

Figure 3. An example output of the recognizer: (a) grey-level image input, (b) edge features, (c) recovered instances.

Positioning of Two-Dimensional Objects", *IEEE Trans. Pat. Anal. and Mach. Intel.*, Vol. 8., No. 1, pp. 44-54, 1986.

2   R. Basri and S. Ullman, "The Alignment of Objects with Smooth Surfaces", *Proc. Second Intl. Conf. on Computer Vision*, pp. 482-488, IEEE Computer Society Press, 1988.

3   J.F. Canny, "A Computational Approach to Edge Detection", *IEEE Trans. Pat. Anal. and Mach. Intel.*, Vol. 8, No. 6, pp. 34-43, 1986.

4   D. Cyganski and J.A. Orr, "Applications of Tensor Theory to Object Recognition and Orientation Determination", *IEEE Trans. on Pat. Anal. and Mach. Intel.*, Vol. PAMI-7, No. 6, pp. 662-673, 1985.

5   M. Dhome, M. Richetin, J.T. LaPreste and G. Rives, "Determination of the Attitute of 3D Objects from a Single Perspective Image", *IEEE Trans. Pat. Anal. and Mach. Intel.*, pp. 1165–1178, Vol. 11, No. 12, 1989.

6   R.O. Duda, and P.E. Hart, *Pattern Classification and Scene Analysis*, Wiley, New York, 1973.

7   M.A. Fischler and R.C. Bolles, "Random Sample Consensus: A Paradigm for Model Fitting with Applications to Image Analysis and Automated Cartography", *Comm. Assoc. Comput. Mach.*, Vol. 24, No. 6., pp. 381-395, 1981.

8   W.E.L. Grimson and T. Lozano-Pérez, "Localizing Overlapping Parts by Searching the Interpretation Tree", *IEEE Trans. Pat. Anal. and Mach. Intel.*, Vol. 9, No. 4, pp. 469-482, 1987.

9   W.E.L. Grimson and D.P. Huttenlocher "On the Sensitivity of the Hough Transform for Object Recognition", *IEEE Trans. on Pattern Analysis and Machine Intelligence*, vol. 12, no. 3, pp. 255-274, 1990.

10   B.K.P Horn and E.J. Weldon, "Filtering Closed Curves", *Proc. Conf. on Computer Vision and Pattern Recognition*, pp. 478-484, 1985.

11   D.P. Huttenlocher and S. Ullman, "Object Recognition Using Alignment", *Proc. First Intl. Conf. on Computer Vision*, pp. 102-111, IEEE Computer Society Press, 1987.

12   D.P. Huttenlocher and S. Ullman, "Recognizing Solid Objects by Alignment", *Intl. Journal of Computer Vision*, , vol. 5, no. 2, pp. 195–212, 1990.

13   T. Kanade and J.R. Kender, "Mapping Image Properties into Shape Constraints: Skewed Symmetry, Affine Transformable Patterns, and the Shape-from-Texture Paradigm", in J. Beck, et. al. (Eds) *Human and Machine Vision*, Academic Press, Orlando, Fla., 1983.

14   Y. Lamdan, J.T. Schwartz, and H.J. Wolfson, "Object Recognition by Affine Invariant Matching", *Proc. IEEE Conf. on Computer Vision and Pattern Recognition*, IEEE Computer Society Press, 1988.

15   S. Linainmaa, D. Harwood, and L.S. Davis, "Pose Determination of a Three-Dimensional Object Using Triangle Pairs", CAR-TR-143, Center for Automation Research, University of Maryland, 1985.

16   D.G. Lowe, "Three-Dimensional Object Recognition from Single Two-Dimensional Images", *Artificial Intelligence J*, Vol. 31, pp. 355-395, 1987.

17   D.G. Lowe, "Organization of Smooth Image Curves at Multiple Scales", *Proc. Second Intl. Conf. on Computer Vision*, IEEE Computer Society Press, 1988.

18   F. Mokhtarian and A. Mackworth, "Scale-Based Description and Recognition of Planar Curves and Two-Dimensional Shapes", *IEEE Trans. Pat. Anal. and Mach. Intel.*, Vol. 8, No. 1, 1986.

19   T. Silberberg, D. Harwood and L.S. Davis, "Object Recognition Using Oriented Model Points," *Computer Vision, Graphics and Image Processing*, Vol. 35, pp. 47-71, 1986.

20 D. Thompson and J.L. Mundy, "Three-Dimensional Model Matching From an Unconstrained Viewpoint", *Proc. IEEE Conf. Robotics and Automation* p. 280, IEEE Computer Society Press, 1987.

21 S. Ullman, "An Approach to Object Recognition: Aligning Pictorial Descriptions", MIT Artificial Intelligence Lab., Memo No. 931, 1987.

Three-Dimensional Object Recognition Systems
A.K. Jain and P.J. Flynn (Editors)
© 1993 Elsevier Science Publishers B.V. All rights reserved.

Representations and Algorithms for 3D Curved Object Recognition[1]

Jean Ponce[a] , David J. Kriegman[b] , Sylvain Petitjean[a], Steven Sullivan[a], Gabriel Taubin[c] , and B. Vijayakumar[b]

[a] Beckman Institute
Department of Computer Science
University of Illinois
Urbana, Illinois 61801 USA

[b] Center for Systems Science
Department of Electrical Engineering
Yale University
New Haven, Connecticut 06520 USA

[c] IBM Thomas J.Watson Research Center
P.O.Box 704
Yorktown Heights, New York 10598 USA

Abstract

We describe a recognition system aimed at recognizing and positioning complex curved 3D objects from monocular image contours. Object models consist of collections of algebraic surfaces and their intersection curves, and the geometric constraints involved in predicting and interpreting the image contours of these models are represented by sets of polynomial constraints. We briefly describe the basic components of the system, including a set of symbolic and numerical tools for solving polynomial constraints and a solid modelling toolkit. We then focus on recent progress in the areas of automatic model construction from sequences of images and monocular feature interpretation.

1. Introduction

Most 3D recognition systems use polyhedra as object models [16, 19, 22, 24, 25, 34, 53, 55]. There are many good reasons for this choice: for example, the geometric constraints associated with the images of polyhedra can often be expressed in a linear fashion [16, 60] and geometric invariants computed from image measurements can be used for indexing [17, 44]. Maybe the most fundamental reason for the popularity of polyhedral models is that the corresponding image features are

[1]This work was supported by the National Science Foundation under Grant IRI-9015749.

simple, and their structure is easily understood. For example, the line-drawing of a polyhedron consists of line segments which are the projection of surface edges and join at points which are either the image of vertices or the t-junctions that occur when an edge is partially occluded by a face. T-junctions are viewpoint-dependent, but the other image features, i.e., the image edges and vertices, can be directly matched to the object features, i.e., the surface edges and vertices [24, 34].

Unfortunately, the world is not simple and polyhedral; it is complex and full of objects which may be curved, smooth or piecewise-smooth, or completely irregular, and the corresponding image features are much more difficult to interpret. For example, the image contours of smooth curved objects are the projection of curves where an imaginary ray passing through the eye grazes the surface. They join at t-junctions but may also terminate (cusp) when the visual ray touches the surface along some special direction. In this case, the image features cannot be matched to viewpoint-independent object features. This challenges us to broaden the scope of recognition systems by using shape representations that are more expressive than polyhedra, but whose image features are still well understood.

As shown in the following, algebraic surfaces are much more flexible than other surface representations such as quadrics [16] or superquadrics [4, 21, 46], but at the same time they allow us to represent the geometric constraints associated with the imaging process in a simple polynomial manner. This is not true of other shape representations such as generalized cylinders [7, 9, 48]. Indeed, algebraic surfaces have recently become the object of active research in the vision community [27, 30, 31, 49, 59, 61, 62].

In this chapter, we address a number of issues raised by the development of a computer vision system aimed at recognizing objects represented by piecewise-smooth collections of algebraic surfaces in a single video image. We will be particularly interested in object modelling and the relationship between models and their images, and we will neglect other important issues such as indexing large object libraries, modelling object classes, and image segmentation.

The rest of the chapter is organized as follows. Section 2 introduces algebraic surfaces more formally and presents a set of basic tools from computer algebra and numerical analysis which can be used to solve the corresponding algebraic constraints. A solid modelling system based on algebraic surfaces is discussed in section 3. Alternatively, we introduce in section 4 several techniques for automatically constructing models by fitting algebraic surfaces to either range data or a sequence of images; a parameterized family of bounded algebraic surfaces is also presented. As shown in section 5, the pose of a modelled object can be estimated from viewpoint-dependent features in a single image. We consider global techniques using point features as well as more accurate local techniques using the entire image contour.

2. Approach

2.1. Algebraic Surfaces

We represent objects by collections of algebraic surfaces and their intersection curves.

A parametric algebraic surface patch is defined by:

$$\mathbf{x}(u,v) = (x(u,v), y(u,v), z(u,v))^t, \quad (u,v) \in I \times J \subset \mathbb{R}^2, \tag{1}$$

where x, y, z are polynomials (or ratios of polynomials) in u, v. Examples include Bézier patches and non-uniform rational B-splines (NURBS).

An implicit algebraic surface of degree d is the set of points $\mathbf{x} = (x, y, z)^T$ where some trivariate polynomial

$$P(\mathbf{x}) = P(x, y, z) = \sum_{i+j+k \leq d} a_{ijk} x^i y^j z^k \tag{2}$$

of degree d vanishes. (The surface is the zero set of P.)

Why use algebraic surfaces? The problem of choosing the right shape representation can be addressed from many points of view [7, 8, 40]; we focus on the expressiveness of the representation and its adequacy to solving vision problems.

Most research in computer vision has focussed on very simple surfaces, namely polyhedra and quadric surfaces [16, 61], and, more recently, on superquadrics [5] or more precisely, superellipsoids [4, 21, 46]. Alone, these surfaces are not expressive enough to represent complex curved objects: for example, a quadric surface has only three shape parameters, and a superquadric surface has only five shape parameters (eight if bending and tapering are allowed). In contrast, a quartic surface has 28 shape parameters! (Each of these surfaces actually has six more coefficients that specify its position and orientation in space, but do not change the surface shape.) Even when more complex objects are modelled by Boolean combinations of these primitives, they cannot be smoothly stitched together; higher degree surface patches are needed. Generalized cylinders [7] form a very general shape representation, but many of them, for example straight homogeneous generalized cylinders [56], can be modelled conveniently by algebraic surfaces [30]. Furthermore, several recent papers have shown how algebraic surfaces could be used for recognizing and locating object models in range and video images [27, 30, 31, 49, 59, 61, 62].

A key property is that, for most vision problems, the geometric constraints associated with algebraic surfaces are themselves algebraic, i.e., they can be expressed by polynomial equalities and inequalities. In turn, this allows us to use extremely powerful methods, such as elimination theory, curve tracing, cell decomposition, and homotopy continuation to solve these problems.

Most of the results reported in this chapter apply to parametric surfaces, but we have chosen to focus on implicit surfaces. Again, there are several reasons for this choice. On the one hand, parametric surfaces are convenient for rendering, since points on a patch are trivially generated by substituting parameter values in (1), and, as noted in [51], parametric polynomial curves and surfaces have received the lion's share in the fitting literature. On the other hand, implicit algebraic surfaces have several key advantages over their parametric counterparts: the implicit function can be used to decide whether a point lies inside or outside the solid bounded by the surface; implicit algebraic surfaces are easily ray traced by solving a univariate polynomial; closed surfaces are naturally represented by implicit surfaces; any parametric algebraic surface can be represented by an implicit algebraic surface but the converse is not true, in other words, implicit surfaces are more general than parametric ones [18, 39]. Also, the computational complexity of many algorithms is related to the degree of the surface, and an implicit representation affords more degrees of freedom at a lower degree than parametric surfaces.

2.2. Basic Algorithms

As stated earlier, the geometric constraints associated with algebraic surfaces are themselves algebraic. This gives us access to an arsenal of computational tools designed to manipulate and solve such constraints. In general, we are given a system $\mathbf{P}(\mathbf{x})$ of n irreducible polynomial equations in m unknowns:

$$\begin{cases} P_1(x_1, \ldots, x_m) = 0, \\ \ldots \\ P_n(x_1, \ldots, x_m) = 0. \end{cases} \tag{3}$$

Homotopy Continuation.

When the system is square ($n = m$), the solutions are normally a discrete set of roots. A number of techniques are available for solving such systems. One approach is to reduce the system to a single univariate equation using resultants (described below), then use some technique for univariate root finding. Alternatively, we use Morgan's homotopy continuation method [43]. Let $\mathbf{Q}(\mathbf{x}) = 0$ be another system of polynomial equations with the same total degree d as $\mathbf{P}(\mathbf{x})$, but with known solutions. A homotopy, parameterized by $t \in [0, 1]$, can be defined between the two systems by:

$$(1 - t)\mathbf{Q}(\mathbf{x}) + t\mathbf{P}(\mathbf{x}) = \mathbf{0}. \tag{4}$$

The d solutions of the target system are found by tracing the d curves (paths) defined in $\mathbb{R}^{m+1}$ by these equations from $t = 0$ to $t = 1$. It can also be shown [43] that for almost any choice of $\mathbf{Q}$, the curve is nonsingular. The paths can be traced independently in parallel. We have implemented this approach for networks of INMOS transputers, networks of Sun SPARCstations, and Intel Hypercubes; in general, we have found a near linear speedup in the number of processors.

Curve Tracing.

When $m = n + 1$, (3) defines a curve in $\mathbb{R}^m$. We have developed an algorithm for characterizing the topology of this curve and tracing a discrete approximation of its regular branches. Briefly, the algorithm is decomposed into four steps (see [32, 50] for details): 1. Compute all extremal points of in some direction, say x_1 (this includes all singular points). 2. Compute all intersections of with the hyperplanes orthogonal to the x_1 axis at the extremal points. 3. For each interval of the x_1 axis delimited by these hyperplanes, intersect and the hyperplane passing through the mid-point of the interval to obtain one sample for each real branch. 4. March numerically from the sample points found in step 3 to the intersection points found in step 2 by predicting new points through Taylor expansion and correcting them through Newton iterations.

Steps 1 to 3 amount to solving a square system of polynomial equations, while step 4 essentially requires solving systems of linear equations. The output of the algorithm is a graph whose nodes are extremal or singular points on and whose arcs are discrete approximations of the smooth curve branches between these points. This is similar to Arnon's s-graph representation of plane curves [2], constructed through cylindrical algebraic decomposition.

Cell Decomposition.

When $n = 1$ and $m = 2$, the curve partitions $\mathbb{R}^2$ into regions. Our curve tracing algorithm can be easily modified into a cell decomposition algorithm by sorting the sample points found in step 3 in increasing x_2 order. The curve branches corresponding to consecutive samples within an interval and the segments of constant x_1 delimited by the points found in step 2 define the boundary of a region (cell). A graph structure whose nodes are the cells and whose arcs are their adjacency relationships can be readily constructed, and the maximal connected components are simply determined. The algorithm is easily extended to the case of several curves by adding the intersection points between curves to the list of singular points. These points can be computed through continuation.

Resultants.

Finally, it may prove necessary to eliminate one or more variables from (3). We use the resultant, a tool from elimination theory [14, 37, 54]. The essential idea is that a necessary and sufficient condition for a system of polynomial equations to have common roots can be expressed as the vanishing of a single polynomial called their resultant. The original variables do not appear in the resultant, hence the name of elimination theory. The resultant itself is a polynomial in the original coefficients. For the work presented here, the Sylvester resultant, which eliminates one variable among two equations, is sufficient and is readily available in most computer algebra systems such as Reduce, Maple, Macsyma and Mathematica.

2.3. Projection Models

The geometry of image formation is modelled as a nonlinear transformation $T : \mathbb{R}^3 \to \mathbb{R}^2$ that associates to the coordinates x of a point in the object frame the coordinates $\hat{p}$ of its projection in the image plane of the camera:

$$\hat{p} = Tx. \tag{5}$$

The transformation T is defined by a set of pose parameters (three for rotation and three for translation) that relate the object coordinate frame and the camera coordinate frame. Conversely, given T and some image point $\hat{p}$, the set of points projecting onto $\hat{p}$ is the line joining the eye to the point $\hat{p}$. It is defined in the object coordinate system by:

$$x = T^*\hat{p} = p + \mu v, \tag{6}$$

where T^* is the "inverse" of T, p is the point $\hat{p}$ expressed in the world coordinate system, v is the direction of the line (the viewing direction), and μ is a scalar representing the unknown depth of the point x along the line. Note that p and v depend only on T^* and $\hat{p}$. In the following, we will alternatively assume perspective and scaled orthographic projection. The viewing direction is a constant in the latter case.

3. A Solid Modelling System

While solid modellers based on algebraic surfaces are becoming common-place, they usually rely on purely numerical methods for representing surface patches and their intersection curves [11, 15, 41], and they lack the symbolic object representation required for vision applications. Here, we describe a solid modeller that overcomes some of these limitations. As in constructive solid geometry (CSG), complex models are constructed through set operations (union, intersection, difference) between simpler solid primitives bounded by algebraic surfaces. As in a boundary representation (BREP), a symbolic description of an object model is created in terms of the topology and defining equations of its boundary elements. More details can be found in [31].

At the heart of the system is the computation of the BREP. First, vertices formed by the intersection of three surfaces are found by solving

$$\begin{cases} P_1(x,y,z) = 0, \\ P_2(x,y,z) = 0, \\ P_3(x,y,z) = 0, \end{cases} \tag{7}$$

for all triplets of surfaces in the model (bounding boxes greatly reduce the number of triplets considered) using homotopy continuation. Next, the intersection curves between all pairs of surfaces are characterized by defining a system of two polynomial equations in three unknowns:

$$\begin{cases} P_1(x,y,z) = 0, \\ P_2(x,y,z) = 0. \end{cases} \tag{8}$$

The topology of the intersection curves can be determined by the numerical curve tracing algorithm [32, 50] described in section 2.2; the previously computed vertices are treated like singular points. A point can be classified as being inside, outside or on the surface of a CSG model; the classification of the sample points found in step 3 can be used to determine if the entire curve branch is part of the BREP.

The line-drawing of a curved object is composed of the projection of intersection curves and occluding contours. The intersection curves are simply rendered by projecting the result of the curve tracing algorithm. The occluding contours of a surface are characterized by the system:

$$\begin{cases} P(x,y,z) = 0, \\ \nabla P(x,y,z) \cdot v = 0, \end{cases} \tag{9}$$

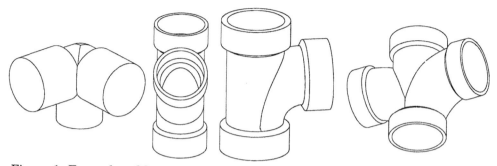

Figure 1. Examples of four pipe fittings including two with quartic surfaces.

where **v** is the viewing direction. The curve tracing algorithm can be used once again to characterize these curves. For an opaque object, the visibility of each point **x** can be determined by tracing a ray in the viewing direction through **x**; this amounts to solving a univariate polynomial equation for each surface in the model [26]. A symbolic line-drawing or image structure graph [38], whose nodes are singular points and whose arcs are the projection of occluding contours or intersection curves, is thus created. The nodes are constructed first by solving square systems of polynomials equations that characterize each type of singular point (cusps, t-junctions, vertices, and curvature-L/3-tangent junctions); these are discussed in more detail in section 5.1. By treating singular points like extrema in the curve tracing algorithm, the arcs are readily created. Furthermore, since contour visibility only changes at singular points, it is sufficient to only ray trace the sample points found in step 3 of the tracing algorithm. Figure 1 shows the drawing of four pipes modelled in our system by a combination of quadratic and quartic surfaces.

4. Fitting Algebraic Surfaces

In the previous section, object models were built interactively using a solid modelling system. Depending on the application, solid models may not always be available (e.g., for natural objects, such as fruits and vegetables, or medical data) or even appropriate (e.g., the representation of the hood of a car by many small bicubic patches is much too detailed and complicated for object recognition–each patch is a surface of degree 18). In this section, we are interested in automatically constructing algebraic surface models from image data. Each model should consist of one or a few algebraic surfaces of moderate degree, capturing the shape of the imaged object with a level of detail appropriate for the targeted application (e.g., object recognition or fast visualization of CT data).

Classical methods for fitting an implicit algebraic surface to 3D data rely on linear least-squares for minimizing the mean squared *algebraic distance* between the surface and the data points [51]. These methods are efficient and guaranteed to find a global optimum. However, they face at least three major problems: (1)

real objects are always compact, but an algebraic surface (e.g., a plane) may be unbounded; (2) an algebraic surface (e.g., a torus) may have holes, and fitting an algebraic surface to 3D data may introduce holes in regions where dense data is in fact available; (3) an algebraic surface may have several disjoint components, and extra sheets may be introduced during fitting where there is no data.

We propose two novel algorithms for fitting implicit algebraic surfaces to 3D data. The first one is based on a new parameterized family of bounded algebraic surfaces, and naturally overcomes problem (1). The second one minimizes the exact geometric distance between the surface and the data points, and effectively overcomes problem (2). See [3, 42] for approaches to problem (3). We also present a novel algorithm for fitting implicit algebraic surfaces to sets of video images by minimizing the exact geometric distance between the surface and the viewing rays joining the eye to the observed contour points. To the best of our knowledge, this is the first time this problem has been attacked. A preliminary implementation of these algorithms has been constructed, and examples are presented.

4.1. Fitting Algebraic Surfaces to 3D Image Data

The classical method for fitting an algebraic surface of degree d to a set of 3D data points x_i, with $i = 1, .., n$, is to minimize the mean squared *algebraic distance* between the surface and the data points, i.e., minimize

$$\sum_{i=0}^{n} P^2(\mathbf{x}_i) = \sum_{i=0}^{n} (\sum_{j+k+l \leq d} a_{jkl} x_i^j y_i^k z_i^l)^2 \tag{10}$$

with respect to the unknown polynomial coefficients a_{jkl}.

The above expression is quadratic in the coefficients a_{jkl}. For a given surface, these coefficients can only be determined up to some scalar factor, and the minimization can be done under the constraint that $\sum a_{jkl}^2 = 1$. This is a linear least-squares problem, whose solution can be found by solving an eigenvalue problem [58].

As remarked earlier, this method is efficient and guaranteed to find a global optimum, but it suffers from significant problems. In particular, the surface fitted to dense range data from a compact connected object without holes may be unbounded, have holes, and extra sheets. We now present two methods that overcome some of these problems.

4.1.1. Bounded-Surface Method

In this section, we introduce a parameterized class of bounded implicit algebraic surfaces [62]. First, we note that any polynomial P of degree d can be written as a sum of homogeneous polynomials, or *forms*, P_i of degree $i \leq d$:

$$P(\mathbf{x}) = \sum_{i=0}^{d} P_i(\mathbf{x}) = \sum_{i=0}^{d} \sum_{j+k+l=i} a_{jkl} x^j y^k z^l. \tag{11}$$

The form P_d of degree d is called the *leading form* of P. The points at infinity of an algebraic surface are non-trivial zeros of its defining polynomial's leading form, so algebraic surfaces whose leading form has no non-trivial zero are bounded. Note that a polynomial whose leading form has non-trivial zeros may also have a

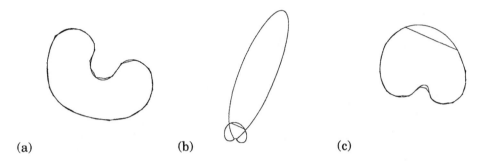

Figure 2. Fitting quartic curves to 2D points (drawn as small circles).

bounded zero set, e.g., the surface $x^4 + y^4 + z^2 - 1 = 0$ is bounded, but its leading form has the non-trivial zero $(0, 0, 1)^T$, corresponding to the isolated point at infinity $(0, 0, 1, 0)^T$. It is easy to show that all surfaces of odd degree are unbounded [62]. Thus, we can restrict our attention to surfaces of even degree.

From Euler's theorem [62, 63] every form P_{2d} of degree $2d$ can be written as a quadratic form $P_{2d}(\mathbf{x}) = \mathbf{X}^T Q \mathbf{X}$ in the vector $\mathbf{X} = (x^d, x^{d-1}y, \ldots, z^d)^T$ formed by the monomials of degree d. Since a quadratic form is positive definite if and only the corresponding symmetric matrix Q can be written as the square of a lower-triangular matrix W [58], any polynomial whose leading form is

$$P_{2d}(\mathbf{x}) = \mathbf{X}^T W^T W \mathbf{X} \tag{12}$$

has a bounded zero set.

Thus, we have a family of polynomials whose leading form is parameterized by the elements of a lower triangular matrix W and whose lower degree forms are simply given as a sum of monomials as in (11). A polynomial in this family is a *non-linear* function of the parameters of its leading form. Unfortunately, this means that local, non-linear minimization procedures must be employed during fitting.

Another property proves extremely useful in practice. It can be shown that the bounded surface associated with the matrix Q is enclosed in a sphere whose radius is inversely proportional to the minimum eigenvalue of Q [62]. This means that the eigenvalues of Q provide an indication of the volume enclosed by the surface. To avoid large surface components that may appear where there are gaps in the data, we have chosen in our fitting experiments to divide the error of fit by the trace of Q, or equivalently the sum of the squares of the eigenvalues of W. This has the effect of maximizing the minimum eigenvalue, thereby minimizing the enclosed volume.

We have implemented the method and used it to fit quartic curves and surfaces to contour and range data. The non-linear minimization algorithm requires a set of initial parameters. In our experiments, we have have either set all initial parameters to 1, or used the quartic $x^4 + y^4 + z^4 = 1$ (suitably translated and scaled to enclose all data points) as an initial guess. Both sets of initial parameters have in general led to identical results at convergence.

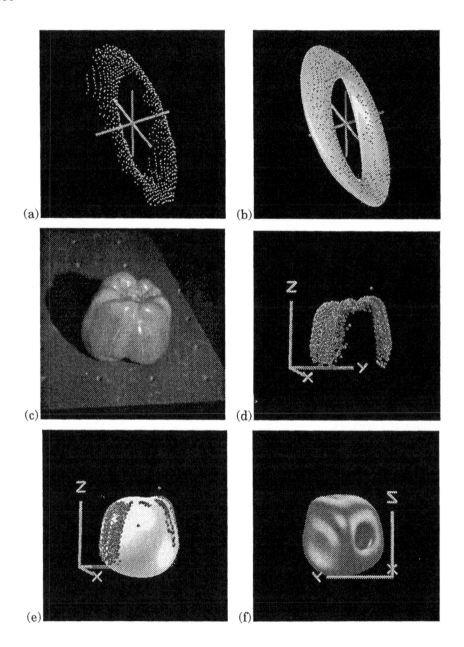

Figure 3. Fitting bounded surfaces to real range data.

Figure 2 shows experiments in curve fitting. Figure 2(a) shows the curve fitted to a polygon entered by hand. A good approximation of the data is obtained. More interestingly, figure 2(b–c) shows another example with a large gap among the data points (once again entered by hand). In figure 2(b), the result of fitting an unconstrained quartic curve is shown: a large component is obtained because of the gap in the data. Figure 2(c) shows the excellent approximation obtained when fitting a bounded, "minimum-area" quartic curve.

Figure 3 shows experiments in fitting bounded quartic surfaces to real range data. Figure 3(a) shows the data from a single range image of a torus, and figure 3(b) shows the fitted surface. Less than half of the surface is visible, but a fairly good approximation of the surface is recovered. Figure 3(c–d) shows a pepper and the corresponding range data obtained by registering and merging three range images. Again, despite noise and large gaps in the data, a reasonable surface model is recovered, as demonstrated by figure 3(e). Unfortunately, there is a hole in the reconstructed surface, as revealed by a second view shown in figure 3(f). This hole actually occurs in a region where the data is dense.

4.1.2. Exact-Distance Method

We just saw that holes in the fitted surface may occur even in regions where dense data is available. An explanation for this phenomenon is that the polynomial P defining the surface is quite shallow in these regions, so the corresponding points have very little influence on the result of the fitting. To remedy this problem, we propose to minimize the *geometric* distance to the surface instead of the *algebraic distance*.

This requires computing the exact geometric distance between the surface and every data point at every iteration of the minimization process. The geometric distance between a point x and a surface is defined as the minimum of the distance between x and all surface points y. It is reached at a point y such that the normal to the surface at y and the line joining x and y are parallel.

In other words, given x, the point y is one of the solutions of the following system of equations:

$$\begin{cases} P(\mathbf{y}) = 0, \\ \nabla P(\mathbf{y}) \times (\mathbf{x} - \mathbf{y}) = 0, \end{cases} \tag{13}$$

where $\nabla P(\mathbf{y})$ denotes the gradient of P at y, and "$\times$" denotes the cross-product operator. The second equation is a vector equation, corresponding to two independent scalar equations. It follows that y can be found by solving the above system of three equations in its three components and then choosing the solution minimizing the distance $|\mathbf{x} - \mathbf{y}|$.

Global numerical methods such as homotopy continuation [43] are available for solving square systems of polynomial equations such as this one. However, they are too slow to be used in practice with hundreds or maybe thousands of data points. We propose the following alternative formulation of the problem. Instead of solving explicitly for y, we minimize the squared distance $D(\mathbf{y}) = |\mathbf{x} - \mathbf{y}|^2$ under the constraint $P(\mathbf{y}) = 0$.

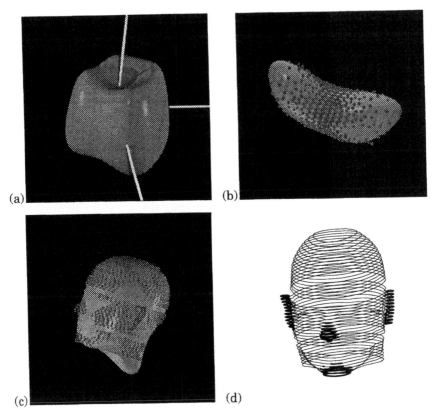

(a) (b)

(c) (d)

Figure 4. Fitting surfaces to real range data using the exact distance.

Writing the first-order necessary conditions of the corresponding Lagrangian $L(\mathbf{y}, \lambda) = D(\mathbf{y}) + \lambda P(\mathbf{y})$, we obtain:

$$\begin{cases} \nabla_{\mathbf{y}} L = \nabla D(\mathbf{y}) + \lambda \nabla P(\mathbf{y}) = 0, \\ \nabla_{\lambda} L = P(\mathbf{y}) = 0. \end{cases} \tag{14}$$

Recursive quadratic programming [36] involves solving the linearized version repeatedly:

$$\begin{bmatrix} \nabla^2 D + \lambda \nabla^2 P & \nabla P \\ \nabla P^T & 0 \end{bmatrix} \begin{pmatrix} d\mathbf{y} \\ d\lambda \end{pmatrix} = - \begin{pmatrix} \nabla D + \lambda \nabla P \\ P \end{pmatrix}. \tag{15}$$

We have implemented this method and present some results. Figure 4(a–b) shows two examples of fitting using the pepper and a banana. This time there is no hole in the surface fitted to the pepper data. Figure 4(c) shows the fit obtained using real CT data of a face. The surface reconstructed is a reasonable approximation but some face features such as the nose, ears and chin, are not well recovered. This is an instance where the exact distance measure proves especially useful since it can be used to automatically segment these parts (figure 4(d)).

4.2. Fitting Algebraic Surfaces to 2D Image Data

In this section, we address the following problem: given a set of 2D views of a given object, can we construct a 3D model of the object? The depth of the contour points is unknown along the contour, but a set of views determines a set of viewing cones and constrains the observed object to lie within the intersection of these cones. Note that non-convex objects can be modelled using this approach.

This is a classical problem in computer vision. Baumgart [6], and later Connolly and Stenstrom [12] have attacked this problem by approximating the visual cones and their intersections by polyhedra; Ahuja and his students [1, 57] have constructed Octree models of the observed objects. Here, we propose a method for directly fitting an implicit algebraic surface to the contour data. Again, we can reformulate the problem of fitting a surface to data points as a problem of minimizing some distance measure between the data points and the surface. The depth of the contour point is not directly observable, but we can minimize the distance between the surface $P(\mathbf{y}) = 0$ and the line $\mathbf{p} + \mu\mathbf{v}$ instead.

The minimum is reached at a point $\mathbf{y}$ on the surface and a point $\mathbf{x} = \mathbf{p} + \mu\mathbf{v}$ on the line such that:

$$\begin{cases} P(\mathbf{y}) = 0, \\ \nabla P(\mathbf{y}) \times (\mathbf{p} + \mu\mathbf{v} - \mathbf{y}) = 0, \\ (\mathbf{p} + \mu\mathbf{v} - \mathbf{y}) \cdot \mathbf{v} = 0. \end{cases} \tag{16}$$

It follows that the points $\mathbf{x}$ and $\mathbf{y}$ can be found by solving the above system of four equations in the three components of $\mathbf{y}$ and in μ, then choosing the solution minimizing the distance $|\mathbf{x} - \mathbf{y}|$. Again, we can minimize instead the squared distance $D(\mathbf{y}) = |\mathbf{p} + \mu\mathbf{v} - \mathbf{y}|^2$ under the constraint $P(\mathbf{y}) = 0$. This leads to a formulation of the Lagrangian conditions analogous to the one derived earlier. We have constructed a preliminary implementation of this method and experimented with synthetic and real data. Figure 5 shows an example.

5. Pose Estimation

We consider the problem of estimating the pose of a piecewise-smooth curved object from a single image. Rather than constructing an intermediate $2\frac{1}{2}$D or 3D representation from the image, and matching it to a 3D object model, we directly match the 2D image features to the 3D model. Similar approaches to recognition and positioning of polyhedra from monocular images have been proposed in [13, 23, 24, 35]; they are based on so-called "rigidity constraints" [16, 20] or "viewpoint consistency constraints" [35]. Feature-matching is simple in the case of polyhedra because most observable image features are the projection of object features (edges and vertices). In contrast, most visible features in the image of a curved object depend on viewpoint and cannot be traced back to particular object features.

We discuss two methods: As shown in [30, 49], elimination thoery can be used to analytically relate the image contours of a curved object to the object geometry and

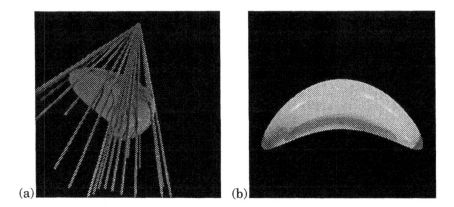

Figure 5. Reconstruction of a bean-shaped surface from its contours. (a) The rays corresponding to one view of the bean. (b) The reconstructed surface overlaid on top of the original surface.

object pose in the form of an implicit algebraic equation. The pose parameters are then estimated by fitting this equation to measured edge points using nonlinear minimization techniques. Unfortunately, these are only local techniques, and we are typically interested in a global minimum. Therefore, a more global approach is needed. In the next section, a minimum number of image features is matched to an object model; global solution techniques are then used to determine the pose. However, the estimated pose will be inherently inaccurate due to measurement noise and the use of only a few measurements. The previously mentioned local method is ideally suited for refining the inaccurate pose and will be discussed in section 5.2. We then show some results in section 5.3, where the pose of an object is first estimated from measured inflections of the image of an edge, and then the entire edge is used to refine the pose. More detail can be found in [33].

5.1. Monocular Image features

Pose estimation from viewpoint-independent image features (vertices) has been discussed extensively elsewhere [13, 24]. In this section, we show how viewpoint-dependent features can be used to determine an object's pose. We use the following superset of Malik's catalogue [38] of point features: t-junctions, cusps, three-tangent junctions, curvature-L junctions, and limb and edge inflections (figure 6). We represent the pose constraints imposed by pairs of surface points projecting onto these image features by square systems of equations in the position of the surface points. For algebraic surfaces, these equations are polynomial, and they can be solved using the techniques described in section 5.1.5. For each pair of points along with the measured contour normal, the parameters of the viewing transformation are then easily calculated from (5). We assume scaled orthographic projection, but the constraints can be extended to perspective.

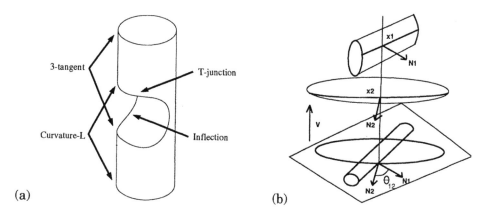

Figure 6. (a) Some viewpoint dependent image features for piecewise-smooth objects. (b) A t-junction and the associated geometry.

5.1.1. T-junctions

Suppose that an observed t-junction is the projection of two limb points x_1, x_2 as shown in figure 6(b). This hypothesis provides the following geometric constraints:

$$
\begin{cases}
P_i(x_i) = 0, & i = 1, 2, \\
(x_1 - x_2) \cdot N_i = 0, \\
N_1 \cdot N_2 = \cos\theta_{12},
\end{cases}
\tag{17}
$$

where $N_i = \nabla P_i(x_i)/|\nabla P_i(x_i)|$ denotes the unit surface normals, and $\cos\theta_{12}$ is the observed angle between the image normals. We have five equations, one observable $\cos\theta_{12}$, and six unknowns (x_1, x_2). In addition, the viewing direction is given by $v = x_1 - x_2$. An extra t-junction hypothesis yields another set of five equations in six unknowns x_3, x_4, plus an additional vector equation $(x_1 - x_2) \times (x_3 - x_4) = 0$, or equivalently two indepedent scalar equations, simply expressing the fact that the viewing direction should be the same for both t-junctions. Thus, two t-junctions and the corresponding hypotheses (i.e., "t-junction one corresponds to patch one and patch two", and "t-junction two corresponds to patch three and patch four") provide us with twelve equations in twelve unknowns. The viewing direction can be computed for each solution of this system, and the other parameters of the viewing transformation are then easily found. Similar constraints are obtained for t-junctions that arise from the projection of edge points by noting that the 3D curve tangent $\nabla P_1(x) \times \nabla P_2(x)$ projects to the tangent of the image contour.

5.1.2. Curvature-L and Three-tangent Junctions

For a piecewise-smooth object, curvature-L or three-tangent junctions occur when a limb terminates at an edge with a common tangent; observe the top and bottom of a coffee cup, or consider figure 6(a). Both feature types have the same local geometry, however one of the edge branches is occluded at a curvature-L junction.

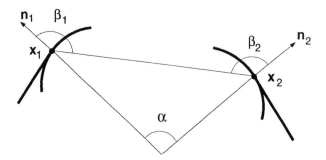

Figure 7. Geometry for pose estimation from three-tangent and curvature-L junctions: the curve branches represents edges, and the straight branches represent limbs.

Consider two points x_1, x_2 on the edges formed by the surfaces P_1, P_2 and P_3, P_4 that project to these junctions; x_1 (resp x_2) is also an occluding contour point for one of the surfaces, say P_1 (resp. P_3). As shown by figure 7, three angles α, β_1, β_2 can be measured using the junctions and the contour normals. Since x_1 and x_2 are limb points, the surface normals are aligned with the measured image normals. Thus, $\cos \alpha = n_1 \cdot n_2 / |n_1||n_2|$, where $n_i = \nabla P_i(x_i)$. Now, define the two vectors $\Delta = x_1 - x_2$ and $\hat{\Delta} = \hat{p}_1 - \hat{p}_2$. Clearly, the angle between the contour normal and $\hat{}$ must equal the angle between n_i and the projection of onto the image plane which is given by $\bar{\Delta} = \Delta - (\Delta \cdot \hat{v})\hat{v}$ where $\hat{v} = n_1 \times n_2 / |n_1 \times n_2|$ is the normalized viewing direction. Noting that $n_i \cdot \hat{v} = 0$, we have $|n_i||\bar{\Delta}| \cos \beta_i = n_i \cdot \Delta$. However, $\bar{}$ is of relatively high degree and a lower degree equation is obtained by taking the ratio of $\cos \beta_i$ and applying the equation for $\cos \alpha$. After squaring and rearrangement, the edge equations (8) along with these equations

$$\begin{cases} (n_1 \cdot n_1)(n_2 \cdot n_2) \cos \alpha - (n_1 \cdot n_2)^2 = 0, \\ \dfrac{\cos \beta_1}{\cos \beta_2}(n_2 \cdot \Delta)(n_1 \cdot n_2) - \cos \alpha (n_2 \cdot n_2)(n_1 \cdot \Delta) = 0, \end{cases} \tag{18}$$

form a system of six polynomial equations in six unknowns. After solving for x_1 and x_2 the pose is then determined from (5).

5.1.3. Inflections

Contour inflections may arise from either limbs or edges. In both cases, observing two inflections is sufficient for determining object pose.

As shown by Koenderink [28] has shown, a limb inflection is the projection of a parabolic point. The parabolic lines of a surface defined implicitly are given by:

$$\begin{aligned} &P_x^2(P_{yy}P_{zz} - P_{yz}^2) + P_y^2(P_{xx}P_{zz} - P_{xz}^2) + P_z^2(P_{xx}P_{yy} - P_{xy}^2) \\ &+2P_xP_y(P_{xz}P_{yz} - P_{zz}P_{xy}) + 2P_yP_z(P_{xy}P_{xz} - P_{xx}P_{yz}) \\ &+2P_xP_z(P_{xy}P_{yz} - P_{yy}P_{xz}) = 0, \end{aligned} \tag{19}$$

where the subscripts indicate partial derivatives. Note that the geometric relationship between the measure image normal and surface normal developed in the

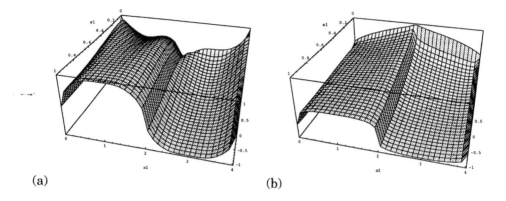

(a)　　　　　　　　　　　　　　(b)

Figure 8. Surface plots of the tables used for fast pose estimation: $\cos \alpha$ and $\cos \beta_1$ are drawn as functions of arc length along the edge curve of the object shown in figure 9. Because of symmetry, only one forth of the table is shown.

previous section and shown in figure 7 applies. Since both points x_1, x_2 are limbs, equation (19) and the surface equation for each point can be added to (18) for measured values of α, β_1 and β_2. This system of six equations in x_1, x_2 can be solved to yield a set of points, and consequently the viewing parameters.

In the case of edges, an image contour inflection corresponds to the projection of an inflection of the space curve itself or a point where the viewing direction is orthogonal to the binormal. Space curve inflections typically occur when the curve is actually planar, and can be treated like viewpoint independent features (vertices). For inflections ocurring when the binormal b_i is orthogonal to the viewing direction, as in figure 6, finding two such points is sufficient for determining pose. It can be shown that the projection of b_i is the image contour normal, and for surfaces defined implicitly (2), the binormal is given by $b = [t^t H(P_1)t]\nabla P_1 - [t^t H(P_2)t]\nabla P_2$ where $H(P_i)$ is the Hessian of P_i, and $t = \nabla P_1 \times \nabla P_2$ is the 3D curve tangent. By including the curve equations (8) with (18) after replacing n_i by b_i, a system of six equations in x_1 and x_2 is obtained. After solving this system for x_1 and x_2, the pose can be readily determined.

5.1.4. Cusps

Observing two contour cusps (terminations) is also sufficient for determining object pose. Cusps occur when the viewing direction is an asymptotic direction at a limb point which can be expressed as $v^t H(P_i)v = 0$ where the viewing direction is $v = \nabla P_1(x_1) \times \nabla P_2(x_2)$. While the image contour tangent is not strictly defined at a cusp (which is a singular point after all), the limit of the tangent as the cusp is approached will be orthogonal to the surface normal. Thus, the cusp and surface equations can be added to the system (18) which is readily solved for x_1 and x_2 followed by pose calculation.

5.1.5. Solving the Systems of Constraints

In the previous sections, square systems of equations were used to characterize the relationship between image measurements and surface points. For algebraic surfaces, these equations are polynomial and the roots can be computed using the techniques described in section 2.2. However, these systems tend to be quite large and since most of the roots are complex and many are at infinity, a faster approach is to restrict our attention to only the real roots. Since the systems may be repeatedly solved during recognition (say, for each hypothesized set of correspondences while searching an interpretation tree), it is advantageous to partition root finding into a more expensive off-line procedure and a faster on-line procedure.

The basic idea is that the equations derived earlier can be viewed as a mapping between measurements, such as the α and β angles in (18), and model points. Thus, we construct off-line a set of tables indexed by discrete pairs of (α, β) values and containing the corresponding surface points. The construction is particularly simple for features lying on specific curves such as parabolic lines or intersection edges: we trace these curves using the algorithm described in section 2.2, and calculate (α, β) for each sample pair of points. Figure 8 shows surface plots of α and β as a function of two edge points on the notched cylinder of figure 9.

We use the tables to find, on-line, the local minima of the difference between measured angle values and stored ones. Each local minimum corresponds to a match hypothesis, and Newton's method [52] can then be used to accurately locate the roots of the corresponding system. The pose is then readily calculated from x_1 and x_2.

In practice, solving the systems of equations using continuation is very expensive and for the examples presented below, nearly 20 hours on a Sun SPARCstation 1 was required. In contrast, the table can be constructed in one hour, yet only two minutes were needed for pose estimation. Clearly, there is a time/space tradeoff here as well. While the table is currently stored as a matrix, there are more compact methods for representing the corresponding mapping, for example quadtrees, splines or perhaps even neural networks.

5.2. Pose Refinement

Determining an object's pose from the minimum number of features is inherently inaccurate, and the effect of measurement noise may be exaggerated by the nonlinear relationship between the measurements and transformation parameters. Naturally, using more data will give a better estimate. When estimating the pose from a set of point features, an explicit correspondence is made between the features and the generating curves and surfaces. This hypothesized match can be carried to the surrounding edge chains which are the projection of either occluding contours or edges.

In [30, 49], it was shown that the relationship between image measurements $\hat{p}$ and object pose T can be expressed by an implicit algebraic equation $C(\hat{p}, T) = 0$ constructed using resultants. Our objective is to determine the transformation T from a set of image points $\hat{p}$ on the occluding contour.

Consider a point $\hat{p}$ which is the image of an edge defined by equation (8). The ray from the focal point to $\hat{p}$ can be written as in equation (6) and substituted into (8) yielding:

$$\begin{cases} P_1(\mathbf{p} + \mu\mathbf{v}) = 0, \\ P_2(\mathbf{p} + \mu\mathbf{v}) = 0. \end{cases} \tag{20}$$

where $\mathbf{p}$ and $\mathbf{v}$ both depend on the pose parameters T and the image measurement $\hat{p}$. The depth μ can be eliminated by computing the resultant of the two equations, yielding a new constraint:

$$C(\hat{p}, T) = 0. \tag{21}$$

Given a set of n contour data points $\hat{p}_i$, the pose T can be recovered by minimizing:

$$\sum_{i=1}^{n} C^2(\hat{p}_i, T) \tag{22}$$

with respect to the pose parameters. When the contours arise from more than one edge or surface as in the example shown in figure 10, the objective function can be simply taken as the sum of individual objective functions given by (22). Because of its simplicity, we have used this function in our implementation. However for greater accuracy, the algebraic distance could be replaced by the exact image distance as in section 4.1.2 or an approximation as in Taubin's formulation [61].

5.3. Implementation and Results

We have implemented pose estimation from features followed by the refinement algorithm. In our experiments with real images, image contour inflections have been used rather than junctions because they are easier to accurately locate in images. However, there has been recent progress in locating junctions as well [45, 47]. The constraint equations are automatically generated from the equations defining an object model using the REDUCE computer algebra system. Both methods described in section 5.1.5 have been used to solve the systems of polynomials; the table method has proven effective in all of our experiments and is significantly faster. Pose refinement from both occluding contours [30, 49] and edges has been implemented; the Levenberg-Marquardt algorithm is used to minimize (22)[52]. In our experiments, the Canny edge detector [10] was applied to the image, edge chains were hand selected, and inflection points were found by fitting cubics to the edgels. Strategies for hypothesizing image–model correspondences such as interpretation trees have not yet been implemented.

Figure 9(a) shows an image of a cylinder with a cylindrical notch and two inflection points. The intersection curve is a fourth-degree algebraic curve; while the binormal of the intersection of two quadrics is generally a fifth degree vector equation in $\mathbf{p}$, it simplifies to a third-degree equation in this case. The arrows indicate the direction of the measured curve normal. The edge constraints of section 5.1.3 lead to a system of six polynomial equations with 1920 roots. However, only two real roots are unique, and figures 9(b) and 9(c) show the corresponding poses. Clearly the pose in figure 9(c) could be easily discounted with additional

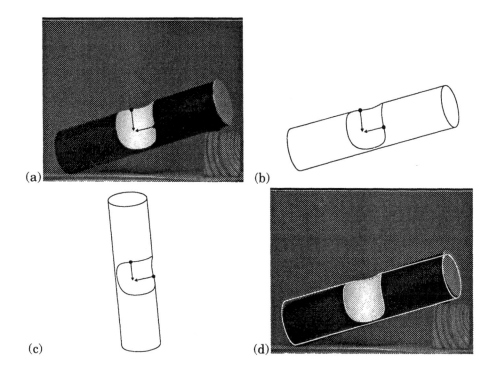

Figure 9. Pose estimation from two inflection points: (a) The original image and the two inflections. (b) & (c) Two recovered poses from the inflections (Note the scale difference in (c)). (d) The refined pose.

image information. As shown in figure 9(d), the pose is further refined using only the edgels from the projection of the intersection curve.

Figure 10.a shows an image of a sphere with two cylindrical notches of different radii and nonintersecting axes (i.e. there is no symmetry). The detected inflections and the curve normals are indicated. Again, the intersection curve is fourth-degree, and the binormals are third-degree. After solving the system of equations described in section 5.1.3 using the table method, six real, non-degenerate solutions were found, and the corresponding poses are shown in figures 10(c–h). The correct pose of 10(c) is further refined and shown in figure 10(b).

6. Conclusions

In this chapter, we have described components of a system for recognizing curved objects in monocular images. The key to our approach is to represent objects by algebraic surfaces, which allows us to formulate the geometric constraints relating 3D models to images as systems of polynomials equations. We have attacked

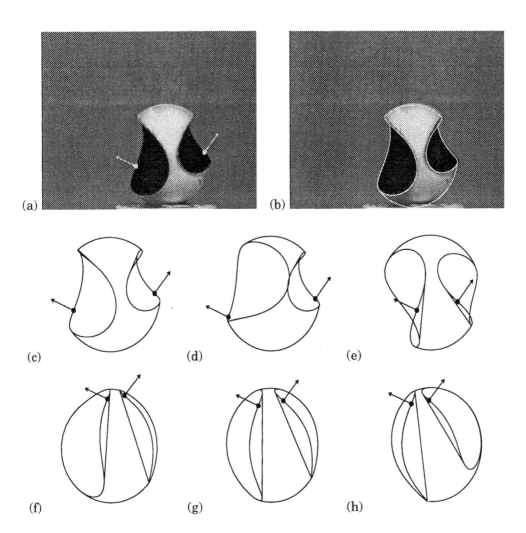

Figure 10. Pose estimation from two inflection points on two curves. (a) Image with two detected inflections and the indicated curve normals. (b) Recovered pose after refinement. (c)-(h) The six poses that are consistent with the two inflections and their normals. They are drawn without hidden line removal and in the correct orientation, but with the scale and translation adjusted to show the pose in detail.

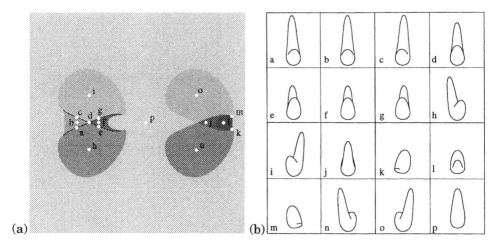

(a)   (b)

Figure 11. Aspect graph of a bean-shaped object similar to the one shown in Fig. 5. (a) The view sphere, drawn in cylindrical coordinates, is partitioned into regions whose stable views are shown in (b).

automatic model construction by developing novel methods for fitting algebraic surfaces to range data and calibrated sequences of video images. We have presented an efficient global method for estimating an object's pose from the minimum number of points features and a local method for refining this estimate using the entire image contour.

Efficient matching strategies are still missing. We are exploring the use of aspect graphs [29] for predicting situations in which certain groups of features are simultaneously observable. The aspect graph is a viewer-centered representation which enumerates all qualitatively distinct line-drawings of an object by partioning the space of viewpoints according to a catalogue of visual events [28]. Figure 11 shows the aspect graph of a bean-shaped object similar to the one in figure 5. See [28, 50] for a discussion of an implemented algorithm for constructing the aspect graph of an algebraic surface.

Beyond that, much work remains to be done. We have not yet addressed the critical issue of segmentation (note that perfect segmentation is not critical for us since we do not use intermediate image representations). Nor have we considered the problem of indexing a large model data base. Certainly classes of objects can be modelled by surfaces whose coefficients are functions of a few shape parameters, and some of our algorithms such as pose estimation can be extended to these models [30]. Finally, many of the ideas presented can be used for other problems in computer vision, and we have just begun to apply these to medical imaging.

REFERENCES

1   N. Ahuja and J. Veenstra. Generating octrees from object silhouettes in ortho-
    graphic views. *IEEE Trans. Patt. Anal. Mach. Intell.*, 11(2):137–149, 1989.
2   D. Arnon. Topologically reliable display of algebraic curves. *Computer Graph-
    ics*, 17(3):219–227, July 1983.
3   C. Bajaj. Electronic skeletons: modeling skeletal structures with piecewise
    algebraic surfaces. In *SPIE Conference on Curves and Surfaces in Computer
    Vision and Graphics II*, pages 230–237, Boston, MA, November 1991.
4   R. Bajcsy and F. Solina. Three-dimensional object representation revisited. In
    Proc. Int. Conf. Comp. Vision, pages 231–240, London, U.K., June 1987.
5   A. Barr. Superquadrics and angle preserving transformations. *IEEE Computer
    Graphics and Applications*, 1:11–23, January 1981.
6   B. Baumgart. Geometric modeling for computer vision. Technical Report AIM-
    249, Stanford University, 1974. Ph.D. Thesis. Department of Computer Science.
7   T. Binford. Visual perception by computer. In *Proc. IEEE Conference on Systems
    and Control*, 1971.
8   J. Brady. Criteria for representations of shape. In J. Beck, B. Hope, and
    A. Rosenfeld, editors, *Human and Machine Vision*, pages 39–84. Academic
    Press, 1983.
9   R. Brooks. Symbolic reasoning among 3-D models and 2-D images. *Artificial
    Intelligence*, 17(1-3):285–348, 1981.
10 J. Canny. A computational approach to edge detection. *IEEE Trans. Patt. Anal.
    Mach. Intell.*, 8(6):679–698, Nov. 1986.
11 M. Casale. Free-form solid modelling with trimmed surface patches. *IEEE
    Computer Graphics and Applications*, 7(1):33–43, January 1987.
12 C. Connolly and J. Stenstrom. 3D scene reconstruction from multiple intensity
    images. In *Proc. IEEE Workshop on Interpretation of 3D Scenes*, pages 124–130,
    Austin, TX, November 1989.
13 M. Dhome, M. Richetin, J. Lapreste, and G. Rives. Determination of the attitude
    of 3-d objects from a single perspective view. *IEEE Trans. Patt. Anal. Mach.
    Intell.*, 11:1265–1278, 1989.
14 A. Dixon. The eliminant of three quantics in two independent variables. *Proc.
    London Mathematical Society, Series 2*, 7:49–69, 1908.
15 R. Farouki. The characterization of parametric surface sections. *Comp. Vis.
    Graph. Im. Proc.*, 33:209–236, 1986.
16 O. Faugeras and M. Hebert. The representation, recognition, and locating of
    3-D objects. *International Journal of Robotics Research*, 5(3):27–52, Fall 1986.
17 P. Flynn and A. Jain. 3D object recognition using invariant feature indexing of
    interpretation tables. *CVGIP: Image Understanding*, 55(2):119–129, 1992.
18 R. Goldman and T. Sederberg. Some applications of resultants to problems in
    computational geometry. *The Visual Computer*, 1:101–107, 1985.
19 W. Grimson and T. Lozano-Pérez. Localizing overlapping parts by searching the
    interpretation tree. *IEEE Trans. Patt. Anal. Mach. Intell.*, 9(4):469–482, 1987.
20 W. E. L. Grimson. *Object Recognition by Computer: The Role of Geometric*

Constraints. MIT Press, 1990.

21 A. Gross and T. Boult. Error of fit measures for recovering parametric solids. In *Proc. Int. Conf. Comp. Vision*, pages 690–694, Tampa, FL, December 1988.

22 M. Hebert and T. Kanade. The 3D profile method for object recognition. In *Proc. IEEE Conf. Comp. Vision Patt. Recog.*, pages 458–463, San Francisco, CA, June 1985.

23 R. Horaud. New methods for matching 3-D objects with single perspective views. *IEEE Trans. Patt. Anal. Mach. Intell.*, 9(3):401–412, 1987.

24 D. Huttenlocher and S. Ullman. Object recognition using alignment. In *Proc. Int. Conf. Comp. Vision*, pages 102–111, London, U.K., June 1987.

25 K. Ikeuchi and T. Kanade. Automatic generation of object recognition programs. *Proceedings of the IEEE*, 76(8):1016–35, August 1988.

26 J. Kajiya. Ray tracing parametric patches. *Computer Graphics*, 16:245–254, July 1982.

27 D. Keren, D. Cooper, and J. Subrahmonia. Describing complicated objects by implicit polynomials. Technical Report LEMS-93, Brown University, 1991.

28 J. Koenderink. *Solid Shape*. MIT Press, Cambridge, MA, 1990.

29 J. Koenderink and A. V. Doorn. The singularities of the visual mapping. *Biological Cybernetics*, 24:51–59, 1976.

30 D. Kriegman and J. Ponce. On recognizing and positioning curved 3D objects from image contours. *IEEE Trans. Patt. Anal. Mach. Intell.*, 12(12):1127–1137, December 1990.

31 D. Kriegman and J. Ponce. Geometric modelling for computer vision. In *SPIE Conference on Curves and Surfaces in Computer Vision and Graphics II*, pages 250–260, Boston, MA, November 1991.

32 D. Kriegman and J. Ponce. A new curve tracing algorithm and some applications. In P. Laurent, A. L. Méhauté, and L. Schumaker, editors, *Curves and Surfaces*, pages 267–270. Academic Press, New York, 1991.

33 D. Kriegman, B. Vijayakumar, and J. Ponce. Strategies and constraints for recognizing and locating curved 3D objects from monocular image features. In *Proc. European Conf. Comp. Vision*, pages 829–833, 1992.

34 D. Lowe. Three-dimensional object recognition from single two-dimensional images. *Artificial Intelligence*, 31(3):355–395, 1987.

35 D. Lowe. The viewpoint consistency constraint. *Int. J. of Comp. Vision*, 1(1):57–72, 1987.

36 D. Luenberger. *Linear and nonlinear programming*. Addison-Wesley, 1984. Second edition.

37 F. Macaulay. *The Algebraic Theory of Modular Systems*. Cambridge University Press, 1916.

38 J. Malik. Interpreting line drawings of curved objects. *Int. J. of Comp. Vision*, 1(1):73–103, 1987.

39 D. Manocha and J. Canny. Algorithm for implicitizing rational parametric surfaces. In *IMA Conf. on Mathematics of Surfaces*, Bath, 1990. To appear in Computer Aided Geometric Design.

40 D. Marr and K. Nishihara. Representation and recognition of the spatial orga-

nization of three-dimensional shapes. *Proc. Royal Society, London*, B-200:269–294, 1978.

41 J. Miller. Sculptured surfaces in solid models: issues and alternative approaches. *IEEE Computer Graphics and Applications*, 6(12):37–48, December 1986.

42 D. Moore and J. Warren. Approximation of dense scattered data using algebraic surfaces. In 24th *Hawaii Intl. Conference on System Sciences*, pages 681–690, Hauai, Hawaii, 1991.

43 A. Morgan. *Solving Polynomial Systems using Continuation for Engineering and Scientific Problems*. Prentice Hall, Englewood Cliffs, NJ, 1987.

44 J. Mundy and A. Zisserman. *Geometric Invariance in Computer Vision*. MIT Press, Cambridge, Mass., 1992.

45 J. Noble. Finding half boundaries and junctions in images. *Image and Vision Computing*, 10(4):219–232, 1992.

46 A. Pentland. Perceptual organization and the representation of natural form. *Artificial Intelligence*, 28:293–331, 1986.

47 P. Perona. Steerable-scalable kernels for edge detection and junction analysis. In *Proc. European Conf. Comp. Vision*, pages 3–18. Springer-Verlag, 1992.

48 J. Ponce, D. Chelberg, and W. Mann. Invariant properties of straight homogeneous generalized cylinders and their contours. *IEEE Trans. Patt. Anal. Mach. Intell.*, 11(9):951–966, September 1989.

49 J. Ponce, A. Hoogs, and D. Kriegman. On using CAD models to compute the pose of curved 3D objects. *CVGIP: Image Understanding*, 55(2):184–197, 1992. Special Issue on Directions in "CAD-based" Vision.

50 J. Ponce, S. Petitjean, and D. Kriegman. Computing exact aspect graphs of curved objects: Algebraic surfaces. In G. Sandini, editor, *Proc. European Conf. Comp. Vision*, volume 588 of *Lecture Notes in Computer Science*, pages 599–614. Springer-Verlag, 1992.

51 V. Pratt. Direct least-squares fitting of algebraic surfaces. *Computer Graphics*, 21:145–152, 1987.

52 W. Press, B.Flannery, S. Teukolsky, and W. Vetterling. *Numerical Recipes in C*. Cambridge University Press, 1988.

53 L. Roberts. Machine perception of three-dimensional solids. In J. T. et al., editor, *Optical and Electro-Optical Information Processing*, pages 159–197. MIT Press, Cambridge, 1965.

54 G. Salmon. *Modern Higher Algebra*. Hodges, Smith, and Co., Dublin, 1866.

55 W. Seales and C. Dyer. Viewpoint from occluding contour. *CVGIP: Image Understanding*, 55(2):198–211, 1992. Special Issue on Directions in "CAD-based" Vision.

56 S. Shafer. *Shadows and Silhouettes in Computer Vision*. Kluwer Academic Publishers, 1985.

57 S. Srivastava and N. Ahuja. Octree generation from object silhouettes in perspective views. *Comp. Vis. Graph. Im. Proc.*, 49(1):68–84, 1990.

58 G. Strang. *Linear algebra and its applications*. Academic Press, Inc., 1980. Second edition.

59 J. Subrahmonia, D. Cooper, and D. Keren. Reliable object recognition using high-dimensional implicit polynomials for 2d curves and 3d surfaces. Technical Report LEMS-94, Brown University, 1991.

60 K. Sugihara. An algebraic approach to the shape-from-image-problem. *Artificial Intelligence*, 23:59–95, 1984.

61 G. Taubin. Estimation of planar curves, surfaces and nonplanar space curves defined by implicit equations, with applications to edge and range image segmentation. *IEEE Trans. Patt. Anal. Mach. Intell.*, 13(11):1115–1138, 1990.

62 G. Taubin, F. Cukierman, S. Sullivan, J. Ponce, and D. Kriegman. Parameterizing and fitting bounded algebraic curves and surfaces. In *Proc. IEEE Conf. Comp. Vision Patt. Recog.*, pages 103–108, Champaign, IL, June 1992. An extended version appears as IBM Computer Science Research report No. RC17659 (#77766), November 1991.

63 R. Walker. *Algebraic Curves*. Princeton University Press, 1950.

Three-Dimensional Object Recognition Systems
A.K. Jain and P.J. Flynn (Editors)
© 1993 Elsevier Science Publishers B.V. All rights reserved.

Structural Indexing: Efficient Three Dimensional Object Recognition[1]

Fridtjof Stein and Gérard Medioni[a]

[a] Institute for Robotics and Intelligent Systems
Department of EE-Systems
University of Southern California
Los Angeles, California 90089 USA

Abstract

We present an approach for the recognition of multiple three dimensional object models from three dimensional scene data. We work on dense data, but neither the models nor the scene data have to be complete. We are addressing the problem in a realistic environment: the viewpoint is arbitrary, the objects vary widely in complexity, and we make no assumptions about the structure of the surface. Our approach is novel in that it uses two different types of primitives for matching: small surface patches, where differential properties can be reliably computed, and lines corresponding to depth or orientation discontinuities. These are represented by *splashes* and 3D curves respectively. We show how both of these primitives can be encoded by a set of super segments, consisting of connected linear segments. These super segments are entered into a table, and provide the essential mechanism for fast retrieval and matching. We address in detail the issues of robustness and stability of our features. The acquisition of the three dimensional models is performed automatically by computing splashes in highly structured areas of the objects, and by using boundary and surface edges for the generation of 3D curves. For every model, all features are recorded in a data base. The scene is screened for highly structured areas, and splashes are computed in these areas and encoded. 3D Curves, corresponding to depth or orientation discontinuities are also encoded. These features are used to retrieve hypotheses from the data base. Clusters of mutually consistent hypotheses represent instances of models. The precise pose of a model instance in the scene is found by applying a least squares match on all corresponding features. We present results with our current system TOSS (Three dimensional Object recognition based on Super Segments) and discuss further extensions.

[1]This research was supported by the Advanced Research Projects Agency of the Department of Defense and was monitored by the Air Force Office of Scientific Research under Contract No. F49620-90-C-0078.

1. Introduction

We present an object recognition system which is able to match general three dimensional objects from partial 3D data in an efficient way by using a method called *structural indexing*. By talking about "general objects" we make very few restrictive assumptions about their shape, so we only exclude statistically defined shapes (e.g. foams) and crumpled objects (e.g. fractals). Matching and recognizing in an "efficient way" is based on a fast indexing and retrieval scheme that has a complexity which grows as $O(kN)$ when N is the number of models, and $k < 1$.

Representing a three dimensional solid object is either possible by using a surface or a volumetric description. Volumetric descriptions from a single view require a difficult inference step to compensate for the unseen part, so we will use descriptions based on visible surfaces instead. The task of object recognition involves identifying a correspondence between a part of one range image and a part of another range image with a particular view of a known object. This requires the ability to match one feature of one range image against a feature of another range image. A feature can be either a surface patch or a general 3D curve. The question is: "How can we represent such features so that they can be matched in an efficient way?"

Reviewing the existing systems, none so far (known to the authors) is able to represent, match, and recognize *general* three dimensional objects. An excellent overview regarding the problem of matching free-formed surfaces can be found in [1]. Most object recognition systems to date either rely on exact, CAD-like models, or make restrictive assumptions on the possible shape of the surface patches. For other related work see Grimson and Lozano Pérez work on interpretation trees [9, 10], Bhanu's work on multiple-view range images [2], Horaud and Bolles [16, 3] 3DPO system, the approach of Faugeras and Hebert in [7], Ikeuchi's method for object recognition in bin-picking tasks [11], Fan's algorithm in [5, 6], the work which is conceptually closest to our approach by Radack and Badler [18], and the 3D-POLY system of Chen and Kak [4]. In the past many systems were developed which can only work with certain objects, such as polygonal shapes, solids of revolution or generalized cylinders. In contrast, we believe that our proposed system TOSS (Three dimensional Object recognition based on Super Segments) is able to recognize rigid objects, whose shapes are not constrained by any simplifying assumptions. Our algorithm uses a combined representation, which captures information about both smooth patches and discontinuity lines.

This chapter is organized as follows: Section 2 presents our approach to recognize objects in range imagery. We focus on the choice of our two features. One, the *3D Curve* is used to represent edges, and the other, the *splash* is used to represent surface patches. In section 3 we describe how we use a table to store our features and retrieve them efficiently for hypotheses generation. Finally in section 4 we present some results on real data, and provide some concluding remarks.

2. Structural Indexing in 3D

In recent years, object recognition in 3D has been either performed based on boundary and edge information (see [1, 13]) or by using surface descriptions (see for example [1, 5, 8, 17]). This creates problems when edges are not well defined, or when the objects cannot be segmented into stable elementary patches.

We present a system which combines both approaches with the following strategy: "Use for the recognition task whatever information is available": We extract edges corresponding to depth and orientation discontinuities, and use them as primitives. These are not sufficient, however, to represent smooth, free form surfaces. So we also compute differential properties, called *splashes*, in smooth areas. We describe both of theses primitives in the following subsections.

2.1 The 3D Curve

For some objects, such as polyhedra, it is natural to use a representation based on edges. For this reason, we extract 3D curves likely to correspond to depth and orientation discontinuities. Edges are in general broken (see e.g. in Figure 1 the surface discontinuities between the fuselage and the wing). We take such inherent limitations into account. Our effort is not to develop a system which can only deal with *perfect* edges. We want to use whatever data current state-of-the-art edge detectors can generate. When we get non-invariant edges (such as limbs, which are viewer-dependent) we treat them just like all other edges. When they are matched against scene edges, they might generate wrong hypotheses, which are then discarded in the verification step. The most stable edges leading to the best matches are the edges which correspond to discontinuities of depth and surface (such as the boundary of the wings in Figure 1).

2.1.1 Curve Extraction

So far we have not discussed the curve extraction process (see Figure 1). We compute the edges of the range image. We use the edge detection algorithm proposed by Saint-Marc *et al.* [5, 6, 19]. It detects surface and depth orientation discontinuities. These features are inferred by examining the zero-crossings and extremal values of the surface curvature. The method uses adaptive smoothing to smooth a range image, which preserves discontinuities and facilitates their detection. This is achieved by repeatedly convolving the image with a very small averaging filter whose weights are a function of the local gradient estimate. In order to extract curvature extrema and zero-crossings, instead of smoothing the original range image R, the original derivatives $P = \frac{\partial R}{\partial x}$ and $Q = \frac{\partial R}{\partial x}$ are computed first. Then the images P and Q are repeatedly smoothed. Finally, the curvature values are computed from the smoothed images P and Q. The curvature extrema and zero-crossings are extracted using hysteresis.

2.1.2. Encoding a 3D curve

Our representation of a general 3D curve is based on a polygonal approximation. We do not rely on any specific feature detection algorithm and we do not explicitly handle distinguished points such as corners or inflection points. Curvature and

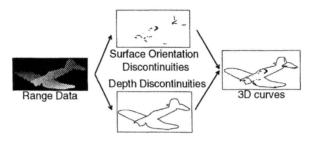

Figure 1. Generation of 3D Curves

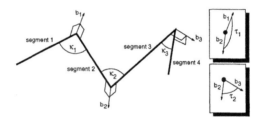

Figure 2. Example of 3D Super Segment of Cardinality 4

torsion are the most important features of a general 3D curve. They are invariant with respect to rotation and translation. By using a polygonal approximation we lose most of the curvature and torsion information, but we approximate it by computing the "curvature" (κ_i) and "torsion" (τ_j) angles between consecutive line segments (see Figure 2).

Obviously, the polygonal approximation for a curve is not unique. Therefore, for the purpose of robustness, we use *several* polygonal approximations with different line fitting tolerances. Since we want to handle occlusion, we do not expect to obtain complete curves in our scenes, but only portions of them. On the other hand, individual segments are too local to be useful as matching primitives. Grouping a fixed number of adjacent segments provides us with our first basic features, the 3D super segments. The 3D super segment is an extension to the 2D super segment which we used for recognition of flat objects in [21]. In accordance to Figure 2, 3D super segments are characterized by their cardinality (number of segments), curvature angles (between consecutive segments), and torsion angles (between consecutive binormals).

As mentioned before, we are mainly interested in the curvature and torsion information implicitly captured by the 3D super segment curvature and torsion angles. This is the reason why we use them to encode a 3D super segment. The curvature (κ_i) and torsion (τ_i) angles are defined in the following way:

$$\kappa_i = \arccos \frac{s_{i+1} \cdot s_i}{|s_{i+1}||s_i|} \quad \text{and} \quad \tau_i = \arccos(b_i \cdot b_{i+1})$$

with the binormals

$$b_i = \frac{s_{i+1} \times s_i}{|s_{i+1}||s_i|}$$

and s_i is the ith segment of the 3D super segment. To encode a 3D super segment ss with cardinality n we use a simple encoding scheme. The list of the quantized curvature and torsion angles values is the code of the 3D super segment ss:

$$\begin{aligned} \text{Code}(ss) \quad = (\quad & \text{Quant}(\kappa_1), \text{Quant}(\kappa_2), ... \text{Quant}(\kappa_{n-1}), \\ & \text{Quant}(\tau_1), \text{Quant}(\tau_2), ... \text{Quant}(\tau_{n-2})) \end{aligned}$$

All the encoded 3D super segments serve as keys into a table (the data base), where we record the corresponding 3D super segments as entries, as explained later.

We now address the following issues:

Polygonal Approximation Because a polygonal approximation with a fixed tolerance is in general not stable, we use multiple line fitting tolerances. A complete analysis is given later. The 3D super segments are built by grouping adjacent segments. Because we cannot assign a specific direction to a 3D super segment, we use both directions for our representation.

Choice of Cardinality Which cardinalities should we use to define the link length of the 3D super segments? To use a fixed cardinality is possible, but it reduces the flexibility of the matching process. But when we have long 3D super segment matches, why not use them? Therefore we compute *all* possible cardinalities: that means that an open curve which is approximated by eight linear segments is represented by two 3D super segments of cardinality 8 (one for each direction), four 3D super segments of cardinality 7, six 3D super segments of cardinality 6, and so on. Higher cardinalities of matched 3D super segments increase the probability of having a *good* match.

2.2. The Splash
2.2.1 Basic Idea

For some objects, however, such as smooth objects, it is impossible to use edges for the representation. Therefore we come up with a new representation based on small surface patches where we can compute differential properties in a reliable way.

Extending the super segment idea is not straightforward. The polygonal approximation of a curve has a property which is crucial, but which is not extendable to higher dimensions: the well defined order of the neighborhood of a linear segment. Every segment on a polygon has two adjacent neighbor segments. Based on this fact, super segments can be generated by grouping adjacent segments together. In surface approximations, however, this ordered neighborhood property does not exist. Polygonal or other segmentations of a surface (or volume) lead in general to patches which can have any number and order of neighbor patches. This is a reason why we decided not to go the path of a polygonal (or higher order) surface

358

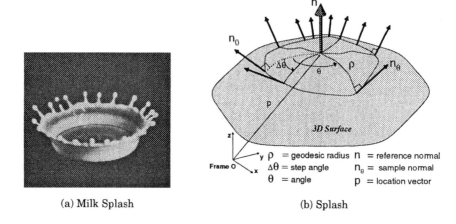

(a) Milk Splash　　　　　　　　　(b) Splash

Figure 3. Splashes

segmentation to obtain a representation for matching and recognition. What are the requirements that a representation for general three dimensional objects has to meet? We want the representation to be

1. translation invariant,

2. rotation invariant,

3. general, in that we do not have to make any assumptions about the shape of the object,

4. local enough, so that we can handle occlusion,

5. robust enough, so that we can handle noise.

In the following, we use lower case letters to describe vectors (n, p...), and upper case letters to describe coordinate frames (N, O...). The basic feature for representing a general surface patch is the *splash*. The name originates from the famous picture of Professor Edgerton (MIT), showing a milk drop falling into milk (see Figure 3 (a)). This picture bears a resemblance to the normals in our basic feature. A splash is best described by Figure 3 (b). At a given location p we determine the surface normal n. We call this normal the *reference normal* of a splash. A circular slice around n with the geodesic radius ρ is computed. Starting at an arbitrary point on this surface circle, a surface normal is determined at every point on the circle. In practice we walk around the reference normal with a θ angle (typically $1° \le \theta \le 15°$) and obtain a set of sample points on the surface circle. The normal at the angle θ is called n_θ. A *super splash* is composed of splashes with

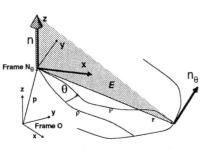

(a) Relationship between n and n_θ

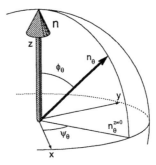

(b) Definition of ϕ_θ and ψ_θ

Figure 4. n and n_θ

different surface radii ρ_i with $i\epsilon\{1,\ldots m\}$, where m is the number of splashes in a super splash.

In other words, the splash is the representation of a surface patch by the Gaussian map in the vicinity of the center of the patch, mapping the tangent with respect to a geodesic distance ρ. One question that is often asked: why do we not use Gaussian curvature for the splash computation (which is invariant to rigid transformations)? Why do we use the tangent information? The answer is straightforward: the computation of curvature requires a higher order derivative than the tangent. This implies that the signal to noise ratio is lower for a curvature based representation than for a tangent based scheme. Therefore from the practical viewpoint we prefer a more reliable representation for the purpose of efficient matching.

We compute a normal in our system by approximating the environment of a normal with triangular patches of small sizes. Every triangle votes for a triangle normal. The average of the three closest triangle normals is the surface normal. This is a very rough method, but the results were always good enough for our approach.

The frame N_θ (see Figure 4 (a)) is defined in the following way:

1. The surface normal n is the z axis.

2. At every location of n_θ, the location of the reference normal p and the tip of the reference normal $n + p$ describe a plane E. The x axis is defined as the vector which is perpendicular to n and lies in the plane E. Furthermore the angle between the x axis and a vector r which is defined between the origin of Frame N_θ and the location of n_θ has to be in the interval $[-90°, 90°]$.

3. The y axis is perpendicular to the x and the z axis in a right handed coordinate system.

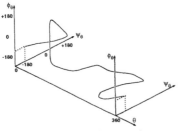
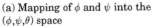
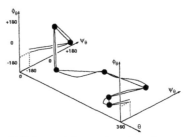

(a) Mapping of ϕ and ψ into the (ϕ,ψ,θ) space

(b) Polygonal Approximation of the Mapping

Figure 5. Vector Mapping $\vec{v}(\theta)$

This frame has the property that the xy-plane always approximates the tangent plane of the surface in p. We represent n_θ in spherical coordinates: we compute the two angles ϕ_θ and ψ_θ:

$$\phi_\theta = \text{angle}(n, n_\theta) \quad \text{and} \quad \psi_\theta = \text{angle}(x, n_\theta^{z=0}).$$

For every sample point of a splash we obtain such a pair. Now we have a two dimensional mapping $\phi(\theta)$ and another one $\psi(\theta)$.

In an earlier implementation (see [22]) we used these two mappings in parallel for our algorithm. But it is in fact possible to combine these two mappings into one compact three dimensional vector mapping $\vec{v}(\theta) = \begin{pmatrix} \phi(\theta) \\ \psi(\theta) \end{pmatrix}$, which describes a curve in the 3D space (ϕ,ψ,θ). By doing so, we are able to use the representation for the general 3D Curve from Section 2.1 for the representation of the mappings of the splash. Drawing a mapping for ϕ and ψ with respect to θ results in a mapping illustrated in Figure 5(a). This mapping has the following properties:

1. Depending on where n_0 is, the mapping is shifted along the θ axis.

2. The mapping is periodic with respect to the θ axis.

3. The variation of the curve represents the structural change in the surface environment around the reference normal n.

   (a) For a splash on a sphere or a plane, the mapping is constant.

   (b) A creased surface results in a curved mapping.

4. Splashes which are located close to each other have a similar shaped mapping. By using the word *similar* we mean similarity in the sense, so that a human would classify them as "pretty much the same". That does not automatically imply that the pairwise difference results in small values (we discuss the issue of robustness in detail in [23]). To be able to compare two mappings, we therefore need a difference measure, introduced below.

2.2.2. Encoding

At this point we have reduced the original question "How do we capture the shape of a general surface patch into a representation?", into the much simpler question "How do we capture the shape of a mapping into a representation?". The solution is straightforward based on our three dimensional approach for representing a general 3D curve (see Section 2.1).

1. For all splashes of a model we compute their mappings. In Section 2.2.3, we discuss the selection of the locations of the splashes.

2. For each splash, the mapping is approximated by polygonal approximations (see Figure 5(b)). It is important to note that the mapping is periodic and therefore the polygon is closed. For the purpose of robustness we use multiple line fitting tolerances. Therefore we get a set of polygons for each mapping.

3. For every polygonal approximation we compute a 3D super segment. The start of the 3D super segment is defined at the point with the maximal distance of the θ axis. This corresponds to the point at which the sample normals have the strongest tilt with respect to the reference normal. If there is more than one global maximum we use one 3D super segment for each of the maxima. With this 3D super segment choice, we obtain rotational invariance in our representation. By starting all 3D super segments at the maximum of the approximation, two shifted polygons with the same shape produce the same 3D super segment.

4. All the obtained 3D super segments are encoded. The encoding works as described in section 2.1. As encodable attributes we take

   (a) the curvature and the torsion angles of a 3D super segment,

   (b) the maximum distance of the mapping from the θ axis,

   (c) the surface radius of the splash.

   Incorporated in the code of the angles of the 3D super segments is also the cardinality (number of segments) of the 3D super segments (by the number of angles). That avoids matching 3D super segments of different cardinality. The encoding of the maximal distance allows to distinguish between different curved surfaces of the same shape (e.g. two spherical surfaces with different sphere radii). The encoding of the radius avoids matches between splashes with different splash radii. In summary, the code for one splash mapping is:
   Code($splash\text{-}mapping$) =
   $$(\text{Quant}(\kappa_1), \text{Quant}(\kappa_2), ...\text{Quant}(\kappa_n),$$
   $$\text{Quant}(\tau_1), \text{Quant}(\tau_2), ...\text{Quant}(\tau_n), \text{Quant}(max), \text{Quant}(radius))$$
   where κ_i is the ith curvature angle of the 3D super segment, τ_i is the ith torsion angle of the 3D super segment, and n is the cardinality of the 3D super segment (n is implicitly encoded in the length of the code).

5. All the encoded 3D super segments serve as keys into a table (the data base), where we record the corresponding splashes as entries, as explained later.

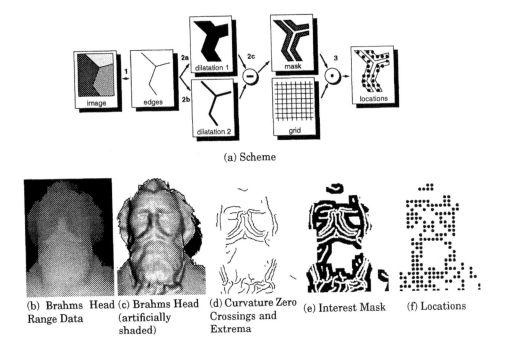

(a) Scheme

(b) Brahms Head Range Data (c) Brahms Head (artificially shaded) (d) Curvature Zero Crossings and Extrema (e) Interest Mask (f) Locations

Figure 6. Interest Operator Scheme on an Example

2.2.3. Interest Operator

One question remains open: at which locations of an object should we compute the splashes? The brute force answer would be: at every pixel (in a range image). A more realistic answer would include the observation that we will not get *structurally rich* splashes at every point, which lead to good and unambiguous matches. Splashes in flat areas result in 3D super segments with extremely low cardinality (e.g. a splash on a plane maps on a 3D super segment consisting of one segment which corresponds to a cardinality of one). Super segments with such low cardinalities are less descriptive than super segments with higher cardinalities, which represent high structured surface patches. Therefore to obtain good and unique matches we are interested in matches of structured patches and high cardinality. These can be found at or near points of high curvature. Our simple selection method works as follows (see Figure 6):

1. To compute the edges (surface and depth discontinuities) we use the algorithm mentioned in Section 2.1.

2. We want to position the splashes in areas where we can expect structured patches on *one* object. This property is not given on a boundary. A boundary edge typically

has the object as one neighborhood and other objects or background information as the other neighborhood. Therefore we use only the "inner object edges" and throw away the boundary edges.

3. For positioning the splashes we are interested in areas around areas of high curvature. Placing a splash on a high curvature point has the disadvantage of an unreliable reference normal. A reliable reference normal is important for a stable splash. Nevertheless we want to capture the structure of the edges in the splash. Therefore the best place for a splash is in the neighborhood of an edge. We get this area in three steps:

   (a) We dilate the edge image by replacing every pixel on the edges by a disc of a certain radius (e.g. $r_1 = 8$ pixels). The resulting image is called *dilatation 1*.

   (b) We dilate the edge image with another radius (e.g. $r_2 = 3$ pixels with $r_1 > r_2$). The resulting image is called *dilatation 2*.

   (c) The subtraction of dilatation 1 and dilatation 2 gives us a mask. This mask describes an area with the above described characteristics. Points in this mask are no high curvature points, but they are close to edges.

4. We compute a grid of splashes on the range image with respect to this mask.

This is obviously not the only way to choose the locations of splashes, but as we will see in the result section, this simple method works quite well. Therefore we have not emphasized this research direction.

3. Recognition

3.1. Object Representation

As mentioned in the previous sections, we want to represent our model (or scene) with super segments for curve representation and splashes for surface patch representation. We want the representation to be compact, fast accessible, and the storage of multiple objects should be possible. We chose for these reasons a table which is implemented as a hash table (more about hashing see [24]). A hash table allows efficient storage (only pointers are recorded), the indexing scheme allows fast access and different features with the same keys can be stored in cellar like buckets.

The representation of an object consists of the following steps (see Figure 7):

1. Compute the features $^m F_i$ of model m with respect to the algorithms described in Section 2.1 for super segments and Section 2.2 for splashes.

2. Encode the features:

   • Encode the curvature and torsion angles for the 3D super segments (Section 2.1).

   • Encode the curvature angles and the other attributes of the mappings of the splashes (Section 2.2.1).

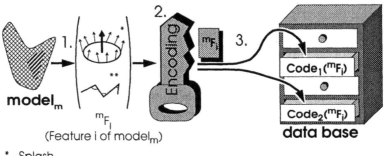

Splash
** Super Segment

Figure 7. Object Representation

3. For every feature ^mF_i several codes $Code_j(^mF_i)$ are computed. They are related to the different line fitting tolerances. Every code of every feature serves as a key for an entry in a table (the data base) where we record the feature.

When we build the data base for more than one object we perform the three steps for every object. The table (data base) grows in size with the number of recorded models. This process of building the data base can be done off line.

3.2. Hypotheses Generation
3.3. Candidate Retrieval

The task of hypotheses generation is the process to establish correspondences between features of stored models and features of the scene. This process results in a set of matching hypotheses which consist of *good* and *false* matches. To separate the good hypotheses from all hypotheses we have to find consistent clusters; this process is discussed in Section 3.5. Our main interest for the candidate retrieval lies in the discriminative power of the hypotheses generation itself. By using indexing, we gain a lot of this power. Several systems of the past (see Section 1) use in this respect very weak features like points or edges. This leaves all the discriminative work to the verifying step. We believe that our indexing mechanism reduces the ratio of false hypotheses to good hypotheses tremendously. One positive side effect is that, in our experiments, models which were not in the scene and therefore provided only false matches had very few hypotheses and could be excluded fairly fast by the verification.

The hypotheses generation consists of the following steps:

1. The scene is preprocessed to generate all the features F_k (splashes and super segments) as explained in Section 3.1.

2. The features are encoded.

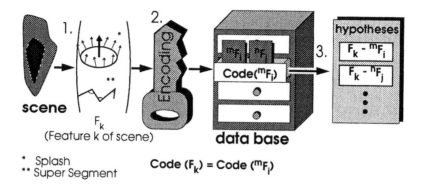

Figure 8. Hypotheses Generation

3. The encoded features are used to retrieve the candidate hypotheses between the features of model and scene.

3.4. Quantization and Cardinality

A crucial parameter for the hypotheses generation is the quantization size for the encoding of the features. Two features match, when both keys are exactly the same. This means that all the pairwise values have to fall into the same quantization intervals. Several questions have to be raised:

- Which quantization (interval size) is the best?

- Should we use different quantizations for different cardinalities?

- Which cardinalities occur in real data and with which frequency?

To answer the first two questions we have to look at the probability of a match, given a certain key length n. We assume equal quantization for all keys. Suppose the range is quantized into intervals of the the same size q. Each value, v, is then assigned a key based on which interval v falls into. If the value v is corrupted by a random additive term bounded by ε, the probability that $v + \varepsilon$ is assigned the same key as v is simply

$$p(k|v) = 1 - p(k-1|v+\varepsilon) - p(k+1|v-\varepsilon) = 1 - \frac{\varepsilon}{q}.$$

When we have a vector of such values v, of length n, the probability that the corrupted vector is assigned the same entry as the original one is

$$p^n(k|v) = (1 - \frac{\varepsilon}{q})^n.$$

Looking at this equation, it is obvious that the same quantization for different keys increases the probability for matches of small cardinality and decreases the probability for matches of large cardinality. In our implementation we try to counteract

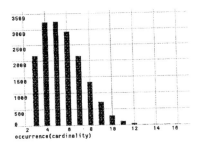

occurrence(cardinality)

Figure 9. Occurrence Distribution with respect to Cardinality

this effect by using larger interval sizes for larger cardinalities. Typical values are (only for the quantization of the curvature and torsion angles):

cardinality	3	4	5	6	7	8	...
number of keys	6	8	10	12	14	16	...
interval size	30	40	45	60	60	90	90

For extremely noisy data we use slightly larger values, for data with little noise we use smaller values.

What kind of splashes do occur in typical range data? Statistics for the three composers (see the results section) and the combined scene gives us the histogram in figure 9. From our experience we get the best matches from the cardinalities 3 to 7. We believe that this is mainly due to the lower probability of matches for larger cardinalities in the equation discussed above.

3.5. Verification

The task of the verification is to distinguish *good* from *bad* hypotheses. Good hypotheses correspond to true matches, bad hypotheses correspond to wrong matches. Good hypotheses have the following properties:

- they correspond to a rigid transformation,
- they can be grouped in geometrically consistent clusters.

Therefore, the verification stage consists of the following steps:

1. We compute all possible matches for the features of the scene with the model features to generate multiple hypotheses. We remove the hypotheses which do not represent a rigid transformation. This can be done for every hypothesis by computing a least squares match between the model feature and the scene feature. The determinant of the resulting rotation matrix should be approximately 1.0 to represent a rigid transformation. Next, we divide the resulting n hypotheses $H = \{h_1, h_2, ... h_n\}$ according to which model the model feature of the hypothesis votes for. We store these into a correspondence table where we have the models m_i as keys and the i_k hypotheses $H_i = \{{}^i h_1, {}^i h_2, ... {}^i h_{i_k}\}$ (with $H_i \subseteq H$) as entries (see Figure 10).

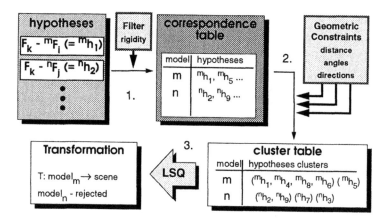

Figure 10. Verification

2. The next step is the formation of consistent clusters. For every model m_i we have to check which hypotheses $^i h_x$ and $^i h_y$ with $^i h_x \neq^i h_y$ are consistent with each other. In theory one matched feature is enough to establish the complete transformation between model and scene. In practice we have to consider several aspects:

- The splash is very local.
- Due to this locality, noise has a lot of influence on the transformation.

Therefore we view, for the geometric analysis, a hypothesis as a match equivalent to a point match. In practice we use the locations of the splashes as the matching point pair. And because three non-collinear point matches define a unique rigid transformation in three dimensions, we adopt the criterion that three consistent hypotheses are sufficient to instantiate a model in the scene. We do not check every hypothesis against every other, instead, if we have three consistent hypotheses $C = \{^i h_r,^i h_s,^i h_t\}$ with $C \subseteq H_i$ for one model m_i, we examine the remaining hypotheses in $H_i \setminus C$ and collect those that are consistent with at least one of the selected three in C. When we have found one instance, represented by $I = C \cup F$, with F the additional found consistent hypotheses, we try to find more instances in the remaining hypotheses $H_i \setminus I$.

But what is meant by consistency? We use the powerful geometrical constraints (distance, direction and angle) introduced by Grimson and Lozano-Pérez [9, 10] to prune efficiently their interpretation trees, and build our clusters. For more detailed discussion of the geometrical constraints, see the appendices in [23]. In the three dimensional domain, these three constraints define the attitude of one feature relative to another since it specifies the five degrees of freedom (three translational for the position and two rotational for the orientation).

3. After this grouping of hypotheses into clusters, we can compute the transformation from the model coordinates to the scene coordinates by applying a least squares calculation on all the matching features. Because of noise, we get in general a good first guess for the transformation but not an exact match. A second least squares match on corresponding corners or segments can refine the result.

3.6. Complexity Analysis

A detailed complexity analysis can be found in [23]. To summarize, the practical complexity of our system is

$$O(n) \leq O_{recognition} \leq O(n^2 \cdot m^3)$$

where n is the number of features in the scene and m the number of models in the data base. In the case of well distinguishable models, such as the first two examples in the results section, the complexity comes close to the above discussed *best case*. An example where the system slows down is shown in the third example of the results section, which presents the results of a cluttered scene, composed of three composer busts. The system detects the correct models and computes the correct locations, but due to noisy data and similar features, the discriminative power is smaller, and the overall recognition process is slower than in the other examples.

4. Results

The recognition mechanism for general three dimensional objects is now illustrated with real data examples. For the presentation of the range data we always display the artificially shaded images. We choose two scenes:

1. a plane and a wagon, which shows that our method works for objects which can be approximated by polygonal surfaces (this input was used by T. J. Fan in [6, 5]),

2. a very complex and cluttered scene with similar objects (busts of composers),

The plane-and-wagon scene is recognized with a data base consisting of 9 objects. The contents of the data base is shown in Figure 11.

4.1. Plane and Wagon

We have four range images, two of the plane from different views and two of the wagon from different views. The range data of all four views was obtained with a laser range finder. One wagon and one plane image serve as models. The scene is composed synthetically by combining the other two range images into the scene image as displayed in Figure 12(a). Figure 12(b) shows the best detected solution. It is interesting to note that the plane was recognized based on the super segment information, whereas the wagon was mainly detected based on splashes (for more details see [23]). This is a good illustration of the system's ability to use whatever information is available: for the wagon the splashes are the best matched features, for the plane the 3D curves.

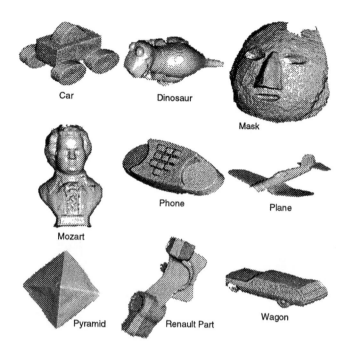

Car Dinosaur

Mask

Mozart Phone Plane

Pyramid Renault Part Wagon

Figure 11. Data Base with 9 Objects

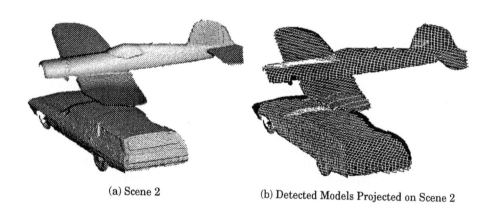

(a) Scene 2 (b) Detected Models Projected on Scene 2

Figure 12. Example Plane and Wagon

4.2. Composers

We obtained three range images from the three busts shown in Figure 13(a-c) with a liquid crystal range finder [20]. These three busts serve as models and are the

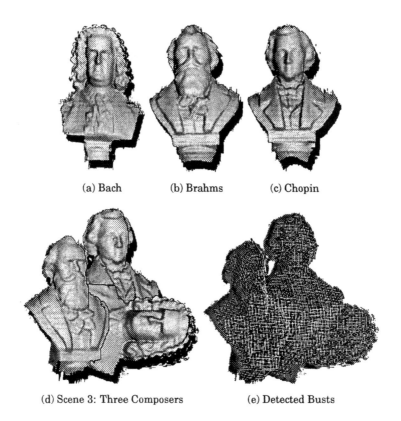

(a) Bach   (b) Brahms   (c) Chopin

(d) Scene 3: Three Composers   (e) Detected Busts

Figure 13. Three Composers

content of the data base. The scene is composed by overlaying three different views of the busts (the range finder's field of view and depth of field are too small to acquire such an amount of data at once). The data is smoothed with the adaptive smoothing algorithm [19] and small holes are closed with bilinear interpolation. The scene is displayed in Figure 13(d). The algorithm finds the correct correspondences and the correct positions despite the fact that the three models are locally very similar (parts of the faces, parts of the clothes). The hypotheses generation step retrieves all possible candidates and the verification step has to unravel the consistent clusters. The projection of the detected models on the scene is shown in Figure 13(e). This example shows the limits of our algorithm with respect to speed. The detection is by far the slowest compared to all the other scenes (see the complexity discussion in [23]). This is mainly due to the similarity of the objects, the noisy data (the liquid crystal range finder provides fairly noisy data due to interreflections and coarse quantization), and the complexity of the scene. It is interesting to note that

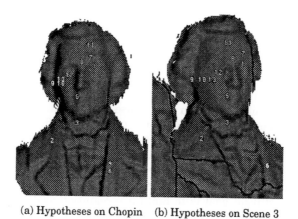

(a) Hypotheses on Chopin   (b) Hypotheses on Scene 3

Figure 14. Example of the Hypotheses found for Chopin (Hypotheses 4 and 6 are overlaid in the Scene)

the algorithm found multiple solutions (e.g. an instance of the Chopin model was also found on Bach in the scene) but they were later rejected based on the rigidity assumption.

In Figure 14 we show as an example the hypotheses which lead to the recognition of Chopin. Hypotheses 11 and 13 illustrate very well the location uncertainty.

4.3. General Observations

We can give some rough numbers about the running time (on a serial Symbolics 3675 Lisp machine).

1. The acquisition of one super splash (consisting of typically 3 splashes with 4 line fitting tolerances) takes about 12 seconds.

2. The plane consists of about 60 splashes, therefore it took about 12 minutes to compute all splashes.

3. The computation of the 3D curves takes less than 3 minutes.

4. The recognition process for the plane and wagon scene takes about 1 minute 30 seconds (retrieval: 15 seconds, verification: 75 seconds).

5. The recognition process for the bust scene takes about 30 minutes (retrieval: 2 minutes, rigidity filter: 7 minutes, verification: 21 minutes).

All these numbers reflect neither the high parallelism which is theoretically possible nor the data redundancy with which we work at the moment. Simple improvements can significantly increase the performance. This is one goal of our future work.

5. Conclusion and Future Work

We showed with our implementation of the TOSS system that structural indexing provides a powerful mechanism for the recognition of general three dimensional objects. We make very few restrictive assumptions about the shape of the objects, and we are able to acquire them automatically. By using two types of primitives, we overcome the problem of recognition in the case where either edge data or surface data provides not enough information for a correct classification. Our encoding scheme allows us to match the primitives and verify the resulting hypotheses in a reasonable time complexity. We are able to handle large object data bases. Our plan for the future is to further exploit the fact that *rich* features such as the splash or the super segment provide enough structural information to recognize objects efficiently. Our long term goal is to build a recognition system which is able to recognize three dimensional models in a two dimensional gray level image.

REFERENCES

1 P. J. Besl. *Machine Vision for Three Dimensional Scenes*, chapter The Free-Form Surface Matching Problem, pages 25–71. Academic Press, 1990. Freeman, H. editor.

2 B. Bhanu. Representation and Shape Matching of 3-D Objects. *IEEE Transactions on Pattern Analysis and Machine Intelligence*, 6(3):340–350, 1984.

3 R. C. Bolles and P. Horaud. 3DPO: A Three-Dimensional Part Orientation System. *International Journal of Robotics Research*, 5(3):3–26, 1986.

4 C. H. Chen and A. C. Kak. A Robot Vision System for Recognizing 3-D Objects in Low-Order Polynomial Time. *IEEE Transactions on Systems, Man, and Cybernetics*, 19(6):1535–1563, 1989.

5 T. J. Fan. *Describing and Recognizing 3-D Objects Using Surface Properties*. Springer Verlag, New York, 1990.

6 T. J. Fan, G. Medioni, and R. Nevatia. Recognizing 3-D Objects Using Surface Descriptions. *IEEE Transactions on Pattern Analysis and Machine Intelligence*, 11(11):1140–1157, 1989.

7 O. Faugeras and Hebert M. The Representation, Recognition, and Locating of 3-D Objects. *International Journal of Robotics Research*, 5(3):27–52, 1986.

8 P. J. Flynn and A. K. Jain. On reliable curvature estimation. In *Proceedings of IEEE Computer Vision and Pattern Recognition*, pages 110–116, San Diego, California, June 1989.

9 W. E. L. Grimson and T. Lozano-Pérez. Model-based recognition and localization from sparse range or tactile data. *International Journal of Robotics Research*, 3(3):3–35, 1984.

10 W. E. L. Grimson and T. Lozano-Pérez. Localizing overlapping parts by searching the interpretation tree. *IEEE Transactions on Pattern Analysis and Machine Intelligence*, 9(4):469–482, 1987.

11 K. Ikeuchi. Precompiling a geometrical model into an interpretation for object recognition in bin-picking tasks. In *Proceedings of the DARPA Image Under-*

standing Workshop, pages 321–339, Los Angeles, California, February 1987. Morgan Kaufmann Publishers, Inc.

12 A. Kalvin, E. Schonberg, J. T. Schwartz, and M. Sharir. Two-Dimensional, Model-Based, Boundary Matching Using Footprints. *International Journal of Robotics Research*, 5(4):38–55, 1986.

13 E. Kishon and T. Hastie. 3-D Curve Matching Using Splines. In *Proceedings of European Conference on Computer Vision*, pages 589–591, Antibes, France, April 1990.

14 Y. Lamdan, J. T. Schwartz, and H. J. Wolfson. On Recognition of 3-D Objects from 2-D Images. In *Proceedings of IEEE International Conference on Robotics and Automation*, April 1988.

15 Y. Lamdan and H. J. Wolfson. Geometric Hashing: A General and Efficient Model-Based Recognition Scheme. In *Proceedings of IEEE International Conference on Computer Vision*, pages 218–249, Tampa, Florida, December 1988.

16 Horaud P. and Bolles R. C. 3DPO's Strategy for Matching Three-Dimensional Objects in Range Data. In *Proceedings of the International Conference on Robotics*, pages 78–85, Atlanta, Georgia, March 1984.

17 B. Parvin and G. Medioni. A constraint satisfaction network for matching 3-D objects. In *Proceedings of the International Conference on Neural Networks*, volume II, pages 281–286, Washington, D.C, June 1989.

18 G. M. Radack and N. I. Badler. Local Matching of Surfaces Using a Boundary-Centered Radial Decomposition. *Computer Vision, Graphics, and Image Processing*, 45:380–396, 1989.

19 P. Saint-Marc, J.-S. Chen, and G. Medioni. Adaptive smoothing: A general tool for early vision. In *Proceedings of the Conference on Computer Vision and Pattern Recognition*, San Diego, California, June 1989.

20 K. Sato and S. Inokuchi. Range-imaging system utilizing nematic liquid crystal mask. In *Proceedings of the IEEE International Conference on Computer Vision*, pages 657–661, June 1987.

21 F. Stein and G. Medioni. Efficient Two Dimensional Object Recognition. In *Proceedings of International Conference on Pattern Recognition*, Atlantic City, New Jersey, June 1990.

22 F. Stein and G. Medioni. TOSS - A System for Efficient Three Dimensional Object Recognition. In *Proceedings of the DARPA Image Understanding Workshop*, Pittsburgh, Pennsylvania, September 1990.

23 F. Stein and G. Medioni. Structural Indexing: Efficient Three Dimensional Object Recognition. *IEEE Transactions on Pattern Analysis and Machine Intelligence*, 14(2):125–146, February 1992.

24 Jeffrey S. Vitter and Wen-Chin Chen. *Design and Analysis of Coalesced Hashing*. Oxford University Press, 1987.

Building a 3-D World Model for Outdoor Scenes from Multiple Sensory Data

Minoru Asada[a]

[a]Department of Mechanical Engineering for
Computer-Controlled Machinery
Osaka University
Osaka 565
Japan

Abstract

This chapter presents a method for building a 3-D world model using range and video images taken from outdoor scenes. The 3-D world model consists of four kinds of maps: a physical sensor map, a virtual sensor map, a local map, and a global map. First, a range image (physical sensor map) is transformed to a tessellated height map representation (virtual sensor map) relative to the observer. Next, the height map is segmented into unexplored, occluded, traversable and obstacle regions from the height information. Moreover, obstacle regions are classified into artificial objects or natural objects according to their geometrical properties such as slope and curvature. A drawback of the height map (recovery of planes vertical to the ground plane) is overcome by using multiple height maps which include the maximum and minimum height points for each tessellated grid on the ground plane. Multiple height maps are useful not only for finding vertical planes but also for mapping obstacle regions into video image for segmentation. Finally, height maps are integrated into a local map by matching geometrical parameters and by updating region labels. We show the results obtained using landscape models and an ALV simulator, and discuss about constructing a global map with local maps.

1. Introduction

The development of an Autonomous Land Vehicle (ALV) is a central problem in artificial intelligence and robotics, and has been extensively studied [1]-[12]. To perform visual navigation, a robot must gather information about its environment through external sensors, interpret the output of these sensors, construct a scene map and a plan sufficient for the task at hand, and then monitor and execute the plan. As a first step, real time visual navigation systems for road following were developed in which simple methods for detecting road edges were applied in simple environments [1, 4, 5, 12]. For even slightly more complicated scenes, the difficulty of the problem increases dramatically, therefore a world model such as a map could be very important for successful navigation through such environments.

This article is based on *"Map building for a mobile robot from sensory data"* by M. Asada, IEEE Trans. on Systems, Man and Cybernetics, Vol. SMC-20, pp. 1326-1336, No. 6, 1990. © 1990 IEEE.

Sometimes, accurate, quantitative maps may be available in advance [13], more often, maps are less descriptive and provide only global information as in a conventional geographical map [14]. In other cases, the robot may try to construct the map from sensory date in unknown environments. The following issues are considered as ideal requirements for map construction.

1. To be able to capture sensor resolution and accuracy.

2. To be able to deal with different kinds of sensors.

3. To be able to easily match and update sensory information.

4. To be able to cope with various kinds of tasks at hand.

5. To be able to adapt to dynamic change in the environment.

We review the existing methods of map construction from the viewpoints of ideal requirements above. These methods are categorized into two types of map representation; one is sensor-based perspective maps [15, 16], and the other is 2-D maps viewed from the top of the robot [17, 18].

Tsuji and Zheng [15] used perspective maps for navigation in which 3-D information obtained by stereo vision is represented in the image coordinate system. Dunlay [16] transformed the range information into the height information from the ground plane and constructed a perspective height image based of the range finder-centered coordinate system. Mapping the road boundaries extracted from the video image onto the perspective height map, he found obstacles inside road regions and determined velocity and steering angle. Since these sensor-based perspective maps can capture sensor resolution and accuracy, they satisfy the first requirement. It seems difficult, however, to integrate perspective maps obtained at different observation locations into one perspective map, and other requirements do not seem to have been considered.

Hebert and Kanade [17] have analyzed ERIM range images and constructed a surface property map represented in a Cartesian coordinate system viewed from top, which yields surface type of each point and its geometric parameters for segmentation of scene map into traversable and obstacle regions. Daily et al. [18] constructed a height map called CEM (Cartesian Elevation Map) which is also represented in a Cartesian coordinate system viewed from top, by transforming the range information into the height information. These robot-centered 2-D maps are easy to deal with different kinds of sensors, and to integrate them into the Cartesian coordinate system, but do not naturally capture sensor resolution and accuracy. They seem to be able to match and update sensory information, but the detailed procedures have not been reported, yet. Therefore, the difference between the robot-centered 2-D maps and the world-centered map obtained by integrating these robot-centered 2-D maps is ambiguous. Kweon et al. [19] proposed an integration scheme of the long sequence of elevation maps. They assumed that the scene include no moving objects, therefore, the statistical method for the whole scene is applicable.

Elfes [20] has developed a sonar-based mapping and navigation system which constructs sonar maps of the environments viewed from the top and updates them with recently acquired sonar information. He proposed a hierarchical representation of sonar map which includes three kinds of axes; an abstract axis, a resolution axis, and a geographic axis. In his system, the outputs of sonar sensors are directly mapped to a 2-D map, therefore, the difference between sensor maps and the 2-D maps is implicit, and other sensory data such as video images and range images seem difficult to be represented in this hierarchy.

In the above works, the fourth and fifth requirements have not been considered. Towards the goal of building a ideal map construction system which satisfies the all requirements, we propose a method for building a hierarchical representation of a 3-D world model for a mobile robot. In this model, we extended and generalized the Elfes's representation so that other kinds of sensor information can be represented, by making the relationships between coordinate systems at different levels explicit. The 3-D world model consists of four kinds of maps: a physical sensor map, a virtual sensor map, a local map, and a global map.

A physical sensor map usually represents sensory data or analyzed data in the sensor-based coordinate system from which the sensory data is taken (e.g. perspective map in [15], or ERIM range image in [17]).

A virtual sensor map represents the sensory data (in the physical sensor map) in the vehicle-centered Cartesian coordinate system. Any other type of coordinate system such as a cylindrical one can be applicable to the virtual sensor map representation in the context of representing sensory data in the vehicle-centered coordinate system. However, the Cartesian mapping seems more suitable for the virtual sensor map representation because the size of the cell on the map (which corresponds to the resolution of the map) is constant everywhere; therefore, fusing the data at the same point but observed from different view points is much easier than other types of mapping such as a cylindrical mapping or Delaunay triangulation which requires a complex algorithm to access data points and their neighborhoods [21]. As one example of a particular instance of the virtual sensor map, we introduce a height map which represents the height information transformed from a range image in the vehicle-centered Cartesian coordinate system. The 2-D map in [17] and CEM (Cartesian Elevation Map) in [18] are also categorized into the virtual sensor map representation. Virtual sensor maps are integrated into a local map which is represented in the object-centered coordinate system. The local map has its own reference (object) on which the integration of the virtual sensor maps is based, therefore, a new local map with a new reference is generated when the current reference cannot be observed as the robot moves. Thus, a number of local maps are generated along with robot navigation. A global map consists of these local maps and the geometrical relationship between them. How to build a global map with relational local maps is proposed by Asada et al. [22]. In this chapter, we focus on the building the physical sensor maps, the virtual sensor maps, and the local map from the video and range data, and have not dealt with how to build the global map.

The features of our representation are;

1. In a hierarchical representation of the world model, we have both sensor-based and observer (robot)-centered maps, each of which can be referred each other when necessary.

2. We can discriminate an object-centered map (a local map) from a robot-cented map (a kind of virtual sensor map) because we can integrate the robot-centered maps observed at different locations into the object-centered map. Through the integration process, stationary and moving objects are separated from the scene, and the moving objects can be tracked.

3. With this model, the mobile robot can derive useful information from a map at adequate level to accomplish various kinds of tasks such as visual navigation, obstacle avoidance, and landmarks and/or objects recognition.

In our system, a range image (one example of the physical sensor map) is transformed to a height map (one example of the virtual sensor map) in the mobile robot centered Cartesian coordinate system. The height is estimated from the assumed ground plane on which the vehicle exists. First, we segment the height map into unexplored, occluded, traversable and obstacle regions from the height information, and then classify obstacle regions into artificial objects or natural objects according to their geometrical properties such as slope and curvature. A drawback of the height map (recovery of planes vertical to the ground plane) is overcome by using a multiple height map which includes the maximum and minimum heights for each point on the ground plane. The multiple height map is useful not only for finding vertical planes but also for mapping obstacle regions into video image (another sensor map) for segmentation. Finally, the system integrates height maps, observed at different locations, into a local map, matching geometrical parameters of obstacle and traversable regions and updating region labels. We show the results applied to landscape models using the ALV simulator of the University of Maryland [4], and discuss construction of a global map with local maps.

2. System Configuration

2.1. Physical Simulation System of ALV And Its Environments

Our ALV physical simulation system was developed in our laboratory [4] for providing a low cost experimental environment for navigation (as opposed to an outdoor vehicle [7, 17]). A range finder based on structured light was added to this system. Planes of light are projected from a rotating mirror controlled by a stepping motor (see [23] for more detail). Recently, we extended the system in two ways. First, we developed a drive simulator program which controls the speed and steering angle of the vehicle (robot arm) during the motion. The camera height and camera tilt to the ground plane are kept constant during the motion through the position feedback of three leg sensors attached to the camera.

Previously, a wooden terrain board, on which a road network was painted, was set vertically to increase the flexibility of camera motion simulated by the robot arm. Due to its vertical setting, it was very difficult to put landscape models such

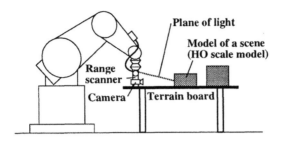

Figure 1. Experimental set up

as trees, bushes, buildings and other vehicles on the board. Thus, we set the terrain board horizontally so that we could place any landscape model without permanently fixing their positions. Fig.1 shows our experimental setup. The robot arm attached with a TV camera and a light-stripe range scanner is set on the board to input a picture and a range image.

2.2. Overall of Map Building System

Fig.2 shows the architecture of our system. In this figure, we omit other modules such as path planner, navigator, pilot and supervisor in [8] in order to concentrate on the map building system.

The 3-D world model for a mobile robot consists of four kinds of maps; a physical sensor map, a virtual sensor map, a local map, and a global map. Elfes [20] proposed a multiple axis representation of a sensor map in his sonar mapping and navigation system (resolution axis, geographical axis and abstraction axis) and adopted three levels (view, local map, global map) in the geographical axis. We extend and generalize these levels so that other sensory data such as range and video data can be represented. Fig.3 shows the geometrical relation between the three coordinate systems of the physical sensor map (sensor-based), the virtual sensor map (vehicle-centered), and the local map (object-centered). The global map is a set of the local maps and discussed in the conclusions. Each sensor has its own coordinate system; for example, an intensity image is represented in the camera-centered coordinate system and a range image in the range-finder-centered coordinate system, both of which are fixed to the robot (vehicle). Here, we assume that the relation between sensor coordinate systems and vehicle coordinate system is known, and that the motion information is available, but not always accurate. A Virtual Sensor Map Builder constructs the virtual sensor maps from the physical sensor map. Stereo matching, which we do not consider in this paper, is one possible strategy of virtual sensor map building for obtaining the depth map (virtual sensor map). Here, we introduce a height map obtained from the range image as a virtual sensor map. The height map is analyzed by the Obstacle Finder and the Obstacle

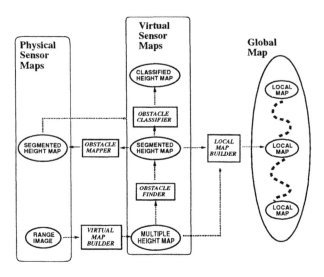

Figure 2. Overview of map building system.

Classifier to segment it into unexplored, occluded, traversable and obstacle regions and then to classify obstacle regions into artificial or natural objects. The result of the height map analysis is mapped onto the intensity image by the Obstacle Mapper in order to segment the intensity image. The Local Map Builder constructs a local map, matching and updating virtual sensor maps at different observing stations in the world coordinate system. In the following, each module is described along with some experimental results.

3. Height Map Analysis

3.1. Virtual Sensor Map Builder (from Range Image to Height Map)

The virtual sensor map builder builds virtual sensor maps from physical sensor maps. Here, we deal with a video image taken by a single camera and a range image obtained from our range finder [23] as physical sensor maps. Even though we use the same camera to take both video and range data, our idea can be applicable to other types of range data such as a range image taken by an ERIM range finder [17]. We discuss the differences between the two types of the range images in the conclusions. Figs.4 show examples of these physical sensor maps. The input scene includes a straight road, T-type intersection, two cabins, one truck, two cars, a mailbox, a stop sign at the intersection, trees and bushes as shown in Figs.4(a),(b), and (c). Figs.4(d),(e), and (f) are the corresponding range image to Fig.4(a),(b) and (c), respectively. All of the images are 512×512, and both the range and the

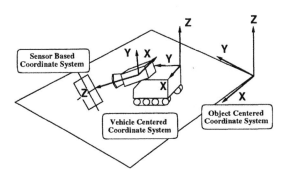

Figure 3. Geometrical relation of coordinate systems between three maps.

intensity values are quantized into 256 levels (8 bits). The brighter points are closer to the range finder and the darker points are farther from it. In the black regions, range information is not available due to inadequate reflection or occlusion. Although actual range finders such as the ERIM range scanner [17] measure the radial distances rather than Cartesian coordinates in both axes, the range image obtained from our range finder has irregular coordinate axes due to its special ranging geometry [23]; the vertical coordinate is radial (scanning angle of the light plane) but the horizontal coordinate is the same as that of the intensity image because range calculation is based on the triangulation with light planes and a single TV camera. Fig.5 shows calibration parameters required to calculate the range.

The range image is transformed to a height map in the vehicle centered coordinate system based on the known height and tilt of the range finder relative to the vehicle. The position of a point in a given coordinate system can be derived from the obtained range R and the direction to that point (it corresponds to the coordinates (x_r, y_r) on the range image). We use the Cartesian coordinate system $O - XYH$ shown in Fig.5, in which case the XY plane corresponds to the ground plane, and the origin O is just below the range finder (the reflecting mirror). The coordinates of a point $P(X_p, Y_p, H_p)$ estimated by the range finder are given by the following equations:

$$X_p = \frac{R_c x_r}{F_x}, \qquad Y_p = R \sin \theta, \qquad H_p = Height - R \cos \theta,$$

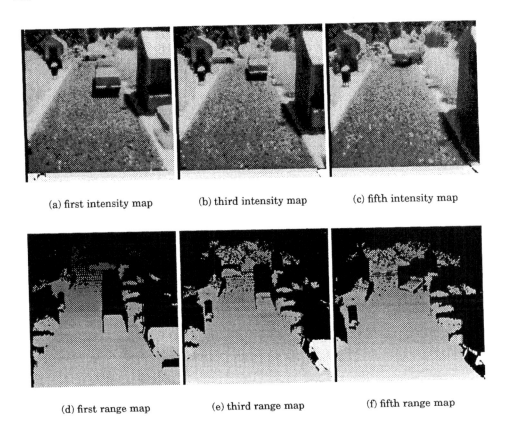

(a) first intensity map   (b) third intensity map   (c) fifth intensity map

(d) first range map   (e) third range map   (f) fifth range map

Figure 4. Sensor maps.

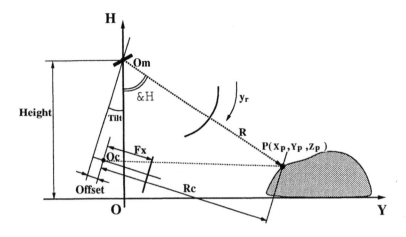

Figure 5. Calculation of the range and height.

where F_x, $Height$, θ, and R_c are the focal length in the horizontal direction of the camera image, the height of the range finder (the reflecting mirror), the vertical scanning angle, and the range in the camera-centered coordinate system which is needed to estimate X-coordinate. The last two parameters are derived from the column position y_r in the range image, the tilt angle of the camera $Tilt$, and the offset parameter $Offset$ by the equations:

$$\theta = a y_r + b, \qquad R_c = R\sin(\theta + Tilt) - \textbf{\textit{Offset}},$$

where a and b are parameters for the transformation from the column position to the vertical scanning angle. An adequate area on the ground plane in front of the vehicle is assigned for a height map and the 3-D coordinates (X_p, Y_p, H_p) are quantized to 8 bits numbers. The height map is a 256×256 image, each pixel corresponds to 1mm^2 on the simulation board. The entire map corresponds to a square of side length 256mm; the scale of the simulation board is 87:1 (HO scale). Gray levels encode the height from the assumed ground plane. Since the range is sparse and noisy at far points, smoothing is necessary. We applied an edge-preserving smoothing method [24] to the height map in order to avoid a mixed pixel problem of high and low points. Fig.6 shows the filtered height map of the input scene (Fig.4(d)); Fig.6(a) shows a gray level image and Fig.6(b) shows its perspective view.

One drawback of the height map is that it is unable to represent vertical planes, especially these under horizontal or sloped planes because the range information corresponding to multiple points in the vertical direction is reduced to one point

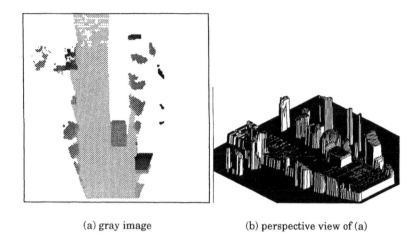

(a) gray image   (b) perspective view of (a)

Figure 6. Height map

in the height map. This is especially undesirable since the range information on the vertical planes is more accurate than that on the horizontal planes. Thus, we compute a multiple height map for one range image which includes the maximum and minimum heights for each point on the height map, and the number of points in the range image which are mapped to one location on the height map. In Fig.6, the maximum height is shown.

3.2. Obstacle Finder (Segmentation of Height Map)

The first step of the height map analysis is to segment the height map into unexplored, occluded, traversable and obstacle regions. The height map consists of two types of regions: those in which the height information is available and those in which it is not. The latter regions are classified into unexplored or occluded regions. Unexplored regions are outside the visual field of the range finder, and therefore are easily detected by using the calibration parameters of the range finder (height, tilt and scanning angle). The remaining regions in this category are labeled as occluded regions. Some regions which are not occluded may be classified into occluded regions if the height information is unavailable due to causes such as inadequate reflection. These regions can be often seen inside bushes or trees with many leaves.

Finding traversable regions is straightforward. First, identify these points close to the assumed ground plane and construct atomic regions for the traversable region. Next, expand the atomic regions by merging other points surrounding them which have low slope and low curvature. The desirable feature of the height map

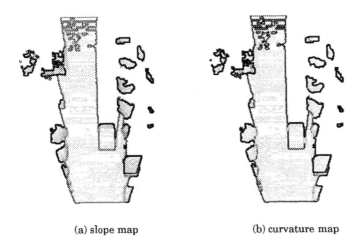

(a) slope map   (b) curvature map

Figure 7. Slope and curvature maps

is that the outputs of the first and second derivatives of the height map correspond
to the magnitudes of the slope and curvature of the surface because the locations of
the points in the height map are represented with Cartesian coordinates. Strictly
speaking, the outputs of the second derivative of height do not directly corresponds
to the curvature of the surface, but at least, it outputs zeros for the sloped plane
such as a roof plane of a house. Figs.7 show the magnitudes of the first (Sobel) and
second derivatives of the height map (Fig.6(a)) whose mask size is 3 by 3 pixels. The
remaining regions are labeled as obstacle regions. Fig.8 shows the final result of
the segmentation of the height map. White, light gray, dark gray and black regions
are obstacle, occluded, traversable and unexplored regions, respectively. We can
see that the boundary of the obstacle regions has high slope and/or high curvature
(see Fig.7).

The result of segmentation of the height map should be useful for path planning
since many path planning algorithm are based on a top view of the configuration of
obstacles and free space [25].

3.3. Obstacle Classifier

The segmented height map constructed by the Obstacle Finder is very useful for
navigation tasks such as avoiding obstacles, but does not contain sufficient explicit
information enough for higher level tasks such as landmark or object recognition.
As a first step in object recognition, we try to classify obstacle regions as artificial objects or parts of natural objects. Many artificial objects such as cabins,
cars, mailboxes and road signs shown in Figs.4 have planar surfaces, which yield

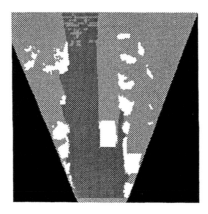

Figure 8. Segmentation of height map.

constant slope and low curvature in the height map and linear features in the intensity image. On the other hand, natural objects such as trees and bushes have fine structures with convex and concave surfaces, which yield various slopes and/or high curvatures in the height map and therefore large variance of brightness in the intensity image (the reverse is not always true).

Thus, utilization of not only the height map but also the intensity image is useful for obstacle classification. In order to use the brightness information in the intensity image, we map the obstacle regions to the intensity image to segment it. The mapping of obstacles to the intensity image seems at first straightforward based on the geometrical relation between the camera and the range finder. However, it is complicated by the need to correctly choose between the maximum and minimum heights associated with each point in the height map. Fig.9 shows a perspective view of a cube on a plane and its height map. If we map only the maximum value of the height of the cube, only a top surface of the cube is cut out from its perspective view and visible side surfaces are left undiscovered (see Fig.9(a)). We should use the minimum height when the object is bounded by traversable regions and use the maximum height when it is occluding other objects behind it (see Fig.9(c)). Classifying the boundary of obstacle regions in the height map and using the multiple height map, the obstacle mapper maps the obstacles to the intensity image as follows.

(1) Classify each boundary point of the obstacle region in the height map according to the geometrical relation between the point, occluded region and traversable region in the segmented height map (Fig.8).

(2) Use the minimum height if that point is adjacent to a traversable region or that point is an occluded boundary (the boundary point is labeled as an occluded

387

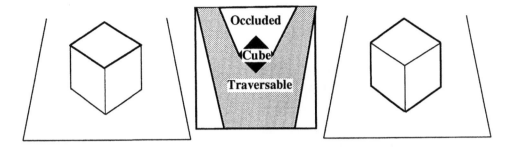

Figure 9. A cube on a plane and its height map.

boundary when the occluded region is between the boundary point and the range finder).

(3) Use the maximum height if that point is occluding boundary (the boundary point is labeled as an occluding boundary when the boundary point is between occluded region and the range finder).

Fig.10 shows the result of this mapping. The truck and the car in front of the cabin on the right side, the mailbox, the stop sign, and the bushes are finely segmented in the intensity image. The car at the intersection is not mapped because its location is outside the height map (far from the viewer). The roof line of the cabin on the left side is incorrect because of the bad range data. The reason that the top roof line of the cabin on the right side drops suddenly to the ground plane is that there is a lower object behind the cabin and the obstacle region in the height map includes both the cabin and the lower object.

The next step is to classify the obstacle using the properties of the height map and the brightness in the intensity image. The obstacle classifier classifies each resegmented region according to the following criteria.

(1) If a region has sufficient size (larger than pre-determined threshold) and constant slope (small variance of slope) and low curvature (low mean curvature and small variance of the curvature), then the region is an artificial object.

(2) If a region has sufficient size and high curvature (high mean curvature and large variance of the curvature) and large variance of the brightness in the intensity image, then the region is a part of a natural object.

(3) Otherwise, the region is regarded as uncertain in the current system.

Fig.10(b) shows an obstacle map, where small regions are labeled as uncertain. The car in front of the cabin on the right side and the truck are correctly interpreted as artificial objects. However, the roofs of two cabins, the mailbox and the stop sign are misinterpreted as natural objects because of the high curvature due to vertical planes and/or insufficient, noisy range data. The region corresponding to the right side cabin includes a roof and the lower object behind it, therefore resegmentation into two regions is necessary ([26]). Uncertain regions and some regions with vertical surfaces would require closer examination for correct interpretation.

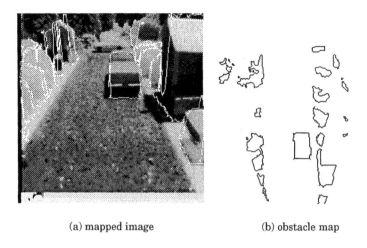

(a) mapped image   (b) obstacle map

Figure 10. Mapping obstacle region into the intensity image.

3.4. Local Map Builder

During the motion of the vehicle, the system produces a sequence of virtual sensor maps constructed at different observing stations. These virtual sensor maps should be integrated into a local map in the world centered coordinate system. The local map builder consists of two parts; the first part matches a new height map and the current local map to determine the correct motion parameters of the vehicle, and the second one updates the description of region properties on the integrated local map. Matching is performed between the (i+1)st height map H_{i+1} and the local map L_i in which the height maps from the first to the ith observing stations are integrated. If $i = 1$, then $L_i = H_i$. The point P_l in the local map L_i can be represented as the point P_h in the height map H_{i+1} by the equation:

$$P_h = R_{lh}P_l + T_{lh},$$

where R_{lh} is a rotation matrix consisting of 3 rotational components, α, β, and θ along each axis, and T_{lh} is a translation vector along each axis, that is, $T_{lh}^{-1} = (X, Y, H)$. The following are matching procedures.

(1) Match the traversable regions between H_{i+1} and L_i. Since the traversable regions are usually the larger planar regions in the segmented map and rough estimates of the motion parameters $(X, Y, H, \alpha, \beta,$ and $\theta)$ are ordinarily available from the internal sensors of the vehicle, this matching is relatively straightforward. The traversable region in H_{i+1} and L_i are on the XY-plane in each coordinate system because we determined the orientation and location of the coordinate system so that

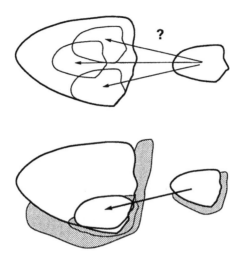

Figure 11. Expanding obstacle region to avoid the matching ambiguity.

the ground plane corresponds to the XY-plane (see Fig.5). Therefore, matching the traversable region between H_{i+1} and L_i is performed by overlaying the XY-plane in L_i onto the XY-plane in H_{i+1}. By this matching, H, α, β are determined (in our experments, all are zero), and the remains are X, Y, and θ; the motion parameters on the XY-plane.

(2) Determine the remaining motion parameters X, Y, and θ by matching the obstacle regions between H_{i+1} and L_i. Search for the motion parameters which take the minimum height difference for each obstacle region, starting the rough motion parameters from the vehicle navigation system as an initial value. We evaluate the following criterion function to find the minimum height difference:

$$H_i^k(x,y,\theta) = \sqrt{\frac{(H_{i+1} - L_i^k(x,y,\theta))^2}{N_i^k}},$$

where X, Y, θ, L_i^k, and N_i^k are estimated translation, rotation, the height of kth obstacle region in L_i, and the number of height points compared between H_i and L_{i+1} for the kth obstacle region, respectively. If we pick up only obstacle region, the matching could be ambiguous. For example, a small obstacle region with a horizontal plane matches at any position of a large region with horizontal plane of the same height (see Fig.11(a)). In order to avoid such an ambiguity, we expand obstacle regions so that they include the height information surrounding them (see Fig.11(b)).

Special care needs to be taken for moving objects because they have different motion parameters from those of the stationary environment. In the current system, we use a heuristic for detecting moving objects. The obstacle surrounded with traversable regions is a candidate moving object because the moving objects should be inside the traversable region (except for flying objects). Three regions, a truck, a mailbox, and a closest bush near the left road boundary are candidate moving objects while only a truck region is actually moving. Although search area for stationary obstacles is ±2mm for translation (X, Y) and ±2° for rotation θ with fine resolution of 1mm and 1°, coarse to fine search algorithm is applied to the candidates for moving objects since moving objects might be outside the search area for the stationary objects.

The algorithm involves estimating motion parameters in a large search area with sparse intervals of translation (±5mm) and rotation (±5°) first, and then, refining it with fine intervals similar to stationary obstacles. In Fig.10, motion parameters for almost of all regions are correctly obtained, and as a result, only a truck region is interpreted as an object moving farther from the viewer (towards the intersection). The motion parameters X, Y, and θ for the stationary objects are 0mm, 15mm, and 0°, respectively, and those for the moving object are 0mm, 45mm, and 0°, respectively. Fig.12 shows the histogram of the motion parameters when $\theta = 0°$, in which the inverse number of the minimum height difference for each obstacle region are accumulated into the corresponding motion parameters (X, Y). Two peaks correspond to the stationary regions (larger) and the moving region (smaller). Velocities of the moving objects are used to predict their locations and orientations in the next height map H_{i+2}.

(3) Integrate the results of the matching into a local map and update region labels. The local map builder overlays all regions in H_{i+1} onto the local map L_i according to their motion parameters. We call the overlaid map a new local map L_{i+1}. For each point on the local map L_{i+1}, the local map builder refines the height information for each point using those in H_{i+1} and L_i. If the height information is available from both, take a mean of both heights with weights (each weight is inversely proportional to the distance from the viewer to that point). When only one height is available, adopt that height; otherwise, retain that point left undetermined. After integration, the obstacle finder segments the integrated local map L_{i+1} into regions using the method described in the previous subsection and assigns a new label for each region. The segmented local map L_{i+1} is used for matching with the next height map H_{i+2}. Thus, the local map is constructed by matching and integrating height maps observed at different locations.

Fig.13 show the local map L_6 integrated with six height maps (from the first to the sixth height maps). Fig.13(b) shows the roof of the cabin on the left side in the first height map H_1 (left) and the local map L_6 (right) from the different view angle from Fig.13(a). In H_1, the roof is eroded and includes noisy heights due to the shallow angle between the light plane from range scanner and the roof plane. While, in L_6, the shape of the roof is clear because the vehicle approached to the cabin, therefore, better height information was obtained. Fig.13(c) shows the shape difference of moving object (truck) between H_6 (left) and L_6 (right). While it is almost invisible due to occlusion by a stop sign in H_6, the region corresponding to the bed of the

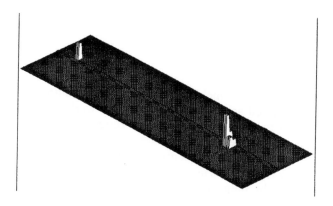

Figure 12. Histogram of motion parameters.

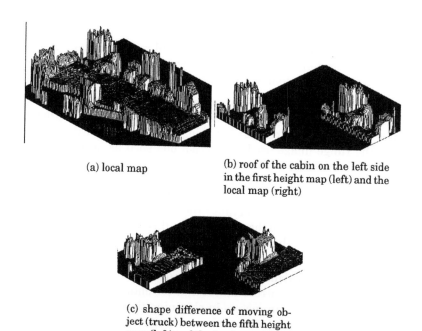

(a) local map

(b) roof of the cabin on the left side in the first height map (left) and the local map (right)

(c) shape difference of moving object (truck) between the fifth height map (left) and the local map (right)

Figure 13. Local map integration with six height maps

truck is visible in L_6 where the system successfully tracks the moving object and update its height information.

4. Conclusion

A map building system for a mobile robot from sensory data has been described. We introduced the virtual sensor map into the 3-D world model construction in order to represent various sensory data such as range and video data and in order to integrate them in a Cartesian coordinate system. We proposed the use of a height map, one example of virtual sensor maps, obtained from the range image for a mobile robot to support various tasks such as path planning and landmark recognition. The height map is easy to recover and calculation of geometrical properties such as slope and curvature is straightforward.

In our experiments, we have dealt with almost flat terrains. In the case of non-flat terrain, the matching process would be more complicated because the orientation of the H-axes in a height map and a local map might be different. One alternative is to adopt the direction of the gravity as the common orientation of the H-axes in the different virtual sensor maps utilizing the output from the orientation sensor in the navigation system. This could make the matching process as simple as in the case of flat terrain.

In the introduction, we have described the ideal requirements for environment map construction system. In the following, we discuss the relationship between the current state of the system and these requirements, and future directions.

1. Capturing sensor resolution and accuracy: In the experiments, we used a range finder based on the structured lighting method which produces dense range images consists of 512×512 image points. The largest difference between the range image used and actual one such as an ERIM range image [17] size of which is 256×64 is the resolution of the range image. Since the size of the height map used was smaller than that of the range image, we referred to the number of range points mapped onto one bucket of the height map as a matching weight because the number of observed range points is inversely proportional to the square of the distance between the observer and the object surface in the perspective projection.

   The matching of height maps are almost successful, but some errors are included in the obtained motion parameters due to the incorrect range data. The main reason for the incorrect data is that the obtained range data is sensitive to image processing parameters in obtaining them when the angle between the light plane and the object surface is shallow. That is, the accuracy of the range information depends on not only the geometrical relationship between the light plane and TV camera but also the angle between the light plane and the object surface. Therefore, it seems difficult to assume the systematic error distribution.

2. Dealing with different kinds of sensor information: In order to deal with both the range and video image data, we proposed a height map representation

defined in a Cartesian coordinate system, where matching and integration are executed. This representation is suitable to cope with spatial sensor information in 3-D space and related one such as range data, video data, and motion parameters of the robot. However, physically different sensor information (e.g. force sensor) does not seem easy to be dealt in our current scheme. Interpretation and representation of multiple sensory information physically different each other is the main issue in sensor fusion.

3. Simple mechanism of matching and integration: Since a height map representation is defined in a Cartesian coordinate system as mentioned above, matching and integration are simple. However, we should develop a method for representing the reliability of the observed and integrated sensor information on which integration (updating) procedure is based.

4. Applicability to various kinds of tasks: The segmentation results of the height map is available for obstacle finding and avoidance. However, more detailed knowledge is needed to recognize landmarks and objects. The current obstacle classifier uses principally the geometrical properties of height information and makes little use of brightness information in identifying the obstacles as artificial or natural objects. The experimental results include some number of errors. Errors due to vertical planes could be corrected by verifying the existence of the vertical plane in the multiple height maps and by obtaining more correct parameters for them from the range image (physical sensor map). Other errors might be corrected by the use of color images since color information is often useful for segmentation of images of outdoor scene [27].

5. Adaptability to dynamic change in the environment: Almost of the existing methods for map construction assume that the environment is stationary. Unlike these methods, our system can deal with moving objects in the scene by assuming that the number of moving objects is not so large and their motions are smooth. If the number of moving objects increased, the system would fail to find correspondence of them between two views. To avoid such situations, we should consider to use not only the geometrical data but also the semantic descriptions for objects in the scene which provide the system with some constraints on the motion of objects.

6. Others: In this paper, we have not dealt with the problem of constructing a global map with a set of local maps. The use of a local map builder which numerically integrates virtual sensor maps using height information could not be applicable to the construction of the global map because a traveling of long distance often results in a significant amount of positional error in the global map and because the robot cannot correct this error unless landmarks in the environment are given in advance [28]. Asada, Fukui, and Tsuji [22] developed a method of representing a global map with relational local maps where the rough description of the geometrical relation between adjacent local maps is described. We could apply their method to our scene. However,

as the map scale becomes larger, the higher level descriptions are generally required to represent a world model instead of numerical representation of it. Our current system has dealt with only region labels (unexplored, occluded, traversable and obstacle) and sublabels (artificial or natural). In order to accomplish higher tasks such as landmark and/or object recognition, knowledge base of the objects expected in the scene and geometrical modeling for them are needed, and the control structure for them should be exploited. These problems are under investigation.

Acknowledgement

The author wishes to thank Prof. Azriel Rosenfeld and Prof. Larry S. Davis for helpful comments and discussions and Mr. Daniel DeMenthon for providing range images at the University of Maryland, and Prof. Saburo Tsuji and Prof. Yoshiaki Shirai for constructive discussions at Osaka University, Japan. He also wishes to thank his son, Ryu Asada, who helped him assemble and paint the landscape models on the simulation board.

REFERENCES

1   A. M. Waxman, J. Le Moigne and B. Srinivasan, "Visual navigation of roadways", *in Proc. of IEEE Int. Cof. Robotics and Automation*, pp.862-867, 1985.
2   S. Tsuji, Y. Yagi, and M. Asada, "Dynamic scene analysis for a mobile robot in man-made environment", *in Proc. of IEEE Int. Conf. Robotics and Automation*, pp.850-855, 1985.
3   S. Tsuji, J. Y. Zheng, and M. Asada, "Stereo vision of a mobile robot: world constraints for image matching and interpretation", *in Proc. of IEEE Int. Conf. Robotics and Automation*, pp.1594-1599, 1986.
4   A. M. Waxman, J. LeMoigne, L. S. Davis, E. Liang, and T. Siddalingaiah, "A visual navigation system", *in Proc of IEEE Int. Conf. Robotics and Automation*, pp.1600-1606, 1986.
5   R. Wallace, K. Matsuzaki, Y. Goto, J. Crisman, J. Webb and T. Kanade, "Progress in robot road-following", *in Proc. of IEEE Int. Conf. Robotics and Automation*, pp.1615-1621, 1986.
6   S. A. Shafer, A. Stentz and C. Thorpe, "An architecture for sensor fusion in a mobile robot", *in Proc. of IEEE Int. Conf. Robotics and Automation*, pp.2002-2011, 1986.
7   C. Thorpe, S. Shafer, T. Kanade et al., "Vision and navigation for the Carnegie Mellon Navlab", *in Proc. of DARPA Image Understanding Workshop*, pp.143-152, 1987.
8   L. S. Davis, D. DeMenthon, R. Gajulapalli, T. R. Kushner, J. Le Moigne and P. Veatch, "Vision-based navigation: a status report", *in Proc. of DARPA Image Understanding Workshop*, pp.153-169, 1987.
9   Y. Goto and A. Stentz, "The CMU system for mobile robot navigation", *in Proc. of IEEE Int. Conf. Robotics and Automation*, pp.99-105, 1987.

10 R. A. Brooks, "A hardware retargetable distributed layered architecture for mobile robot control", *in Proc. of IEEE Int. Conf. Robotics and Automation*, pp.106-110, 1987.

11 R. S. Wallace, "Robot road following by adaptive color classification and shape trucking", *in Proc. of IEEE Int. Conf. Robotics and Automation*, pp.258-263, 1987.

12 M. A. Turk, D. G. Morgenthaler, K. D. Gremban and M. Marra, "Video road-following for the autonomous land vehicle", *in Proc. of IEEE Int. Conf. Robotics and Automation*, pp.273-280, 1987.

13 S. Tsuji, "Monitoring of a building environment by a mobile robot", *in Proc. of 2nd Int. Symp. on Robotics Research*, pp.349-365, 1985.

14 D. Lawton, T. S. Levvit, C. C. McConnell, P. C. Nelson and J. Glicksman, "Environmental modeling and recognition for an autonomous land vehicle", *in Proc. of DARPA Image Understanding Workshop*, pp.107-121, 1987.

15 S. Tsuji and J. Y. Zheng, "Visual path planning by a mobile robot", *in Proc. of 10th Int. Joint Conf. Artificial Intell.*, pp.1127-1130, 1987.

16 R. T. Dunlay, "Obstacle avoidance perception processing for the autonomous land vehicle," *in Proc. of IEEE Int. Conf. Robotics and Automation*, pp.912-917, 1988.

17 C. Thorpe, M. H. Hebert, T. Kanade, and S. A. Shafer, "Vision and navigation for the Carnegie-Mellon Navlab," *IEEE Trans. Pattern Anal. Machine Intell.*, vol. PAMI-10, pp.362-373, 1988.

18 M. Daily et al., "Autonomous cross-country navigation with ALV", *in Proc. of IEEE Int. Conf. Robotics and Automation*, pp.718-726, 1988.

19 M. Hebert, T. Kanade, and I. Kwen, "3-D vision techniques for autonomous vehicles", *Technical Report* CMU-RI-TR-88-12, Robotics Institute, Carnegie Mellon University, 1988.

20 A. Elfes, "A sonar-based real world mapping and navigation", *IEEE Journal of Robotics and Automation*, vol. RA-3, pp.249-265, 1987.

21 D. J. Orser and M. Roche, "The extraction of topographic features in support of autonomous underwater vehicle navigation", *in Proc. of Fifth Int. Symp. on Unmanned Untethered Submersible*, 1987.

22 M. Asada, Y. Fukui, and S.Tsuji, "Representing global world of a mobile robot with relational local maps," *IEEE Trans. on System, Man, and Cybernetics*, vol.SMC-20, no.6, pp.1456-1461, 1990.

23 D. DeMenthon, T. Siddalingaiah and L. S. Davis, "Production of dense range images with the CVL light-stripe range scanner", *Center for Automation Research Technical Report* CAR-TR-337, University of Maryland, 1987.

24 M. Nagao and T. Matsuyama, "Edge preserving smoothing", *Computer Graphics Image Processing*, vol.9, pp.394-407, 1979.

25 S. Puri and L. S. Davis, "Two dimensional path planning with obstacles and shadows", *Center for Automation Research Technical Report* CAR-TR-255, CS-TR-1760, University of Maryland, January 1987.

26 M. Asada, "Building 3-D world model for a mobile robot from sensory data", *Center for Automation Research Technical Report* CAR-TR-332, CS-TR-1936,

University of Maryland, 1987.

27 Y. Ohta, T. Kanade and T. Sakai, "Color information for region segmentation", *Computer Graphics Image Processing*, vol.13, pp.222-241, 1980.

28 R. A. Brooks, "Visual map making for a mobile robot", *in Readings in Computer Vision*, eds. M. A. Fischler and O. Firscein, pp.438-443, Los Altos, Calif.: Morgan Kaufman.

Three-Dimensional Object Recognition Systems
A.K. Jain and P.J. Flynn (Editors)
© 1993 Elsevier Science Publishers B.V. All rights reserved.
397

Understanding Object Configurations

Jeff L. DeCurtins, Prasanna G. Mulgaonkar, and Cregg K. Cowan [a]

[a]Advanced Automation Technology Center
SRI International
Menlo Park, California 94025 USA

Abstract

This paper describes our continuing work toward discovery and utilization of the knowledge we use to complete our understanding of a scene about which only partial visual information is available (i.e., the knowledge we use to understand that all the cookies are there even though we see only some of them). We describe our range image interpretation system developed for analyzing scenes containing piles of objects. In particular, we concentrate on two aspects of the system: The first, the use of shape models and sensor geometry to form hypotheses about the complete geometry of partially visible but unknown objects in a scene; second, the use of free-body analysis to evaluate the physical stability of the hypothesized objects in the presence of Coulomb friction.

1. Introduction

Analyzing the contents of a pile of unknown objects using noncontact sensing is an area of great interest because of the large and important set of material-handling applications in unstructured or partially structured environments, such as mailpiece singulation (the separation of individual pieces from a pile), feeding random or mixed parts, scavenging, and other similar tasks. An important characteristic of this problem domain is that complete information about the object shapes in the pile cannot be obtained — the interior of the pile will always be occluded, even if multiple viewpoints are used. For such applications, the key problem is to infer the size and shape of the objects and their spatial relationships to each other and to any unseen objects that may be underneath them: in other words, to estimate the object shapes and their configuration in the pile. These estimates can then be used to determine the required manipulation of the objects. For example, an autonomous robot operating in a damaged area must form hypotheses about piles of objects blocking its path and estimate the properties of the piles, such as inter-object forces and contacts, in order to manipulate an object without having other objects (i.e., those supported by the moved object) fall on the robot.

1.1. Prior Work

Most prior work in analysis of three-dimensional images has concentrated on two topics. Model-based techniques have concentrated on approaches to represent and recognize instances of known objects, such as, the work of Bolles, et al. [1], of Bhanu [2], and of Faugeras and Hebert [3]. Other work has concentrated on techniques to generate robust descriptions of surface shape, e.g., the work of Fan, Medioni and Nevatia [4], of Brady et al. [5], and of Besl and Jain [6]. Shape descriptions from multiple range images, covering all sides of an unknown object, have been used to develop geometric models or maps, including the work by Potmesil [7], by Dane and Bajcsy [8], and by Connolly [9]. As the amount of significant prior work in these areas is very large, we refer the interested reader to a survey of the various techniques [10].

In recent years a few researchers have investigated techniques to generate complete object descriptions, in the form of superquadric shape descriptors, from incomplete range data. Techniques to describe a range image as a collection of parts are described by Pentland [11]. The method uses minimal-length encoding to generate the simplest explanation of the range data (simplest in terms of the superquadric shape primitives). Boult and Gross [12] describe a system using least-squares minimization to describe positive or negative superquadrics in single-object images. Range images of single objects are also used in the work by Solina and Bajcsy [13], which describes the object as a superquadric with parametric distortions, e.g., tapering, bending and cavities.

1.2. Overview

We begin with three basic pieces of knowledge: a range image of a scene, the geometry of the sensor used to acquire the image, and the expectation that the scene consists of objects whose geometries can be represented by a set of generic shapes (e.g., boxes and cylinders).

The range image contains the three-dimensional location of visible points on the surfaces of unknown objects in a pile (see Figure 1). These surfaces occlude the volumes of the objects within the scene; these are the hidden volumes we wish to explore.

The range image can be acquired using different sensors, each of which uses a camera and some form of a light source. Because we know the geometry of each sensor, we can use the three-dimensional relationships among the camera, the light source, and the acquired image data to reason about the scene. For example, we know that the space between an imaged surface and the camera is free of any other object; we cannot make the same assertion for the space *behind* that surface.

We expect our scene to consist of objects whose shapes are familiar but whose dimensions are unknown. In the work reported here we restrict these shapes to cylinders and boxes. If the dimensions were also known, we would have precise geometric models of the objects and could have used object-recognition techniques to hypothesize the occluded portions with certainty, once we had identified instances of particular models from data in the image.

One object that is always present is the support surface on which the pile of objects rests. To simplify the data interpretation, we assume that this surface is planar and that it extends beyond the boundaries of the camera's field of view. The range image of this surface is easily segmented from the entire image, because the

Figure 1. Range Image of Three Boxes and a Cylinder

camera geometry is usually calibrated to this surface.

Given the information about the image, the sensor, and the expected shapes, we look for surface patches consistent with our expected shapes. For boxes and cylinders, these patches are either planar or cylindrical and, if planar, are either polygonal or circular. We look for adjacency relationships among these surface patches to detect any of our generic-shape models, e.g., two adjacent rectangular patches at right angles imply a box. These surface segmentation and adjacency extraction techniques have been described elsewhere [14].

At this point we form an initial hypothesis about the configuration of the objects in the scene. This first guess will consist of all matched objects. For all surfaces not matched to a generic shape, we add a three-dimensional object formed by treating the unmatched surface as a paper-thin entity. Figure 2 shows a typical example.

Except in rare cases, this initial hypothesis is unrealistic, because the region of contact between objects and the support surface is occluded by the upper surfaces of the objects themselves. The hypothesis suggests that the objects are floating in space. The question is: How do we fix this hypothesis?

In Section 2 below, we describe our technique for extending objects of the initial hypothesis to form a final hypothesis of the complete configuration of the pile. From the discussion on sensor geometry we know that these extensions must be directed into the occluded volumes of the imaged space. What may not be immediately apparent is the fact that the order in which these extensions are performed affects the final hypothesis. Figure 3 provides a simple example of this order dependence. Thus, there are many final hypotheses, one for each ordering of the extensions. In this enhancement of our previous work, we examine all possible extension orderings.

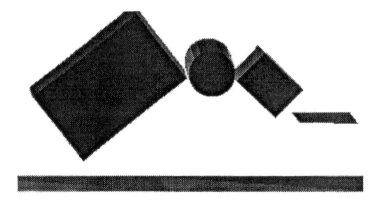

Figure 2. Initial Hypothesis for Image in Figure 1

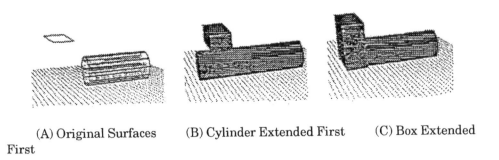

(A) Original Surfaces   (B) Cylinder Extended First   (C) Box Extended First

Figure 3. Scene Description is Dependent on Order of Extension

In Section 3 we discuss our free-body analysis technique for evaluation of each of the final hypotheses. This technique involves a quick, iterative solution of the free-body diagram (FBD) for each of the objects in a final hypothesis. The resulting set of computed forces and torques at the contact points between objects allows us to determine if the hypothesis is physically stable. Since we know the actual configuration of objects is stable, this technique provides a method to reject the unreasonable final hypotheses and select the true representation.

In Section 4 we conclude with suggestions for further research.

1.3. Sensor Description

The techniques described in this paper were developed for use on data from five different sensors. Three of the sensors were based on triangulation, one using a simple plane of light through which the piles of objects were transported on a conveyor, and two observing stationary piles of objects [15, 16]. The other two sensors used amplitude-modulated lidar [17], one for observing a static scene, the other for objects transported on a conveyor. The range data from each sensor is encoded as X, Y, and Z images, so that the data in each image are a function of column (U) and row (V) coordinates.

A fundamental aspect of our technique is to exploit knowledge of the sensor geometry. One important use of sensor geometry is the explicit representation of known free space, as described below in Section 2.1. A second use is to evaluate hypothesized data with respect to the observed data. Figure 5 illustrates the evaluation of a hypothesized 3-D point, P, with respect to a range image, R, from the sensor depicted in Figure 4. The evaluation calculates the camera image pixel at which P would have been recorded and compares it to the camera image actually obtained there. If the (x,y,z) of P_i matches that of any image data within a neighborhood of 3 pixels, then P is considered visible in the range image (positive evidence: score +1). If P_i is closer to the light source than the image data, then P should have been visible but was not found, implying that P is transparent (negative evidence: score -1). If P_i if farther from the light source, then P was occluded (neutral evidence: score 0). We refer to this evaluation procedure as the *Visibility Test*.

2. Hypothesis Generation

A complete configuration hypothesis is one in which every object from our initial (surface patch) hypothesis has been extended to its maximum dimensions within the occluded volume of space above the support surface.

Two factors physically limit these maximum dimensions. First, the hypothesized extension of an object's dimension must not imply that this additional portion of the object would have been visible to the sensor. We call this the *visibility criterion*. Second, the hypothesized extension must not enter volume already occupied by another object.

2.1. Representation of Space

The volume of space explored by the sensor consists of two separate pyramids, one with its apex at the camera and the other with its apex at the light source. These

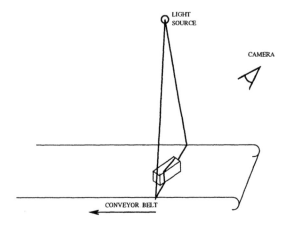

Figure 4. Range Sensor Configuration

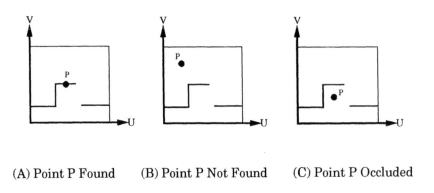

(A) Point P Found   (B) Point P Not Found   (C) Point P Occluded

Figure 5. The Visibility Test of a 3-D Point P

pyramids share a common base, namely the support surface. The only surfaces that can be imaged are those within the intersection of the two pyramids, where object surfaces may be illuminated by the light source *and* viewed by the camera.

Both factors limiting object extension are directly related to spatial occupancy. To enforce the limitations, we choose a voxel representation that encodes the occupancy status of the imaged volume. This representation consists of an array of elements each of which corresponds to a small volume of the imaged space. The size of this elemental volume may be chosen according to the desired resolution of the objects in the final configuration hypothesis. The spatial resolution of the sensor itself forms a lower limit on the voxel element size. In our experiments, the sensor spatial resolution was approximately 0.07 inches. Typical object sizes were on the order of 5 to 10 inches. We chose a voxel cube of 0.1 inch on a side. The occupancy of each of these elemental volumes is denoted by one of three integer values: *empty*, K (i.e., occupied by object K), and *freespace*.

Freespace corresponds to the space between visible data points and the camera or the light source. Freespace must be free of any objects; otherwise, the data points would not be visible to the sensor. K corresponds to the space enclosed by the surface of the *Kth* object in the initial hypothesis. The motivation for marking the voxel with the object number (as opposed to simply marking it *occupied*) will become clear below. The third integer value, *Empty*, corresponds to regions within the imaged space that are neither freespace nor occupied by an object.

The total number of different values a voxel must store is the maximum number of objects expected in the scene plus two (one value to designate *empty* and one value to designate *freespace*). We have used one byte (eight bits) per element, permitting as many as 254 objects to be represented in the scene.

Initialization of the voxel array proceeds as follows:

1. Mark Empty: Set the value of all voxels to *empty*.

2. Mark Freespace: For each data point in the image, find the line segment from the 3-D location of that point to the 3-D location of the camera. Set to *freespace* the value of each voxel through which the line segment passes. Repeat this process for the line segment from the data point location to the light source location.

3. Mark Objects: For each *Kth* object in our initial hypothesis, produce a sampling of points on its surface. The resolution of this sampling must be finer than that of the voxel array, to ensure that no voxel corresponding to location on the object surface is left unmarked. Set to K the value of voxels bounding the sample points from the *Kth* object.

When this initialization is complete, we have a data structure that can answer the question "Can object K be extended into the space near the point at (x,y,z)?" The answer is "yes" only if the voxel corresponding to (x,y,z) is marked as *empty*.

2.2. Enumerating Possible Extensions

Given a set of N objects in our initial hypothesis, we would like to explore all possible extensions of these initial shapes into occluded regions within the imaged space. How many possible extensions are there? If our set of generic shapes consists only of boxes and cylinders, we might make the following first estimate:

- A box is extensible in the six directions of its surface normals.

- A cylinder is extensible in two directions corresponding to the surface normals of the two circular end faces.

The total number of different extensions is $6 \times B + 2 \times C$, where B is the number of boxes and C is the number of cylinders. Because the order in which these extensions are made affects the final configuration hypothesis, we should, theoretically, examine all $(6 \times B + 2 \times C)!$ cases. The values for B and C in any complex scene make examination of all possible cases impractical. For example, the scene shown in Figure 1 gives $B = 3$, $C = 1$, giving $6 \times 3 + 2 \times 1 = 19$ possible directions and 19! possible final hypotheses.

To reduce the permutations to a manageable set, we perform a preliminary test on each possible direction to see if it can be extended. The technique for this test is identical to the extension mechanism described in Section 2.3. This test labels each direction as extensible or nonextensible and results in the elimination of at least one and usually most of the possible directions for each of the initial objects that were hypothesized from visible surface patches. On extension, at least one surface on each of these objects immediately collides with the known freespace between the sensor and the data points that gave rise to the patches. See, for example, Figure 6A. A similar condition eliminates many of the side faces of the initial objects, as illustrated in Figure 6B.

When this preliminary test is performed on the initial hypothesis of the scene in Figure 2, both extensions of the cylinder are eliminated. Only the extension away from the sensor is possible for each of the two tilted boxes. Two extensions remain for the thin horizontal box—one away from the sensor and the other horizontally underneath the tilted boxes and the cylinder (see Figure 7). This leaves us with only 4! = 24 extension permutations.

We had hoped to further trim the number of possible extensions by dividing them into independent subsets. Extensions in each subset would have no effect on extensions from any other subset. Unfortunately, in most cases these independent subsets either do not exist or have very few members, because objects in a pile tend to lean on each other.

We are now ready to enumerate all possible final hypotheses for the configuration of a pile of N objects in a given scene. The first ordering of extensions is simply the list as it is constructed, beginning with the first possible extension of object 1 and ending with the last extension of object N. As described below, we perform these extensions in order. This generates the first final hypothesis. The entries in the extension list are then permuted and the next final hypothesis is generated. This cycle continues until all permutations have been tried.

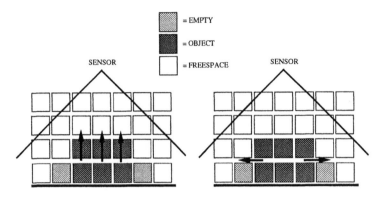

(A) Blocked Extension Toward the Sensor   (B) Blocked Lateral Extension

Figure 6. Extensions Limited by Freespace

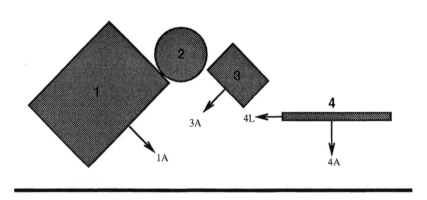

Figure 7. Possible Extensions for Scene with Three Boxes and Cylinder

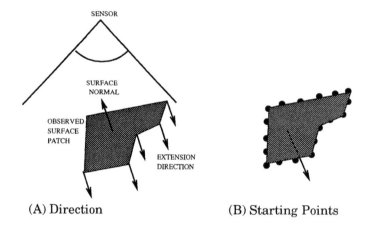

(A) Direction   (B) Starting Points

Figure 8. Axial Extension

2.3. Performing Extensions

We distinguish between two extension types: *Axial*, those that extend an object in a direction opposite the sensor along the normal to a surface patch detected in the image, and *Lateral*, those that extend an object perpendicular to the normal of the detected patch.

2.3.1. Axial Extensions

An axial extension occupies the space behind a visible surface patch (see Figure 8A). Such an extension does not change the shape of the surface patch; only its hypothesized depth in the occluded region may change.

To begin an extension, we determine the direction for the extension and a sampling of points from which to start it. The extension from each sample point stops when that extension collides with another object or with freespace.

The direction of an axial extension of an object face is opposite to the normal of the visible face (see Figure 8A). To generate starting points for our extension, we should, in principle, sample the entire face at a finer resolution than was used to create the freespace/object voxel array. This would ensure that no occupied voxel would be bypassed as the extension proceeded. In practice, however, we have found it sufficient to test the extension from the center point and a sample of points along the boundary of the face (see Figure 8B).

The size of the extension is then determined by the following algorithm, repeated for each starting point in our sample:

1. Scale the direction vector so that its maximum component equals the resolution of the voxel array. This will permit us to move through the space corresponding to the array without skipping voxels or testing the same voxel

Figure 9. Axial Extension of Large Tilted Box from Figure 2

twice. The starting point becomes the first test point.

2. Move the location of the test point along one unit of the direction vector. Find the voxel corresponding to the new location of the test point.

3. Check for collision of the above voxel with any occupied space. If we are extending Object K, a collision occurs if the voxel contains any value except *empty* or K. The value of K is allowed because the granularity of the voxel array may already have caused the neighborhood of the starting point to be marked as occupied by K.

4. If a collision is detected, stop and report the distance of the final test point from the starting point; otherwise, continue at Step 2.

The size of the extension is the minimum distance returned by the above algorithm over all starting points. The resulting extension of the largest boxlike object in the initial hypothesis of Figure 2 is shown in Figure 9. The box was extended downward (away from the sensor) until it collided with the volume occupied by the support surface.

2.3.2. Lateral Extensions

A lateral extension changes the shape of a visible boundary by extending its edges into nonviewable regions (see Figure 10). Typically, such a situation arises if a portion of the object is occluded from the camera or the light source by another object. In these situations we assume that any hypothesized extension of a boundary edge must result in a shape that is consistent with one of the generic shapes we expect to see in the scene. Thus, the hidden end of a partially visible box face should not

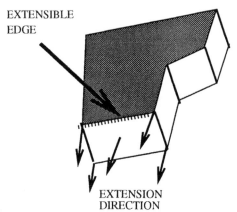

EXTENSIBLE
EDGE

EXTENSION
DIRECTION

Figure 10. Lateral Extension

be hypothesized to be round. The visible portion of a surface boundary is used as a hint of the actual configuration in the occluded region.

Note that if the entire *visible* surface boundary is composed of occluded edges, no extension direction is implied and thus the lateral extension cannot be performed. We expect to address this situation in future work.

The algorithm for lateral extensions is identical to that for axial extensions except in the selection of direction vectors for the extension. Currently we explore lateral extensions for rectangular boundaries only. Two possible cases arise, one along the parallel edges of the rectangle, the other at a corner.

Note that each edge of a polyhedron belongs to two intersecting surfaces. An edge of one surface's boundary is labeled extensible or nonextensible according to the extensibility (as determined above in Section 2.2) of the other surface to which the edge also belongs.

If two nonextensible edges on a rectangular boundary are parallel, we are extending a box along one of its dimensions (see Figure 11). If two nonextensible edges form a right angle, we are extending the visible corner of a box (see Figure 12). In both cases we choose extension directions along the edge directions. In the case of parallel edges, the extension direction is single. For a box corner, extension is in two orthogonal directions.

To generate starting points, we sample the boundary and center of the extensible face of the box in the same manner as for axial extensions.

For extension in a single direction, the extension algorithm outlined above is used to find the minimum noncolliding distance. This minimum distance fully determines the object extension. For extension in a double direction we apply the extension algorithm to each direction separately. This results in two minimum extension distances corresponding to the two extremes of the object's rectangular boundary (see Figure 12). Because we know that the object's face is rectangular but we do not know its dimensions, the extension technique is not adequate to deter-

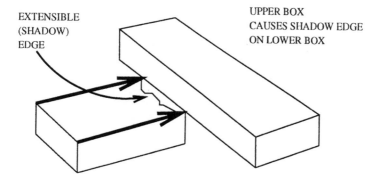

Figure 11. Box Extension Along Parallel Edges

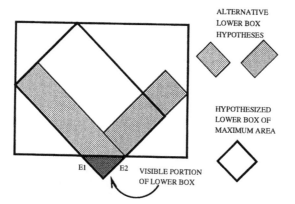

Figure 12. Ambiguity in Extension of Box Corner

Figure 13. Lateral Extension of Horizontal Box from Figure 2

mine which shape we should choose in the continuum between the two extremes. Some help may come from parameterizing the free-body analysis discussed below, so as to restrict the continuum to a set of rectangles that would provide stability. This parameterization has not yet been implemented. Currently, we choose the rectangular shape of maximum area.

The result of the single-direction, parallel-edge lateral extension of the horizontal box in Figure 2 is shown in Figure 13.

2.3.3. Incorporating Extensions

When an extension succeeds, i.e., when the minimum collision distance is greater than zero, the object representation is enlarged to match the predicted extension. The surface of this enlarged object is then sampled at the voxel resolution and the voxels corresponding to the sample points are marked as occupied by that object. Note that it is necessary to sample only the surface of an object, not its volume, for marking the voxel array, because any extension by another object must collide with surface voxels before it collides with interior voxels.

At this point, the updated hypothesis is examined to see if any smaller objects appear to be portions of the newly enlarged one. This occurs if the surface segmentation process has interpreted a single surface as two distinct surfaces in the initial hypothesis.

In a similar fashion, the updated hypothesis may now imply that two objects could be different pieces of the same object. This happens, for example, when an object occludes the middle portion of a cylinder, leaving the two ends visible as distinct cylinders. If one of the ends is extended until it collides with the other, the

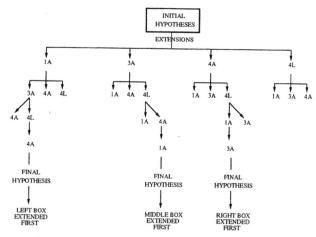

Hypothesis Permutation Tree

(A) Left Box Extended First   (B) Middle Box Extended First   (C) Right Box Extended First

Figure 14. Effect of Extension Order on Final Hypothesis

two coaxial cylinders of equal radii that are now present in the hypothesis can be linked into a single object. Details on both these procedures for inclusion of smaller objects have been described by Mulgaonkar et al. [14].

When all extensions in the current permutation have been performed, we are left with a final hypothesis for the configuration of the pile. This hypothesis will be consistent with the sensor geometry and all acquired image data. It is then evaluated, using the free-body technique described below, and is saved if rated best of all final hypotheses produced so far.

The extension order is then permuted, the voxel array and hypothesis returned to their initial states, and the extension process repeated. Figure 14 shows the complete set of final hypotheses for the objects in the initial hypothesis of Figure 2. These fall into three groups, depending upon which object was extended first.

3. Hypothesis Evaluation

Each final hypothesis for the configuration of the pile accounts for all the visible data. Thus, from an image-processing standpoint, there is no way to discriminate among these alternative configurations. We therefore turn to a non-visual approach, using the physical consequences of each proposed configuration to rule out impossible arrangements and to rank the remaining, plausible hypotheses.

The use of physical properties to aid image understanding is not new. The class of shape-from-x techniques uses a priori physical knowledge to determine various geometric properties of the scene acquired by the sensor. Here we reverse this process by computing a physical property from the geometry inferred by the image and using that property to further the interpretation of the image. That physical property is *stability*. Stability is tested by first computing a free-body diagram (FBD) for each object in a final hypothesis. The free-body diagram denotes the direction and magnitude of all forces and torques acting on an object from any source external to the object. An object is stable when the forces and torques in its FBD sum to zero.

3.1 Preliminary Computations

For each final hypothesis, several preliminary computations must be performed before computing the FBD. These include determination of the contact points between objects, the directions of normal forces at the contact points, the values of the coefficient of friction at these contact points, the identification of objects that give no support to other objects, and an efficient ordering of the objects for the iterative FBD computation.

We first compute the contact points between one object and another and between any object and the support surface. As in the hypothesis extension algorithm, we use a voxel approach because it is easy to implement and is fast for this application. To find the contact points, the bounding boxes of each pair of objects are computed and divided into voxels. These voxels are initialized to *empty*. Next, all voxels containing a point on the surface of the first object in the pair are marked as *occupied*. If the voxel corresponding to any point on the surface of the second object in the pair is already marked as *occupied*, the location corresponding to the center of the voxel is recorded as a contact point.

Depending on the configuration of the object pair and the voxel resolution chosen (we use three times the resolution of the extension voxels), this process typically yields several dozen contact points. If possible, this collection is simplified to a point, a line, or polygonal convex hull around the set of contact points. For rigid objects, it is sufficient to determine the forces at the points in this simplified set [18]. The simplification process uses line and plane fitting algorithms to determine if the contact set is amenable to reduction.

The maximum magnitude of a static frictional force at a point of contact between two objects is:

$$|\mathbf{F}| = \mu |\mathbf{N}| \tag{1}$$

where N is the normal force at the contact point and μ is the coefficient of friction.

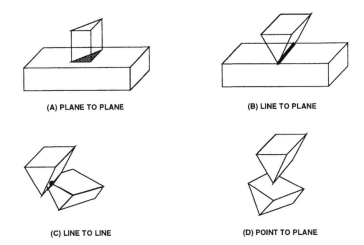

(A) PLANE TO PLANE

(B) LINE TO PLANE

(C) LINE TO LINE

(D) POINT TO PLANE

Figure 15. Types of Object Contact

Although the free-body technique to be described accommodates different coefficients of friction at the various contact points, all objects in our test scenes were paper-wrapped boxes and cylinders typical of a mail room environment. Thus, a constant coefficient was used throughout.

The object geometry at the point of contact determines whether the surface normal is well defined or not. Figure 15 shows the four types of contact we typically encounter. For plane-to-plane contact, we simply use the normal of either plane. For line-to-plane and point-to-plane contact, we use the normal direction of the single planar surface. For line-to-line contact, the surface normal is degenerate. We therefore choose the normal at such a point as the direction mutually perpendicular to the two edges.

The range of possible resultant force vectors (i.e., the sum of the normal and frictional forces) at each contact point is described by the *friction cone* (see Figure 16). The apex of the cone is located at the contact point, and the radius of the base per unit height is μ. The cone is directed *into* the object because we have allowed only pushing forces in our hypotheses. A cone directed outward would imply the existence of glue.

To speed the computation of forces, it helps to note which objects do not support any other object. These *nonsupporting* objects are typically found at the top of a pile. Once the contact points have been found, an object is labeled nonsupporting if no vector within any of its friction cones has a negative Z component, i.e., is directed against gravity. The force-torque balance for these objects can then be determined once and will not be affected by the iterations of the balancing algorithm.

The force-torque balance of each object in a pile is intertwined with the balance of other objects. Thus, balancing one object may mean unbalancing another. We

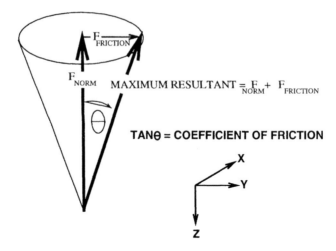

Figure 16. Friction Cone Definition

want to minimize this see-saw effect within the algorithm by choosing an optimal ordering of objects to which we apply the balancing act. We apply Fahlman's method [19], ordering the objects by the height of their centroids, but with a slight modification: we place all nonsupporting objects at the front of the list, regardless of their heights.

3.2. Balancing Forces and Torques

Following the preliminary computations, we analyze the balance of forces and torques within each hypothesized configuration of the pile. Treating each object within the hypothesized configuration of the pile as a free body, this process evaluates all the gravitational and contact forces acting on the object. We attempt to find a static solution, i.e., we try to assign plausible forces to each contact point such that the vector sum of all forces and the vector sum of all torques on the object each add to zero. The solution of the free-body problem for any one object is related to all others in the pile in accordance with Newton's First Law: the force exerted on Object A by Object B at a given contact point is equal and opposite to the force exerted on Object B by Object A at the same point.

For a given object, the algorithm iteratively selects contact points and updates forces at these points in an attempt to bring the force and torque sums below their respective thresholds. When this is achieved, the object balance is declared to be verified, and the algorithm repeats this computation for the next object in the ordered list.

The thresholds for force balance and torque balance must be selected with some care. If a postcard leans against a heavy package, a force-balance threshold of, say, ±10 pounds might be adequate for the heavy package but not for the postcard.

Furthermore, the package's balance cannot be calculated to within the milligrams required for a postcard, because such a threshold is much smaller than the errors in modeling the weight and dimensions of the package. Fortunately, piles are usually composed of objects of similar weight and dimensions.

3.2.1. Balance Thresholds

We have set the force-balance threshold to one percent of the average object weight in the initial (unextended) hypothesis. This threshold is tighter than the same calculation would yield on any of the final (extended) hypotheses, because the dimensions in the initial hypothesis are at their minimum. In our experiments we have assumed a constant density for all objects, so that their weight is directly proportional to their volume. Note, however, that the free-body analysis described below does not depend on this assumption.

To determine the torque balance threshold, we find the largest radial arm for each object. (A *radial arm* is a vector from the center of an object to a point on the object's surface). At the endpoint of the maximal arm of each object, we calculate a torque,

$$\mathbf{L} = \mathbf{R} \times \mathbf{F} \tag{2}$$

where $\mathbf{R}$ is the largest radial arm vector of the object and $\mathbf{F}$ is the force-balance threshold. We take one percent of the average of these computed torques as our torque balance threshold. Both the force- and torque-balance thresholds are constant during the evaluation of all final hypotheses.

3.2.2. Balance Algorithm

The free-body algorithm begins with the first object on the ordered list. (Typically, this is a nonsupporting object initially subject only to gravity, such as the cylinder in Figure 14B.) The vector sum of all forces acting on that object is computed. The data structure describing a contact point between Object A and Object B maintains the current value of the force applied to Object A at that point. This force is equal in magnitude and opposite in direction to the force Object A applies to Object B. Summing the forces on a given object consists of initializing the sum to the object's weight and adding the force at each contact point on the given object.

The next step involves selecting a candidate subset of the object's contact points at which the forces will be changed. These changes will counter a portion of the current sum of forces on the object. In this way we hope to eventually reduce the current sum to zero. The complete set of contact points for a given object may not be available for force modification. (See the long, slanted box of Figure 14B.)

Permanently excluded from modification are points of contact with nonsupporting objects. The forces at these points of contact are fixed by the free-body algorithm's operation on the nonsupporting objects themselves, and by definition, are only a function of the nonsupporting object's weight (e.g., the force exerted on the long slanted box by the cylinder). In addition, all contact points with objects to which the free-body algorithm has already been applied are initially excluded from consideration (e.g., the force exerted on the long slanted box by the large slanted box).

If the current object can be balanced without changing the forces applied to previously balanced objects, then previous efforts will not be undone. If balancing fails for an object, these initially excluded points will be brought into the set of candidate changeable forces.

A given object's contact points (and their associated friction cones) provide the complete set of possible applicable forces. This finite-dimensional set differs from the usual vector basis in several ways. First, although only one force vector exists at each contact point, the number of vectors to choose from within the friction cone is infinite. Second, the resultant force at a contact must remain within the cone (see Figure 16). Finally, the changes made to the contact forces also affect the sum of the torques on the object.

Rather than attempt a coordinated solution of all components of the balance problem at once (i.e., a linear programming approach), we simply compute a change in the contact forces sufficient to counter the maximum component of the current total force or torque on an object. This countering of the maximum component is repeated until the sum of forces and torques on the object is brought below threshold.

To counter the maximum component of the current total force on an object, we find, within our candidate set of contacts, a subset whose friction cones contain vectors that can oppose that maximum component. Typically, a friction cone that contains one such vector will contain many such vectors. Which one should be chosen as the new contact force? If the nonfrictional force vector (i.e., the cone axis) is among the possibilities, we choose it. This reflects the fact that frictional forces arise only when necessary. For instance, a block sitting on a flat, level table uses no frictional force at its contact points, even though the coefficient of friction there is greater than zero (see Figure 17A). This preference for nonfrictional forces has some built-in fuzziness. The optimal counterforce must be far enough off the axis of the friction cone for us to choose a frictional force over a frictionless force (see Figure 18). Our threshold value for far enough off axis is half the width of the friction cone. If frictional forces are present within the pile, the tightness of this threshold affects the speed at which the algorithm converges to a balance solution and the magnitude of non-necessary frictional forces present in that solution (see Figure 17B).

The contacts selected as candidates are further filtered to remove those whose contribution toward countering the resultant force is less than half the average contribution of all candidates.

Once our final candidate set of contacts has been established for a given iteration on a given object, we calculate the force contributed by each contact to counter the maximum component of the current sum of forces. To do this, we find the value for k in the linear combination

$$S_{max} = -k \sum C^i_{max} \tag{3}$$

where S_{max} is the value of the maximum component of the current sum of forces and C^i_{max} is the corresponding component of the normalized candidate vector from

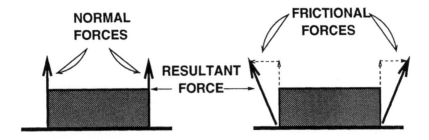

(A) No Frictional Forces   (B) Frictional Forces Present

Figure 17. Two Solutions with $\sum F = 0$ and $\sum L = 0$

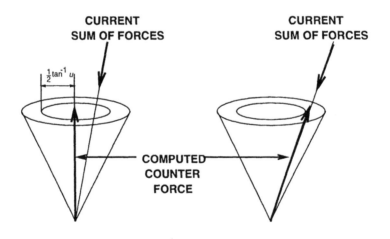

(A) Counter Without Friction   (B) Counter Using Friction

Figure 18. Choosing When to Apply Frictional Force at Contact Point

contact point i. The force to be added to the existing force at contact i is then kC^i_{max}. We form the new resultant force at contact point i:

$$\mathbf{F}^i_{new} = \mathbf{F}^i_{current} + k\mathbf{C}^i_{max} \tag{4}$$

and check to ensure that it lies within the friction cone at point i. If it does not, we reduce the value of k for this contact point until the resultant force is within the cone. This reduction in k, if it is necessary, means our counter of the maximum component of the sum of forces will only be approximate. Further counterbalance is left to the next iteration over this object.

After countering the maximum component of the current sum of forces, the sum of torques is computed and the sequence of candidate contact selection and component balancing is applied to the torques. The only difference between force and torque balancing is that when a torque update is postulated, the force that gives rise to the update must be computed. This is necessary because contacts are maintained in terms of force rather than torque. The magnitude of the force corresponding to the desired torque $\mathbf{L}$ is

$$|\mathbf{F}| = \frac{|\mathbf{L}|}{|\mathbf{R}| \sin \theta} \tag{5}$$

where $|\mathbf{R}|$ is the length of the radial arm from object center to contact point, and $\sin \theta$ is the angle between $\mathbf{F}$ and $\mathbf{R}$. The $\sin \theta$ term means we have a range of possible magnitudes and directions for $\mathbf{F}$. We choose $\mathbf{F}$'s direction so as to minimize $|\mathbf{F}|$.

When the torque pass is complete, the new sums of forces and torques are computed. If the magnitude of *both* of these sums is below their respective thresholds, the object is declared to be balanced. Otherwise, a test is made to see if the sums are converging toward zero. If they are, but have not yet reached threshold, the balance process is repeated. For a stable, symmetrically supported object, one or two passes are generally sufficient to obtain sums within threshold. Figure 19 shows the friction cones and the computed force at each contact point for the horizontal box of Figure 14B.

Note that the lower four contact points are the reduced set determined from the convex hull of the plane-to-plane contact between the box and the support surface. Similarly, the upper two contact points are the reduced set from the line-to-plane contact between the box and the long, slanted box of Figure 14B.

Sometimes no candidate vectors can be found to counter the current sum of forces or torques on an object, or the sums are not converging after several cycles. If either of these conditions occurs, the contact points that were initially excluded from candidacy because their forces had already been used in balancing another object are now made available for selection and change. The balance algorithm is repeated for the current object with the newly enlarged set of candidate contact points. If the force at one of these contact points is changed, the previously balanced (other) object to which it belonged must be reanalyzed. If again we find no candidate

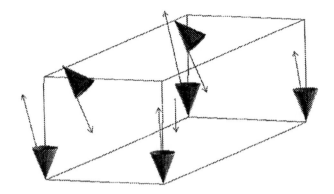

Figure 19. Free-Body Forces on Horizontal Box of Figure 14(B)

vectors or a nonconverging sequence, the current object is declared to be unbalanced and the process moves on to the next unbalanced object in the ordered list.

The balancing algorithm may be summarized as follows:

1. Select the next unbalanced object in the ordered list.

2. If no unbalanced objects are left, rate the hypothesis as stable and exit.

3. Compute the vector sum of forces and torques on the object.

4. If both the force and torque sums are below their respective thresholds, mark the object as balanced and go to Step 1.

5. Find a candidate set of contact forces such that a combination of these forces will yield a force to counterbalance the maximum component of the current sum of forces on the object.

6. Update each contact force according to its coefficient in the balancing combination of Step 5.

7. Compute the new sum of torques using the force updates from the previous step.

8. Find a candidate set of contact torques that, in combination, will yield a torque that counterbalances the maximum component of the current sum of torques on the object.

9. Derive the forces corresponding to the combined, counterbalancing torques in Step 8. Update each corresponding contact force.

10. If both force and torque sums are not converging, or no candidate set could be found for either the force or torque sum, enlarge the set of candidate contacts to include those from previously balanced (but not nonsupporting) objects. If this has already been tried, mark the object as unbalanced and go to Step 1. Otherwise, continue at Step 3.

3.3. Stability Rating

The free-body analysis procedure outlined above partitions the objects of an hypothesis into two classes: balanced and unbalanced. For the hypothesis corresponding to the true configuration of a pile, we expect all objects to be in the balanced class.

We establish a stability rating for each hypothesis as the number of balanced objects divided by the total number of objects. Thus, a completely balanced hypothesis has a rating of 1.0, while all others are rated at a lower value.

The three final hypotheses for our example are shown in Figure 14B. The stability of the left-box-first and right-box-first hypotheses are rated 0.25, while the middle-box-first hypothesis is rated 1.0. This stability evaluation identifies the interpretation that matches the actual configuration of the pile.

In addition, knowledge of the contact force relationships yielded by the stability analysis permits the immediate use of intelligent grasping techniques and paves the way for future areas of research described below.

4. Conclusions

In this paper we have described our techniques for the generation and evaluation of alternate hypotheses for the configuration of a pile of objects. These techniques have proved to be useful in predicting the size and shape of the hidden portions of generically shaped objects in a scene and also in estimating the force relationships among the objects. These techniques, like the hidden regions they hypothesize, seem to be relatively unexplored in image-understanding research.

One immediate possibility for future work is the inclusion of a larger set of generic shapes. This would not only make the process more robust but would also help define its limits. In addition, the larger shape set would allow the exploration of assumptions about the coefficient of friction as a function of shape and size.

In the longer term, free-body analysis could be pushed forward in the image understanding process. It could be used to suggest the appropriateness of one extension over another, in order to reduce the explosive number of alternative hypotheses. Beyond that, it might be used to suggest the existence of an entirely unseen objects underneath the visible portion of a pile.

ACKNOWLEDGEMENTS

Research work leading to this paper was funded by the Office of Naval Research under contract N00014-90-C-0063. The range images used in the research were provided by the USPS Office of Advanced Technology.

REFERENCES

1 R.C. Bolles, P. Horaud, and M. J. Hannah. 3DPO: A three-dimensional part orientation system. In *Proc. 8th IJCAI*, Karlsruhe, West Germany, August 1983. MIT Press, Cambridge, Massachusetts.

2 B. Bhanu. Representation and shape matching of 3-d objects. *IEEE Transactions on Pattern Analysis and Machine Intelligence*, PAMI-6(3):340–350, May 1984.

3 O.D. Faugeras and M. Hebert. A 3-d recognition and positioning algorithm using geometrical matching between primitive surfaces. In *Proc. 7th IJCAI*, August 1981.

4 T.J. Fan, G. Medioni, and R. Nevatia. Segmented descriptions of 3-d surfaces. *Robotics and Automation*, pages 527–538, December 1987.

5 M. Brady, J. Ponce, A. Yuille, and H. Asada. Describing surfaces. *CVGIP*, 32:1–28, 1985.

6 P. Besl and R. Jain. Segmentation through symbolic surface description. In *Proc. IEEE Conf. on Computer Vision and Pattern Recognition*, pages 77–85. IEEE Computer Society, 1986.

7 M. Potmesil. Generating models of solid objects by matching 3d surface segments. In *Proc. 8th IJCAI*, pages 1089–1093, Karlsruhe, West Germany, August 1983. MIT Press, Cambridge, Massachusetts.

8 C. Dane and R. Bajcsy. An object-centered three-dimensional model builder. In *Proc. 6th ICPR*, pages 348–350, Munich, West Germany, October 1982. MIT Press, Cambridge, Massachusetts.

9 C. I. Connolly. The determination of next best views. In *Proc. International Conference on Robotics and Automation*, pages 432–435, St. Louis, Missouri, March 1985. IEEE.

10 P. Besl and R. Jain. Three dimensional object recognition. *Computing Surveys*, 17(1):75–145, March 1986.

11 A. P. Pentland. Recognition by parts. Technical Report 406, SRI International, Menlo Park, California, December 1986.

12 T. E. Boult and A. D. Gross. Recovery of superquadrics from depth information. In *Proc. AAAI Workshop on Spatial Reasoning and Multi-Sensor Fusion*, St. Charles, Illinois, October 1987. Morgan Kaufmann Publishers, Inc., Los Altos, California.

13 F. Solina and R. Bajcsy. Recovery of parametric models from range images: The case for superquadrics with global deformations. *IEEE Trans. on PAMI*, 12(2):131–147, February 1990.

14 P. G. Mulgaonkar, C. K. Cowan, and J.L. DeCurtins. A system to understand object configurations using range images. Technical Report ITAD-8842-TN-91-6, SRI International, Menlo Park, California, January 1991. Accepted for publication in IEEE PAMI Special Issue on Interpretation of 3-D Scenes, Part II. The issue is expected in December, 1991.

15 G. Pierce, K.C. Nitz, and M. McDowell. Sri three dimensional range image acquisition system. In *Proc. 3rd. USPS Advanced Technology Conference*, pages

32–42. United States Postal Service, May 1988.

16 J.P. Rosenfeld and C. Tsikos. High speed space encoding projector for 3d imaging. *SPIE Proceedings*, 728:146–151, October 1986.

17 C. Jacobus, A. Riggs, L.M. Tomko, and K.G. Wesolowicz. Laser radar range imaging sensor for postal applications. In *Proc. 3rd. USPS Advanced Technology Conference*, pages 6–31. United States Postal Service, May 1988.

18 R. Featherstone. The dynamics of rigid body systems with multiple concurrent contacts. In *Proc. Third International Symposium on Robotics Research*, pages 24–31, Gouvieux, France, October 1985.

19 S.E. Fahlman. *A Planning System for Robot Construction Tasks*. PhD thesis, Massachusetts Institute of Technology, Cambridge, Massachusetts, May 1973.

Three-Dimensional Object Recognition Systems
A.K. Jain and P.J. Flynn (Editors)
© 1993 Elsevier Science Publishers B.V. All rights reserved.

Modal Descriptions for Modeling, Recognition, and Tracking

Alex Pentland, Stan Sclaroff, Bradley Horowitz, and Irfan Essa[a]

[a]Vision and Modeling Group
The Media Laboratory
Massachusetts Institute of Technology
Cambridge, Massachusetts 02139 USA

Abstract

An efficient and reliable method for recovering, recognizing, and tracking 3-D solid models from 2-D and 3-D measurements is presented. Because the modal representation is based on the finite element method, the dynamics of the observed object can be accurately modeled. As a consequence, optimal estimates of object motion and shape can be made, and physical predictions and simulations can be made directly from recovered models. The modal representation is unique except for rotational symmetries, and in typical situations the solution is overconstrained. Consequently recognition and database search using the recovered 3-D models is both efficient and stable. The performance of the method is evaluated using synthetic range data with various signal-to-noise ratios, laser rangefinder data, X-ray imagery, and normal intensity imagery.

1. Introduction

Vision research has a long tradition of trying to go from collections of low-level measurements to higher-level "part" descriptions such as generalized cylinders [3, 12, 13], deformed superquadrics [14, 16, 20, 23], or geons [2, 6], and then of attempting to perform object recognition. The general idea is to use part-level modeling primitives to bridge the gap between image features (points, edges, or corners) and the symbolic, parts-and-connections descriptions useful for recognition and reasoning.

Recently, several researchers have successfully addressed the first of these problems — that of recovering part descriptions — using deformable models. There have been two classes of such deformable models: those based on parametric solid modeling primitives, beginning with our own work on superquadrics [14], and those based on mesh-like surface models, such as employed by Terzopoulos, Witkin, and Kass [22]. In the case of parametric modeling, fitting has been performed using the modeling primitive's "inside-outside" function [14, 20, 16], while in the mesh surface models a physically-motivated energy function has been employed [22].

The description of shape by use of a parametric solid with orthogonal parameters has the advantage that it can produce a unique, compact description that is well-suited for recognition and database search, but has the disadvantage that it may not have enough degrees of freedom to account for fine surface details. The deformable mesh approach, in contrast, is very good for describing shape details, but produces descriptions that are neither unique nor compact, and consequently cannot be used for recognition or database search without additional layers of processing. Both of these approaches share the disadvantage that shape recovery is relatively slow, requiring dozens of iterations in the case of the parametric formulation, and up to hundreds of iterations in the case of the physically-based mesh formulation.

We have improved on these previous techniques by adopting an approach based on both the finite element method (FEM) and on parametric solid modeling. This provides both the convenience of the physically-based mesh formulation, and the canonical representation of the parametric approach. Because the representation employed is based on the finite element method, the dynamics of the observed object can be accurately modeled. As a consequence, optimal estimates of object motion and shape can be made even in non-stationary enviroments, and physical predictions and simulations can be made directly from recovered models [15, 16].

More importantly, however, we have been able to obtain develop a formulation whose degrees of freedom are orthogonal, and thus **decoupled**, by posing the dynamic equations in terms of the FEM equations' eigenvectors. These eigenvectors are known as the object's *free vibration* or *deformation modes*, and together form a frequency-ordered orthonormal basis set analogous to the Fourier transform.

By decoupling the degrees of freedom we achieve substantial advantages:

- The fitting problem has a simple, efficient, closed-form solution.

- The model's intrinsic complexity can be adjusted to match the number of degrees of freedom in the data measurements. Consequently the solution can always be made overconstrained, and thus stable and noise-resistant.

- The shape representation is unique, except for rotational symmetries and degenerate conditions. Thus the solution is well-suited for recognition and database tasks.

The plan of this paper is to first review our representation, and to show how it may be used to obtain closed-form solutions to the shape recovery problem. We will then show how recovered object models may be efficiently compared to stored models to achieve accurate object recognition using either 2-D or 3-D data. Finally, we will demonstrate how the representation can be used to construct Kalman filters that estimate shape and motion estimation from image sequences.

2. Background: The Representation

Our representation describes objects using the force-and-process metaphor of modeling clay: shape is thought of as the result of pushing, pinching, and pulling on a lump of elastic material such as clay [14, 16]. The mathematical formulation is based on the finite element method (FEM), which is the standard engineering technique for simulating the dynamic behavior of an object.

In the FEM, energy functionals are formulated in terms of nodal displacements U, and iterated to solve for the nodal displacements as a function of impinging loads R:

$$M\ddot{U} + C\dot{U} + KU = R \tag{1}$$

This equation is known as the FEM **governing equation**, where U is a $3n \times 1$ vector of the (x, y, z) displacements of the n nodal points relative to the object's center of mass, M, C and K are $3n$ by $3n$ matrices describing the mass, damping, and material stiffness between each point within the body, and R is a $3n \times 1$ vector describing the x, y, and z components of the forces acting on the nodes.

When a constant load is applied to a body it will, over time, come to an equilibrium condition described by

$$KU = R \tag{2}$$

This equation is known as the **equilibrium governing equation**. The solution of the equilibrium equation for the nodal displacements U is the most common objective of finite element analyses.

In the type of shape modeling done in computer vision, sensor measurements are used to define virtual forces which deform the object to fit the data points. The equilibrium displacements U constitute the recovered shape. For additional detail, see reference [17].

2.1 Modal Analysis

To obtain an equilibrium solution U, one integrates Equation 1 using an iterative numerical procedure at a cost proportional to the stiffness matrices' bandwidth. To reduce this cost we can transform the problem from the original nodal coordinate system to a new coordinate system whose basis vectors are the columns of an $n \times n$ matrix P. In this new coordinate system the nodal displacements U become **generalized displacements** $\tilde{U}$:

$$U = P\tilde{U} \tag{3}$$

Substituting Equation 3 into Equation 1 and premultiplying by P^T transforms the governing equation into the coordinate system defined by the basis P:

$$\tilde{M}\ddot{\tilde{U}} + \tilde{C}\dot{\tilde{U}} + \tilde{K}\tilde{U} = \tilde{R} \tag{4}$$

where

$$\tilde{M} = P^T M P; \quad \tilde{C} = P^T C P; \quad \tilde{K} = P^T K P; \quad \tilde{R} = P^T R \tag{5}$$

With this transformation of basis, a new system of stiffness, mass, and damping matrices can be obtained which has a smaller bandwidth then the original system.

The optimal basis Φ has columns that are the eigenvectors of both M and K [1]. These eigenvectors are also known as the system's *free vibration modes*. Using this transformation matrix we have

$$\Phi^T K \Phi = \Omega^2, \qquad \Phi^T M \Phi = I \qquad (6)$$

where the diagonal elements of Ω^2 are the eigenvalues of $M^{-1}K$ and remaining elements are zero. When the damping matrix C is restricted to be *Rayleigh damping*, then it is also diagonalized by this transformation.

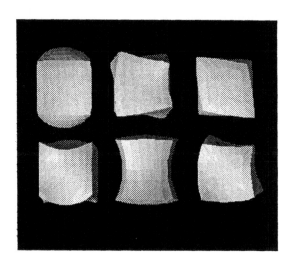

Figure 1. A few of the vibrations mode shapes of a 27 node isoparametric element.

The lowest frequency modes are always the rigid-body modes of translation and rotation. The next-lowest frequency modes are smooth, whole-body deformations that leave the center of mass and rotation fixed. Compact bodies — solid objects like cylinders, boxes, or heads, whose dimensions are within the same order of magnitude — normally have low-order modes which are intuitive to humans: bending, pinching, tapering, scaling, twisting, and shearing. Some of the low-order mode shapes for a cube are shown in Figure 1. Bodies with very dissimilar dimensions, or which have holes, etc., can have very complex low-frequency modes.

2.2. Modeling Using Implicit Functions

It is important to have a unified representation for both geometric and physical modeling. Our approach is to combine the modal shape deformations defined above with an implicit function surface such as a sphere or cube. This combination gives us the advantage of being able to accurately and simply describe physical deformations, and yet to be able to use the implicit function representation's *inside-outside* function for solving intersection problems and model fitting [21, 15, 7].

In object-centered coordinates $\mathbf{r} = [r, s, t]^T$, the implicit equation of a spherical surface is

$$f(\mathbf{r}) \;=\; f(r, s, t) \;=\; r^2 \;+\; s^2 \;+\; t^2 \;-\; 1.0 \;=\; 0.0 \tag{7}$$

This equation is also referred to as the surface's **inside-outside function**, because to detect intersection between a point $\mathbf{X}_p = [X_p, Y_p, Z_p]^T$ and the volume bounded by this surface, one simply substitutes the coordinates of $\mathbf{X}$ into the function f. If the result is negative, then the point is inside the surface. Generalizations of this basic operation may be used to find line-surface intersections or surface-surface intersections.

A solid defined in this way can be easily positioned and oriented in global space, by transforming the implicit function to global coordinates, $\mathbf{X} = [X, Y, Z]^T$ we get [21]:

$$\mathbf{X} = \mathcal{R}\mathbf{r} + \mathbf{b} \tag{8}$$

where $\mathcal{R}$ is a rotation matrix, and $\mathbf{b}$ is a translation vector. The implicit function's positioned and oriented (rigid) inside-outside function becomes (using Equation 8):

$$f(\mathbf{r}) = f(\mathcal{R}^{-1}(\mathbf{X} - \mathbf{b})). \tag{9}$$

Any set of implicit shape functions can be generalized by combining them with a set of global deformations $\mathcal{D}$ with parameters m. For particular values of m the new deformed surface is defined using a deformation matrix $\mathcal{D}_m$:

$$\mathbf{X} = \mathcal{R}\mathcal{D}_m\mathbf{r} + \mathbf{b} \tag{10}$$

In our system the deformations used are the **modal shape polynomial functions**, defined by transforming the original finite element shape functions to the modal coordinate system (see [17] and Appendix A of this paper). These polynomials are a function of r, Equation 10 becomes:

$$\mathbf{X} = \mathcal{R}\mathcal{D}_m(\mathbf{r})\mathbf{r} + \mathbf{b} \tag{11}$$

The inside-outside function, with nonrigid deformations becomes (using Equation 11):

$$f(\mathbf{r}) = f(\mathcal{D}_m^{-1}(\mathbf{r})\mathcal{R}^{-1}(\mathbf{X} - \mathbf{b})) \tag{12}$$

This inside-outside function is valid as long as the inverse polynomial mapping $\mathcal{D}_m^{-1}(\mathbf{r})$ exists. In cases where a set of deformations has no closed-form inverse mapping, Newton-Raphson and other numerical iterative techniques have to be used.

This method of defining geometry, therefore, provides an inherently more efficient mathematical formulation for intersection detection than geometric representations such as polygons or splines. See Pentland and Williams [15] and Sclaroff and Pentland [21] for a discussion of the computational complexity of intersection detection algorithms.

2.3. Idealized Modes

For applications that do not require accurate physical modeling, such as object recognition, we have found that it is adequate to use a single set of particularly simple deformations derived using idealized elasticity properties. Moreover, because the elastic properties of the model are of no concern, it is sufficient to set $\tilde{K}$ to be the identity matrix, except for rigid-body modes which have zero stiffness.

The entries of the idealized deformation matrix $\mathcal{D}_m$ for these idealized modes are as follows,

$$
\begin{aligned}
d_{00} &= m_6 + sm_{12} + tm_{15} - (m_{13} + m_{16})sgn(r) - m_{14} - m_{17} \\
d_{01} &= m_{11} + 2s(m_{13} + sgn(r)m_{14}) \\
d_{02} &= m_{10} + 2t(m_{16} + sgn(r)m_{17}) \\
d_{10} &= m_{11} + 2r(m_{19} + sgn(s)m_{20}) \\
d_{11} &= m_7 + rm_{18} + tm_{21} - (m_{19} + m_{22})sgn(s) - m_{20} - m_{23} \\
d_{12} &= m_9 + 2t(m_{22} + sgn(s)m_{23}) \\
d_{20} &= m_{10} + 2r(m_{25} + sgn(t)m_{26}) \\
d_{21} &= m_9 + 2s(m_{28} + sgn(t)m_{29}) \\
d_{22} &= m_8 + rm_{24} + sm_{27} - (m_{25} + m_{28})sgn(t) - m_{26} - m_{29}
\end{aligned}
\tag{13}
$$

where $\mathbf{m} = [m_0, m_1, \ldots, m_{p-1}]^T$ is a p x 1 vector of the *modal amplitudes*, and $\mathbf{r} = [r, s, t]^T =$ is the coordinate of a point in undeformed space.

The modal amplitudes m_i formulated in this way have an intuitive meaning. Modal amplitudes m_0 - m_5 are the rigid body modes of translation and rotation, m_6 - m_8 are the x, y, and z sizes, m_9 - m_{11} are shears about the x, y, and z axes and the rest are bends, tapers and pinches in various axes. Figure 2 illustrates a few of these idealized deformation modes for a cube; the reader should compare this figure to Figure 1.

3. Recovering 3-D Models

Let us assume that we are given m three-dimensional sensor measurements (in the global coordinate system) that originate from the surface of a single object

$$
\mathbf{X}^w = [x_1^w, y_1^w, z_1^w, \cdots, x_m^w, y_m^w, z_m^w]^T
\tag{14}
$$

We then attach virtual springs between these sensor measurement points and particular nodes on our deformable model. This defines an equilibrium equation whose solution U is the desired fit to the sensor data. Consequently, for m nodes with corresponding sensor measurements, we can calculate the virtual loads R exerted on the undeformed object while fitting it to the sensor measurements. For node k these loads are simply

$$
[r_{3k}, r_{3k+1}, r_{3k+2}]^T = [x_k^w, y_k^w, z_k^w]^T - [x_k, y_k, z_k]^T
\tag{15}
$$

where

$$
\mathbf{X} = [x_1, y_1, z_1, \cdots, x_n, y_n, z_n]^T
\tag{16}
$$

Figure 2. A few of the vibrations mode shapes of a cube, using idealized deformations

are the nodal coordinates of the undeformed object in the object's coordinate frame. When sensor measurements do not correspond exactly with existing nodes, the loads can be distributed to surrounding nodes using the interpolation functions used to define the finite element model, as described in [17].

Thus to fit a deformable solid to the measured data we solve the following equilibrium equation:

$$KU = R \tag{17}$$

where the loads R are as above, the material stiffness matrix K is as described above and in [17], and the equilibrium displacements U are to be solved for. The solution to the fitting problem is simply

$$U = K^{-1}R \tag{18}$$

The difficulty in calculating this solution is the large dimensionality of K, so that iterative solution techniques are normally employed.

However a closed-form solution is available simply by converting this equation to the modal coodinate system. This is accomplished by substituting $U = \Phi\tilde{U}$ and premultiplying by Φ^T, so that the equilibrium equation becomes

$$\Phi^T K \Phi \tilde{U} = \Phi^T R \tag{19}$$

or equivalently

$$\tilde{K}\tilde{U} = \tilde{R} \tag{20}$$

where $\tilde{R} = \Phi^T R$ and $\tilde{K} = \Phi^T K \Phi$ is a *diagonal* matrix. Again, note that the calculation of Φ needs to be performed only once as a precomputation, and then stored for all future applications. Further, it is normally not desirable to use all of the eigenvectors (as explained below), so that the matrix remains of managable size even when using large numbers of nodes. In our implementation we normally use between 30 and 81 modes.

The solution to the fitting problem, therefore, is obtained by inverting the diagonal matrix $\tilde{K}$:

$$\tilde{U} = \tilde{K}^{-1}\tilde{R} \tag{21}$$

Note, however, that as this formulation is posed in the object's coordinate system the rigid body modes have zero eigenvalues, and must therefore be solved for separately by setting $\tilde{u}_i = \tilde{r}_i$, $1 \le i \le 6$. The complete solution may be written in the original nodal coordinate system, as follows

$$U = \Phi(\tilde{K}_6)^{-1}\Phi^T R \tag{22}$$

where $\tilde{K}_6 = (\tilde{K} + I_6)$ is a matrix whose first six diagonal elements are ones, and remaining elements are zero.[1]

The major difficulty in calculating this solution occurs when there are fewer degrees of freedom in sensor measurements than in the nodal positions — as is normally the case in computer vision applications. Previous researchers have suggested adopting heuristics such as smoothness and symmetry to obtain a well-behaved solution; however in many cases the observed objects are neither smooth nor symmetric, and so an alternative method is desirable.

A better method is to discard some of the high-frequency modes, so that the number of degrees of freedom in $\tilde{U}$ is equal to or less than the number of degrees of freedom in the sensor measurements. To accomplish this, one simply row and column reduces $\tilde{K}$, and column reduces Φ so that their rank is less than or equal to the number of available sensor measurement degrees of freedom. The motivation for this strategy is that:

- When there are fewer degrees of freedom in the sensor measurements than in the model, the high-frequency modes cannot in any sense be accurate, as there is insufficient data to constrain them. Their value primarily reflects the smoothness heuristic employed.

- While the high-frequency modes will not contain information, they are the *dominant* factor determining the cost of the solution, as they are both numerous and require the use of very small time steps [15].

- In any case, high-frequency modes typically have very little energy, and even less effect on the overall object shape. This is because (for a given excitatory

[1]Inclusion of the matrix I_6 into Equation 22 may also be interpreted as adding an external force that constrains the solution to have no residual translational or rotational stresses.

energy) the displacement amplitude for each mode is ***inversely*** proportional to the ***square*** of the mode's resonance frequency, and because damping is proportional to a mode's frequency.

Perhaps the most interesting consequence of discarding some of the high-frequency modes, however, is that it allows Equation 22 to provide a generically ***overconstrained*** estimate of object shape. Note that discarding high-frequency modes is **not** equivalent to a smoothness assumption, as sharp corners, creases, etc., can still be obtained. What we cannot do with a reduced-basis modal representation is place many creases or spikes close together.

3.1. 3-D Shape from 2-D Contours

In the case where we are given only 2-D contour information we can still employ the same equations to estimate shape, however we must generalize Equation 22 to reflect the uncertainty we have about the z coordinate of each contour point. This can be accomplished by altering Equation 22 to reflect the fact that some sensor measurements are more certain than others. We accomplish this by introducing a $3n$ x $3n$ diagonal weighting matrix $\mathbf{W}$:

$$\tilde{\mathbf{U}} = \tilde{\mathbf{K}}_6^{-1}(\mathbf{W}\boldsymbol{\Phi})^T\mathbf{R} \tag{23}$$

The diagonal entries of $\mathbf{W}$ are inversely proportional to the uncertainty (variance) of the data associated with each of the nodal coordinates. The effect of $\mathbf{W}$ is to make the strength of the virtual springs associated with each data point reflect the uncertainty of the measurement.

3.2. Determining Spring Attachment Points

Attaching a virtual spring between a data measurement and a deformable object implicitly specifies a correspondence between some of the model's nodes and the sensor measurements. In most situations this correspondence is not given, and so must be determined in parallel with the equilibrium solution. In our experience this attachment problem is the most problematic aspect of the physically-based modeling paradigm. This is not surprising, however, as the attachment problem is similar to the correspondence problems found in many other vision applications.

Our approach to the spring attachment problem is similar to that adopted by other researchers [20, 22]. Given a segmentation of the data into objects or "blobs," the first step is to define an ellipsoidal coordinate system by examination of the data's moments of inertia. These estimates of position, orientation, and size define an elliptical coordinate system. Data points are then projected onto this ellipsoid in order to determine the spring attachment points.

For simple objects (e.g., cubes, bananas, cylinders) our experimental results show that this method of establishing spring attachment points yields accurate, stable object descriptions. It should be noted, however, that because $\boldsymbol{\Phi}$ linearizes object rotation it is important that the elliptical coordinate system established by the moment method be sufficiently close to the object's true coordinate system. We have found that as long as our initial estimate of orientation is within ± 15 degrees, we can still achieve a stable, accurate estimate of shape.

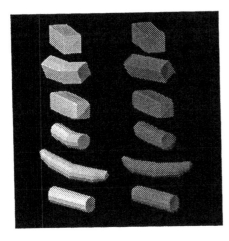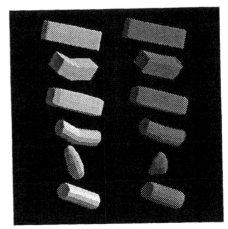

Figure 3. Front (a) and side (b) views of fitting a rectangular box, cylinder, banana, etc., using sparse 3-D point data with 5% noise. The original models are shown in white, and the recovered models are shown in darker grey. Given the positions of some of the object's visible surface points (as shown in (a)), we can recover the full 3-D model as is shown in the side view (b).

For complex objects, however, the spring attachment problem is sufficiently non-linear that we have found it necessary to establish attachment positions iteratively. We accomplish this by first fitting a solid model as above, and then use that derived model to more accurately determine the spring attachment locations. In our experiments we have found that two or three iterations of this procedure are normally sufficient to produce a good solution to both the spring attachment and the fitting problems.

3.3. Examples of Recovering 3-D Models

Figure 3 shows an example using very sparse synthetic range information. White objects are the original shapes that the range data was drawn from, and the grey objects are the recovered 3-D models. For each white object in Figure 3(a) the position of visible corners, edges, and face centers was measured, resulting in between 11 and 18 data points. These data points were then corrupted by adding 5 mm of uniformly distributed noise to each data point's x, y, and z coordinates; the maximum dimension of each object is approximately 100 mm. Note that only the *front-facing* 3-D points, e.g., those visible in Figure 3(a) at the left, were used in this example. Total execution time on a standard Sun 4/330 is approximately 0.1 seconds.

Despite the rather large amount of noise and a complete lack of information about the back side of the object, it can be seen that Equation 22 does a good job

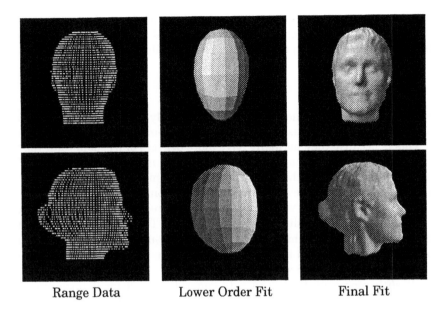

| Range Data | Lower Order Fit | Final Fit |

Figure 4. Fitting laser rangefinder data of a human face. Left column: original range data, Middle column: recovered 3-D model using only low-order modes, Left Column: full recovered model.

of recovering the entire 3-D model. This is especially apparent in the side view, shown in Figure 3(b), where we can see that even the back side of the recovered models (grey) are very similar to the originals (white). This accuracy despite lack of 360° data reflects the fact that Equation 22 provides the shape estimate with the least internal strain energy, so that symmetric and mirror symmetric solutions are preferred.

A second example uses 360° laser rangefinder data of a human head, as shown in the left-hand image of Figure 4. There are about 2500 data points. Equation 22 was then used to estimate the shape, using only the low-frequency 30 modes. The low-order recovered model is shown in the middle column; because of the large number of data points execution time on a Sun 4/330 was approximately 3 seconds. It can be seen that the low-order modes provide a sort of qualitative description of the overall head shape.

A full-dimensionality recovered model is shown in the right-hand image of 4. In the ThingWorld system [15, 16], rather than describing high-frequency surface details using a finite element model with as many degrees of freedom as there are data points, we normally augment a low-order finite element model with a spline description of the surface details. This provides us with a two-layered representation (low-order finite element model + surface detail spline description = final model) that we find to be both more efficient to recover and more useful in recognition, simulation, and visualization tasks than a fully-detailed finite element model.

4. Object Recognition

Perhaps the major drawback of previous physically-based modeling techniques is that they have not been useful for recognition, comparision, or other database tasks. This is because they normally have more degrees of freedom than there are sensor measurements, so that the recovery process is underconstrained. Therefore, although heuristics such as smoothness or symmetry can be used to obtain a solution, they do not produce a stable, unique solution.

The major problem is that when the model has more degrees of freedom than the data, the model's nodes can slip about on the surface. The result is that there are an infinite number of valid combinations of nodal positions for any particular surface. This difficulty is common to all spline and piecewise polynomial representations, and is known as the **knot problem**.

For all such representations, the only general method for determining if two surfaces are equivalent is to generate a number of sample points at corresponding positions on the two surfaces, and observe the distances between the two sets of sample points. Not only is this a clumsy and costly way to determine if two surfaces are equivalent, but when the two surfaces have very different parameterizations it can also be quite difficult to generate sample points at "corresponding locations" on the two surfaces.

The modal representation, assuming that all modes are employed, decouples the degrees of freedom within the non-rigid dynamic system, but it does not by itself reduce the total number of degrees of freedom. Thus a complete modal representation suffers from the same problems as all of the other representations.

The obvious solution to the problem of non-uniqueness is to discard enough of the high-frequency modes that we can obtain an overconstrained estimate of shape, as was done for the shape recovery problem above. Moreover, the reduced-basis modal representation is a **unique** representation of shape because the modes (eigenvectors) form an orthonormal basis set. Therefore, there is only one way to represent an object in canonical position.

Further, because the modal representation is frequency-ordered, it has stability properties that are similar to those of a Fourier decomposition. Just as with the Fourier decomposition, an exact subsampling of the data points points does not change the low-frequency modes. Similarly, irregularities in local sampling and measurement noise tend to primarily affect the high-frequency modes, leaving the low-frequency modes relatively unchanged.[2]

---

[2]Note that when there are many more data points than degrees of freedom in the finite element model, the interpolation functions act as filters to bandlimit the sensor data, thus reducing aliasing phenomina. If, however, the number of modes used is much smaller than the number of degrees of freedom in the finite element

The primary limitation of this uniqueness property stems from the finite element method's linearization of rotation. Because the rotations are linearized, it is impossible to uniquely determine an object's rotation state. Thus object symmetries can lead to multiple descriptions, and errors in measuring object orientation will cause commensurate errors in shape desription.

Thus by employing a reduced-basis modal representation we can obtain overconstrained, canonical shape estimates. To compare objects l and k with known mode values $\tilde{U}^l$ and $\tilde{U}^k$, one simply compares the two vectors of mode values:

$$\varepsilon = \frac{\tilde{U}^l \cdot \tilde{U}^k}{\|\tilde{U}^l\|\|\tilde{U}^k\|} \tag{24}$$

Vector norms other than the dot product can also be employed; in our experience all give roughly the same recognition accuracy.

To recognize a recovered model with estimated mode values $\tilde{U}$, one compares the recovered mode values to the mode values of all of the p known models:

$$\varepsilon_k = \frac{\tilde{U} \cdot \tilde{U}^k}{\|\tilde{U}\|\|\tilde{U}^k\|} \qquad k = 1, 2, \ldots, p \tag{25}$$

The known model k with the maximum dot product ε_k is the model best matching the recovered model, and thus declared to be the model recognized. Note that for each known model k, only the vector of mode values $\tilde{U}^k$ needs to be stored.

The first six entries of $\tilde{\Phi}$ are the rigid-body modes (translation and rotation), which are normally irrelevant for object recognition. Similarly, the seventh mode (overall volume) is sometimes irrelevent for object recognition, as many machine vision techniques recover shape only up to an overall scale factor. Thus rather than computing the dot product with all of the modes $\tilde{U}$, we typically use only modes number eight and higher, e.g.,

$$\varepsilon_k = \frac{\sum_{i=8}^{i=m} \tilde{u}_i \tilde{u}_i^k}{\sqrt{\sum_{i=8}^{i=m} \tilde{u}_i^2}\sqrt{\sum_{i=8}^{i=m} (\tilde{u}_i^k)^2}} \qquad k = 1, 2, \ldots, p \tag{26}$$

where m is the total number of modes employed. By use of this formula we obtain translation, rotation, and scale-invariant matching.

The ability to compare the shapes of even complex objects by a simple dot product makes the modal representation well suited to recognition, comparison, and other database tasks. In the following section we will evaluate the reliablity of the combined shape recovery/recognition process using both synthetic and laser rangefinder data.

4.1. Recognition Accuracy using 3-D Measurements

To assess accuracy, we conducted an experiment to recover and recognize face models from range data generated by a laser range finder. In this experiment we obtained laser rangefinder data of eight people's heads from a five different viewing

---

model, then it is possible to have significant aliasing.

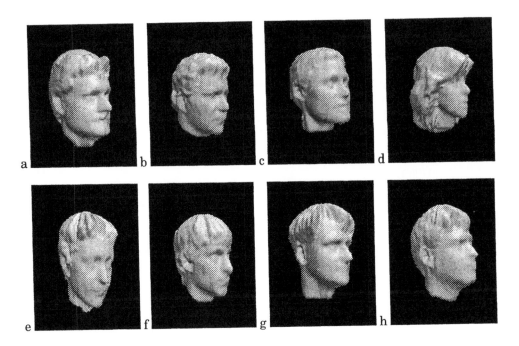

Figure 5. Eight heads used in our recognition experiment. Note that in some cases there is significant missing data.

directions: the right side ($-90°$), halfway between right and front ($-45°$), front ($0°$), halfway between front and left ($45°$), and the left side ($90°$). We have found that people's heads are only approximately symmetric, so that the $\pm45°$ and $\pm90°$ degree views of each head have quite different detailed shape. In each case the range data was from the forward-facing, visible surface only.

Data from a 360° scan around each head was then used to form the stored model of each head that was later used for recognition. Full-detail versions of these eight reference models are shown in Figure 5; note that in some cases a significant amount of the data is missing. As previously, only the low order 30 deformation modes were used in the shape extraction and recognition procedure. Because the low order modes provide a coarse, qualitative summary of the object shape (see the middle column of Figure 4) they can be expected to be the most stable with respect to noise and viewpoint change. Total execution time on a standard Sun 4/330 averaged approximately 5 seconds per fitting and recognition trial.

Recognition was accomplished by first recovering a 3-D model from the visible-surface range data, and then comparing the recovered mode values to the mode values stored for each of the three known head models using Equation 26. The known model producing the largest dot product was declared to be the recognized object. The first seven modes were not employed, so that the recognition process

	−90°	−45°	0°	45°	90°
a	0.16	-0.19	-0.24	-0.19	-0.12
b	0.26	0.28	0.35	0.37	0.30
c	**0.88**	**0.99**	**0.99**	**0.98**	**0.99**
d	0.03	0.15	0.21	0.21	0.11
e	0.06	-0.13	-0.10	-0.04	-0.06
f	0.58	0.44	0.46	0.46	0.50
g	0.53	0.53	0.52	0.50	0.58
h	0.42	0.47	0.53	0.49	0.50

Figure 6. Recognizing faces from five different points of view.

was translation, rotation, and scale invariant.

Figure 6 illustrates typical results from this experiment. The top row of Figure 6 illustrates the five models recovered from range data from the front, visible surface using viewpoints of −90°, −45°, 0°, 45°, and 90°. Each of these recovered head models look similar, and more importantly have approximately the same deformation mode values $\bar{U}$, despite the wide variations in input data. Modes 8 through 30 of these recovered models were then compared to each of the stored head models. The dot products obtained are shown below each recovered head model.

In Figure 6 all of the input data was views of Kim (depicted as "head c" in the tables). As can be seen, the dot products between recovered 3-D model and known model are quite large for Kim's head model. In fact, in this example the smallest correct dot product is almost three times the magnitude of any of the incorrect dot products; the same was also true for range data of the other subjects.

In this experiment 92.5% accurate recognition was obtained. That is, we successfully recovered 3-D models and recognized each of the eight test subjects from each of the five different views with only three errors. Analysis of the recognition results showed that, while the average dot product between different reference

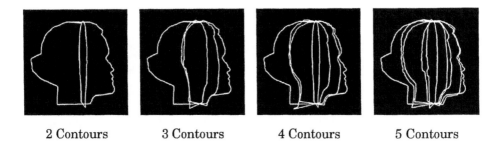

| 2 Contours | 3 Contours | 4 Contours | 5 Contours |

Figure 7. Set of 2-D contour groups used in head recognition (contours are illustrated as lying in planes, however no depth information along the contour was employed). These contours were taken from the same head depicted in Figure 4 (**Head c** in Figure 5).

models was 0.31 (72°), the average dot product between models recovered from different views of the same person was 0.95 (18°). Thus recognition was typically extremely certain. All three errors were from front-facing views, where relatively few discriminating features are visible; remember that only overall head shape, and not details of surface shape, were available to the recognition procedure as only 30 modes were employed.

4.2. Recognition Accuracy Using 2-D Contours

In a similar head recovery and recognition experiment, we used a few 2-D head contours instead of full range data to see how well our techniques performed in the case of sparse silhouette data. In this experiment, we recovered heads from 2, 3, 4, and finally 5 contours, in order to approximate the information available in an active vision scenario. In each trial, the contours were spaced evenly in rotation. An example of the contours used in this experiment is shown in Figure 7; these contours were taken from the same head depicted in Figure 4 ("head c" in Figure 5). Note that although the contours in this figure are illustrated as lying in planes, the actual depth of each contour point was unknown.

As in the previous experiment, the recovered heads were compared against the full-detail versions of the reference heads shown in Figure 5, and the model producing the largest dot product was declared to be the recognized object. Heads were compared using the scale, rotation, and translation invariant matching of Equation 26.

In our experiments with two contours, recognition accuracy averaged 93.75%. Accuracy improved as more contours were added until 96.875 percent of the heads were correctly identified when 5 contours were used. The results were not as good if the contours did not include the traditional side "silhouette," and performed best when the data's spring attachment was smoothed out more across the surface.

Total execution times were slightly greater than those for the full data experiment, averaging 5 seconds per fitting and recognition trial on a Sun 4/330. The greater execution time is attributable to the more careful distribution and smoothing of spring attachment between contours and the underlying deformable model.

5. Recovery of Rigid and Non-Rigid Motion

In the previous sections we have addressed static shape estimation and recognition. For sequences, however, it is necessary to also consider the *dynamic* properties of the body and of the data measurements. The Kalman filter is the standard technique for obtaining estimates of the state vectors of dynamic models, and for predicting the state vectors at some later time. Outputs from the Kalman filter are the optimal (weighted) least-squares estimate for non-Gaussian noises.

Kalman filtering has been used in many motion estimation applications [4, 8], but normally it is both too expensive and requires too much storage to apply to the large number of variables typical of a whole-body finite element model. However, because the modal representation allows us to summarize the dynamic state of the body with only a small number of parameters, development of a Kalman filter is straightforward.

The following section outlines the development of the Kalman filter equations. Readers interested in additional detail are referred to references [18, 19].

5.1 The Kalman Filter

Let us define a dynamic process

$$\dot{\mathbf{X}} = \mathbf{AX} + \mathbf{B}a \tag{27}$$

and observations

$$\mathbf{Y} = \mathbf{CX} + n \tag{28}$$

where a and n are white noise processes having known spectral density matrices. Then the optimal estimate $\hat{\mathbf{X}}$ of $\mathbf{X}$ is given by the following *Kalman filter*

$$\dot{\hat{\mathbf{X}}} = \mathbf{AX} + \mathbf{K}_f(\mathbf{Y} - \mathbf{CX}) \tag{29}$$

Assuming that the cross-variance between the system excitation noise a and the observation noise n is zero, then the Kalman gain matrix $\mathbf{K}_f = \mathbf{PC}^T \mathcal{N}^{-1}$, with $\mathbf{P}$ being the solution to the Riccati equation.

5.1.1 Estimation of Displacement and Velocity

In the current application we are primarily interested in estimation of the modal amplitudes $\tilde{\mathbf{U}}$ and their velocities $\tilde{\mathbf{V}} = \dot{\tilde{\mathbf{U}}}$. In state-space notation our system of equations is

$$\begin{bmatrix} \dot{\tilde{\mathbf{U}}} \\ \dot{\tilde{\mathbf{V}}} \end{bmatrix} = \begin{bmatrix} \mathbf{0} & \mathbf{I} \\ \mathbf{0} & \mathbf{0} \end{bmatrix} \begin{bmatrix} \tilde{\mathbf{U}} \\ \tilde{\mathbf{V}} \end{bmatrix} + \begin{bmatrix} \mathbf{0} \\ \mathbf{I} \end{bmatrix} a \tag{30}$$

where a is a noise vector due to nodal accelerations. The observed variable will be the estimated 3-D modal amplitudes $\tilde{\mathbf{U}}_o$:

$$\tilde{\mathbf{U}}_o = \tilde{\mathbf{U}} + n \qquad (31)$$

where n is a vector of the observation noise.

We will assume that n and a originate from independent noise with standard deviations n and a respectively, so that $\mathcal{N} = n^2\mathbf{I}$, $\mathcal{A} = a^2\mathbf{I}$. Experimentally, this model of $\mathcal{N}$ has been shown to be approximately correct. Given $\mathcal{N}$ and $\mathcal{A}$ we may then determine the Kalman gain matrix, and thus the Kalman filter equation:

$$\begin{bmatrix} \dot{\tilde{\mathbf{U}}} \\ \ddot{\tilde{\mathbf{U}}} \end{bmatrix} = \begin{bmatrix} \dot{\tilde{\mathbf{U}}} + (\frac{2a}{n})^{1/2} \left(\tilde{\mathbf{U}}_o - \tilde{\mathbf{U}}\right) \\ (\frac{a}{n}) \left(\tilde{\mathbf{U}}_o - \tilde{\mathbf{U}}\right) \end{bmatrix} \qquad (32)$$

Each mode is independent within this system of equations, and so we may write the Kalman filter separately for each of the modes. For instance, the displacement prediction for mode i at time $t + 1$ is:

$$\tilde{u}_i^{t+1} = \tilde{u}_i^t + \dot{\tilde{u}}_i^t + ((a/n) + (2a/n)^{1/2}) \left(\tilde{u}_{o,i} - \tilde{u}_i^t\right) \qquad (33)$$

6. Examples Using Real Data

Figure 8(a) shows one frame from a sequence of X-ray images, with the zero-crossing edge contours overlayed. From these contours the 3-D shape was estimated using Equation 23. The resulting shape is shown in Figure 8(b) as a 3-D wireframe overlayed on the original X-ray data. Figure 8(c) shows the recovered model from the side. Because only bounding contour information was available, the shape estimated along the z axis (shown in Figure 8(c)) is determined by finding the minimum stress state that still fits the bounding contour. The z-axis shape cannot, therefore, be regarded as accurate but only as plausible and consistent. Note that use of a minimum stress criterion for solution means that symmetric and mirror symmetric shapes are preferred. Execution time was approximately one second on a standard Sun 4/330. For additional detail see Pentland and Sclaroff [17].

Figure 9 shows an example of recovering non-rigid motion from contour information. The 3-D shape and motion of the heart ventricle was tracked over time using the contour information shown at the top within each box of Figure 9. For each frame Equation 23 was used to obtain an estimate of 3-D shape from the contour information (see Figure 8). These shape estimates were then integrated in a Kalman filter formulation, as described in Pentland and Horowitz [18]. As in the single-image case, deformations along the z axis are determined by finding the minimum stress state that still fits the bounding contour. The z-axis deformations cannot, therefore, be regarded as accurate but only as plausible. Execution time was approximately one second per frame on a standard Sun 4/330.

Figure 10 illustrates a more complex example of tracking rigid and non-rigid motion. This figure shows three frames from a twelve image sequence of a well-known tin woodsman caught in the act of jumping. Despite the limited range of

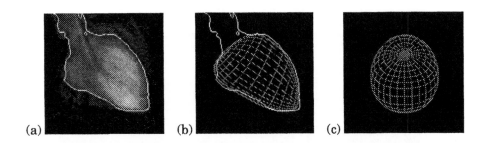

Figure 8. An example showing the use of a 2-D image contour to recover a 3-D deformable solid model. The original image and contour are shown in (a). The model is recovered from the contour as shown in (b). An orthogonal side view is shown in (c).

motion, this example is a difficult one because of the poor quality optical flow, due to pronounced highlights on thighs and other parts of the body.

In this example an articulated 3-D model was constructed by hand, with spring-like attachment constraints inserted between the various body parts. In this manner the combined behavior of the various parts were constrained to be consistent with the physics of the situation: parts must stay connected, movement by one part causes an equal but opposite reactions among the other parts, and inertia is conserved.

We then calculated optical flow by use of a block-wise Horn-Schunk algorithm, and then our Kalman filter formulation used to estimate the motions of the various parts. The between-part attachment constraints then introduce additional forces that enforce the conservation of force and inertia. All of these forces are then integrated to produce a final physically-consistent estimate of the overall motion at each instant in time.

The estimated motions for this sequence are illustrated by the bottom row of Figure 10. As can be seen by comparing the 3-D motion of the model with that in the original image, the resulting tracking is reasonably accurate. For additional information, see references [18, 19].

7. Summary

We have described an efficient method for recovering, recognizing, and tracking physically-based 3-D solid models using either 2-D or 3-D sensor measurements. In our current implementation we typically use between 30 and 81 deformation modes, so that the solution can be overconstrained using as few as 11 independent 3-D sensor measurements. The representation is unique except for rotational symmetries and degenerate conditions.

Because the recovered 3-D shape description is canonical, we may efficiently measure the similarity of different shapes by simply calculating normalized dot

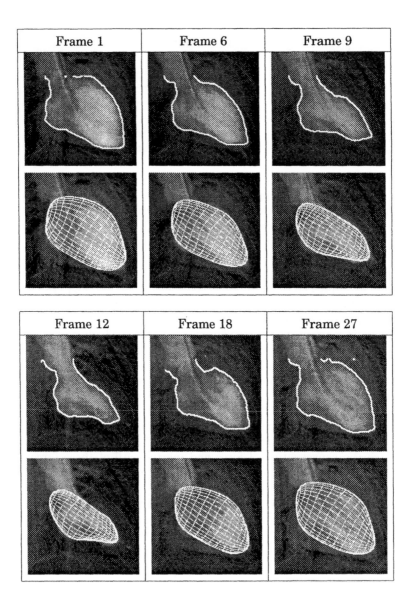

Figure 9. A heart's nonrigid motion as it was recovered from contours taken from a motion sequence. The contours, extracted via a simple threshold and zero crossing scheme, are shown along the top. The deformable model recovered from these contours is shown in wireframe.

Figure 10. Three frames from an image sequence showing tracking of a jumping man using an articulated, physically-based model. Note that despite poor quality optical flow (due to pronounced highlights on thighs and other parts of the body) the overall tracking is reasonably accurate.

products between the mode values Ū of various objects. Such comparisons may be made position, orientation and/or size independent by simply excluding the first seven mode amplitudes. Thus the modal representation seems likely to be useful for applications such as object recognition and spatial database search.

The major weaknesses of our current method are

- The need to estimate initial object orientation to within ±15°, as in our formulation rotational variation has been linearized.

- The need to segment data into simple, approximately convex "blobs" in a stable, viewpoint-invariant manner.

In the current implementation of our system we use a standard method-of-moments to obtain initial estimates of object orientation, and a minimum description length technique to produce segmentations [5]. We are now working to integrate the segmentation procedure of Dickinson, Pentland, and Rosenfeld [6] (described elsewhere in this volume) to produce an integrated segmentation-fitting-recognition system.

REFERENCES

1   K. Bathe. *Finite Element Procedures in Engineering Analysis*. Prentice-Hall, 1982.
2   I. Biederman. Recognition-by-Components: A Theory of Human Image Understanding. *Psychological Review*, 94(2):115–147, 1987.
3   T. Binford. Visual Perception by Computer. *Presented at The IEEE Conference on Systems and Control*, December 1971.
4   T. J. Broida and R. Chellappa. Estimation of Object Motion Parameters from Noisy Images, *IEEE Trans. Pattern Analysis and Machine Intelligence*, 8(1):90–99. January 1986.
5   T. Darrell, S. Sclaroff, and A. Pentland. Segmentation by Minimal Description. In *Proc. Third International Conference on Computer Vision*, December 1990.
6   Dickinson, S., Pentland, A., and Rosenfeld, A., (1992) From Volumes to Views: An Approach to 3-D Object Recognition, *CVGIP: Image Understanding*. Vol. 55, No. 2, March, pp. 130-154.
7   I. Essa. *Contact Detection, Collision Forces and Friction for Physically-Based Virtual World Modeling*. Master's thesis, Dept. of Civil Engineering, M.I.T., 1990.
8   O. D. Faugeras, N. Ayache, and B. Faverjon Building Visual Maps by Combining Noisy Stereo Measurements, *Proc. IEEE Conf. on Robotics and Automation*, San Francisco, CA., April 1986.
9   B. Friedland. *Control System Design*. McGraw-Hill, 1986.
10 D. Hoffman and W. Richards. Parts of Recognition. In *From Pixels to Predicates*, A. Pentland (ed.), 1985.
11 M. Leyton. Perceptual Organization as Nested Control. *Biological Cybernetics*, 51:141–153, 1984.

12 D. Marr and K. Nishihara. Representation and Recognition of the Spatial Organization of Three-dimensional Shapes. In *Proc. of the Royal Society - London B*, 1978.

13 R. Mohan and R. Nevatia. Using Perceptual Organization to Extract 3D Structures. *IEEE Trans. Pattern Analysis and Machine Intelligence*, 11(11):1121–1139, November 1989.

14 A. Pentland. Perceptual Organization and Representation of Natural Form. *Artificial Intelligence*, 28(3):293–331, 1986.

15 A. Pentland and J. Williams. Good Vibrations : Modal Dynamics for Graphics and Animation. *Computer Graphics*, 23(4):215–222, 1989.

16 A. Pentland. Automatic Extraction of Deformable Part Models. *International Journal of Computer Vision*, 107–126, 1990.

17 Alex Pentland and Stan Sclaroff. Closed form solutions for physically based shape modeling and recovery. *IEEE Trans. Pattern Analysis and Machine Intelligence*, 13(7):715–729, July 1991.

18 Alex Pentland and Bradley Horowitz. Recovery of nonrigid motion and structure. *IEEE Trans. Pattern Analysis and Machine Intelligence*, 13(7):730–742, July 1r991.

19 Alex Pentland, Bradley Horowitz, and Stan Sclaroff. Non-rigid motion and structure from contour. In *IEEE Workshop on Visual Motion*, pages 288–293. IEEE Computer Society, 1991.

20 F. Solina and R. Bajcsy. Recovery of Parametric Models from Range Images: The Case for Superquadrics with Global Deformations. *IEEE Trans. Pattern Analysis and Machine Intelligence*, 12(2):131–147, 1990.

21 Stan Sclaroff and Alex Pentland. Generalized implicit functions for computer graphics. *Computer Graphics*, 25(4):247–250, 1991.

22 D. Terzopoulos, A. Witkin, and M. Kass. Symmetry-Seeking Models for 3-D Object Reconstruction. In *Proc. First Conference on Computer Vision*, pages 269–276, London, England, December 1987.

23 Demetri Terzopoulos and Dimitri Metaxas. Dynamic 3d models with local and global deformations: Deformable superquadrics. *IEEE Trans. Pattern Analysis and Machine Intelligence*, 13(7):703–714, July 1991.

Three-Dimensional Object Recognition Systems
A.K. Jain and P.J. Flynn (Editors)
© 1993 Elsevier Science Publishers B.V. All rights reserved.   447

Function-Based Generic Recognition for Multiple Object Categories[1]

Melanie Sutton, Louise Stark and Kevin Bowyer[a]

[a] Department of Computer Science and Engineering
University of South Florida
Tampa, Florida 33620 USA

Abstract

In *function-based* object recognition, an object category is represented by knowledge about object function. Function-based approaches are important because they provide a principled means of constructing *generic* recognition systems. Our work concentrates specifically on the relation between shape and function of rigid 3-D objects. Recognition of an observed shape is performed by reasoning about the function that it might serve. Previous efforts have dealt with only a single basic level object category. A number of important issues arise in extending this approach to deal with multiple basic level categories. One issue is whether the knowledge about object function can be organized into general primitive chunks that are re-usable across different categories. Another issue is how to efficiently index the knowledge base so as to avoid exhaustive testing of an object shape against each known category. In order to better explore these issues, we have implemented a system whose domain of competence is a number of different object categories within the superordinate categories furniture and dishes. The performance of this system has been evaluated on a database of over 400 shapes.

1. Introduction

Most work to date on the problem of object recognition has assumed that a geometric model is made available, a priori, for each object that the system will need to recognize. Something more is clearly needed if the goal of visual perception for "autonomous" and "real world" systems is to be achieved. Even in principle, it is simply not possible to give a real world system a geometric model of each object that it will need to recognize. A real world system must be able to deal with objects for which it has *no explicit prior shape model*. A function-based model does not specify any explicit geometric or structural plan. Instead, an object category is defined in terms of knowledge about what is necessary in order to function as an instance of the category. The underlying representation is in the form of primitive chunks of

[1]This research was supported by AFOSR grant F49620-92-J-0223, NSF grants IRI-9120895 and CDA-91-00898 (Research Experiences for Undergraduates), and a NASA Florida Space Grant Consortium graduate fellowship.

knowledge about shape, physics and causation that may be used as primitives in building definitions of required function.

The *form and function* concept is intuitively appealing and has been discussed by a number of researchers in AI and computer vision [1, 3, 5, 7, 8, 12, 13]. Brady and co-workers implemented a system which performs a type of function-based reasoning based on a 2-D outline of the object in its typical pose [4], and also outlined goals for a more ambitious system [3]. Di Manzo *et al.* [5] described a design for a system to recognize shapes of chairs in 3-D. Vaina and Jaulent [12] investigated the use of both conceptual and structural object description for functional recognition. Kitahashi *et al.* [7] brings the user, or "functant," into the representation to help define prototypical shapes. Stark and Bowyer reported on a system that analyzes 3-D shapes in arbitrary initial orientation to determine if they could fulfill the function of a chair [10] and later described an extension of the system for the superordinate category furniture [9].

The work described in this paper explores the issues that must be addressed in generalizing the function-based approach to deal with multiple basic-level categories. Our ideas have been realized in an implemented system and experimental results are reported. Section 2 outlines the structure and operation of the GRUFF-3 system (Generic Representation Using Form and Function). The complexity analysis in section 3 shows that recognition is at worst case polynomial time. Section 4 describes an evaluation of the system's competence in recognition. Section 5 presents a function-based category indexing scheme that leads to greater efficiency. Finally, section 6 summarizes the current state of the project and suggests directions for continued research.

2. The Function-Based Recognition System

This section details the operation of a function-based recognition system for the superordinate category *furniture*, comprised of the basic level categories *chair, table, bench, bed* and *bookshelf*, and the superordinate category *dishes*, comprised of the basic level categories *cup/glass, bowl, plate, pot/pan*, and *pitcher*. We first describe the six *knowledge primitives* that are the basis of the function-based definitions of the object categories. Next we show how a sequence of invocations of the primitives is used to define a *functional property* that is specific to some particular category. We then present the overall definition of the basic level categories and their subordinate categories. The analysis of an example shape is traced to provide a better understanding of how recognition occurs. During the recognition process an *association measure* is accumulated to reflect the level of confidence that the system holds in support of the object belonging to the hypothesized category. This section ends with a description of how this association measure is accrued throughout the evaluation process.

2.1. The "Knowledge Primitives"
A knowledge primitive can be thought of as a chunk of parameterized procedural knowledge that implements some basic concept about shape, physics or causation.

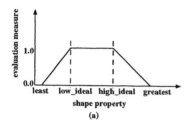 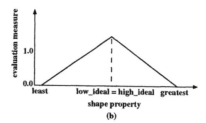

Figure 1. Using Shape Properties to Calculate Evaluation Measure

Each knowledge primitive takes some (specified portions of a) 3-D shape as its input, along with parameter values for the primitive, and returns an *evaluation measure* between zero and one. (Individual evaluation measures from a series of primitive invocations are accumulated into an overall association measure for the functionality of the shape.) The six knowledge primitives used by the system are as follows.

- **relative_orientation (normal_one, normal_two, range_parameters)**

This primitive determines if the angle between the normals for two surfaces (normal_one and normal_two) falls within a desired range. There are four range_parameters: least, low_ideal, high_ideal, and greatest (see Figure 1 a). These parameters are used to calculate a value for the evaluation measure based on where the relative orientation falls within the specified range. Any value between low_ideal and high_ideal results in a measure of one. Values outside this range but between least and greatest fall off linearly to zero. (Low_ideal and high_ideal could be set to the same value, resulting in a "triangle rule" for calculating the measure value (see Figure 1 b).) This primitive can also be used to calculate the transformation which would position the shape in the desired orientation. For example, a common operation is to find the transformation of the shape which would orient a specified surface parallel to the support plane. When relative orientation is used in this manner, a measure of one is always returned.

- **dimensions (shape_element, dimension_type, range_parameters)**

This primitive can be used to determine, for example, if the width or depth of a surface lies within a specified range. The evaluation measure is calculated using four parameters, in much the same manner as for relative orientation.

- **proximity (shape_element_one, shape_element_two, range_parameters)**

This primitive can be used to check qualitative relations between shape_elements, such as *above*, *below* and *close to*. Again, the measure is calculated as described above.

- **clearance (object_description, clearance_volume)**

This primitive can be used to check that there is a specified volume of unobstructed

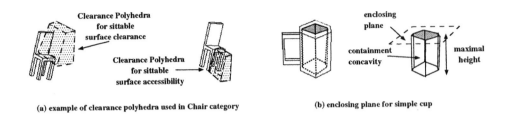

(a) example of clearance polyhedra used in Chair category   (b) enclosing plane for simple cup

Figure 2. Examples of Knowledge Primitive Tests Performed

free space in a particular location relative to a particular part of the shape. The volume is specified by a *clearance polyhedron* (see Figure 2(a)), implemented as a rectangular volume of space specified by a set of six faces and eight vertices. This primitive is generally used for the purpose of checking the accessibility of a potential functional element of an object shape. The evaluation measure is one if the volume specified is unobstructed, or zero if it is obstructed.

- **stability (shape, orientation, applied_force)**

This primitive can be used to check that a given shape is stable when placed on a supporting plane in a given orientation and with a (possibly zero) force applied. It is assumed that the object has homogeneous density, so that the center of mass may be calculated directly from the shape description. This primitive returns an evaluation measure of one if the shape is stable in the specified orientation, or zero if it is not.

- **enclosure (concavity, orientation, enclosing_plane)**

This primitive is used to determine if there exists a concavity in the shape which can be "enclosed" by a single plane introduced parallel to the support plane with the shape in a given orientation. For a cup, in its normal upright orientation, a plane which encloses the concavity can be introduced parallel to the support plane (see Figure 2(b)).

Actually, an infinite number of planes could be introduced at different *levels*, each enclosing a different volume. We are interested in finding the maximal volume that can be enclosed for a given concavity. If the invocation finds an enclosable concavity, then it returns a measure of one along with a description of the concavity. If the invocation does not find an enclosable concavity, then it returns a measure of zero.

All of the system's reasoning about 3-D shapes is constructed out of these six building blocks. The exact set of "primitives" is not so important– it would be easy to create a different set that allowed the construction of a system with equivalent capabilities. However, one of the underlying assumptions of our work is that a "small" number of primitives suffices to define a "large" number of categories, and that the number of primitives required grows "slowly" as a function of the number of categories. This represents one of the fundamental advantages of a function-based system over a more traditional model-based system; that is, that a small amount of a priori knowledge can establish a large domain of competence. Primitives 1 through 5 above were used to construct function-based recognition systems for the single category chair [10] and for a collection of basic level categories in the superordinate category furniture [9]. One of the contributions of this work is to demonstrate that only a single additional primitive suffices to allow a much larger domain.

2.2. Functional Properties Defined By Knowledge Primitives

An object category is defined in terms of its functional plan, specified as a set of *functional properties*. Each functional property is defined by a sequence of invocations of the knowledge primitives, the basic building blocks in the representation. Because functional properties tend to be specific to particular categories, there are far too many to cover each in detail here. Considering one in detail should give the flavor of the approach. The subcategory *king_size_bed* has the following definition:

king_size_bed ::= *provides_king_size_sleeping_surface* + *provides_stability*

indicating that it is defined as a shape placed in an orientation that satisfies the conjunction of the two properties *provides_king_size_sleeping_surface* and *provides_stability*. Note that since the system reasons only on the basis of static 3-D shape, we do not consider properties such as "softness" that may be very important in a complete function-based definition. Sensing of such properties would obviously require more than just static shape as input.

A pseudo-code description of the particular sequence of primitive invocations that constitutes the definition of *provides_king_size_sleeping_surface* is given in Figure 3(b), along with the actual measures returned during the evaluation. The order of the primitives is chosen to help eliminate surfaces from consideration as early as possible using computationally simple tests. For example, the first invocation of dimensions simply tests to see if a surface is within the proper range of area. This is a very simple test, since the area of each face is calculated prior to hypothesis generation. However, the area test does not ensure that the surface is of the proper dimensions in width and depth. These dimension tests require slightly more processing and are only invoked for surfaces which have already passed the area test. For the shape in Figure 3, only two surfaces survive the initial area test, and these are the only surfaces tested by the subsequent primitives of the functional property. Throughout the analysis, the evaluation measures returned from each primitive are combined to reflect an overall *association measure*. The calculation of the association measure is covered in a later section. At the end of the *provides_king_size_sleeping_surface* evaluation, there are two candidate surfaces, each with a different orientation for the shape and a different association measure.

To complete evaluation of the subcategory king size bed, stability has to be confirmed (see Figure 3(c)). The candidate surface corresponding to the bottom of the bed was eliminated by the *provides_stability* functional property, due to the shape not being stable when this surface is oriented parallel to the support plane. Thus there is only one orientation which allows the necessary function, and the overall association measure of this shape as a king size bed is 0.70 (see Figure 3(d)).

initial orientation
of object shape

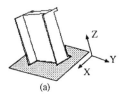

(a)

primitives implementing provides_king_size_sleeping_surface	aggregation of evaluation measure at each step			
dimensions (surface, area, least_1, low_1, high_1, greatest_1) /* check for appropriate total sleeping surface area */		primitive=0.73 aggregate=0.73		primitive=0.97 aggregate=0.97
dimensions (surface, width, least_2, low_2, high_2, greatest_2) /* check for appropriate width of the sleeping surface area */		primitive=1.0 aggregate=0.73		primitive=1.0 aggregate=0.97
dimensions (surface, depth, least_3, low_3, high_3, greatest_3) /* check for appropriate depth of the sleeping surface area */		primitive=1.0 aggregate=0.73		primitive=1.0 aggregate=0.97
dimensions (surface, contiguous_area, least_4, low_4, high_4, greatest_4) /* check for appropriate degree of contiguity of face */		primitive=1.0 aggregate=0.73		primitive=1.0 aggregate=0.97
dimensions (surface, height, least_5, low_5, high_5, greatest_5) /* check for surface being within the appropriate height range */		primitive=0.96 aggregate=0.70		primitive=0.58 aggregate=0.56
relative_orientation (surface, ground, least_6, low_6, high_6, greatest_6) /* find orientation to position surface parallel to the ground */		primitive=1.0 aggregate=0.70		primitive=1.0 aggregate=0.56
clearance (object, 3d_volume_above_sleeping_surface) /* check for appropriate clearance above the sleeping surface */		primitive=1.0 aggregate=0.70		primitive=1.0 aggregate=0.56
clearance (object, 3d_volume_to_left_side_of_sleeping_surface) /*check for appropriate accessibility from the left side */		primitive=1.0 aggregate=0.70		primitive=1.0 aggregate=0.56
clearance (object, 3d_volume_to_right_side_of_sleeping_surface) /* check for appropriate accessibility from the right side */		primitive=1.0 aggregate=0.70		primitive=1.0 aggregate=0.56

(b)

knowledge primitives implementing provides_stable_support	aggregation of evaluation measure at each step			
stability (object shape, given_orientation, self_stable) /* check for stability with no external forces except gravity */		primitive=1.0 aggregate=0.70		primitive=0.0 aggregate=0.0
stability (object_shape, given_orientation, forces_on_surface) /* check for stability of object with external forces applied */		primitive=1.0 aggregate =0.70		

(c)

Evaluation Result:
Recognition for King Size Bed Subcategory,
association measure: = 0.70

provides king size
sleeping surface

provides stable
support

(d)

Figure 3. Knowledge Primitives Defining King Size Bed Subcategory

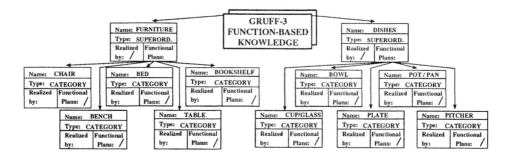

Figure 4. Representation of **GRUFF**-3 Function-Based Knowledge

2.3. Basic Level, Subordinate and Superordinate Categories

The system's knowledge about its domain of competence can be considered as a frame system organized into a tree structure. Each node of the tree is represented by a frame that has four fields: *Name, Type, Realized_By,* and *Functional_Plans.* Nodes are of one of four Types: *Superordinate, Category, Subcategory* or *Function_Label.* Each *Superordinate* frame has associated with it one or more *Category* frames which are linked to the *Functional_Plans* field.

Specifically in our current system, the knowledge base consists of the superordinate category *furniture,* with five immediate children representing the basic level categories *chair, table, bench, bed* and *bookshelf* and the superordinate category *dishes* with five immediate children representing the basic level categories *cup/glass, bowl, plate, pot/pan,* and *pitcher* (see Figure 4). At this level, the branches of the tree represent non-exclusive disjunction. That is, any one input shape may possibly be recognized as belonging to more than one category. Of course, there may be a different preferred orientation of the shape and a different association measure for each category. Each basic level category acts as a root node for the functional plans defined within that category. Figure 5 depicts the defining structures for two of the basic level categories known to the current system. The arcs to the immediate descendants of the basic level category nodes also represent a type of disjunction. The different subcategory nodes at this level represent functional variations of the parent basic level category that are important enough to be separately named, such as "lounge chair" and "Balans chair."

Below any immediate descendant of a basic level category, the nature of the representation changes somewhat. The essential functional plan is fixed, and is represented by the conjunction of the functional properties listed in the *Realized_By* field of the subcategory node. However, there may be different levels of refinement or elaboration of the essential functional plan that are important enough to be separately named. These would be listed in the *Functional_Plans* field of the same sub-

454

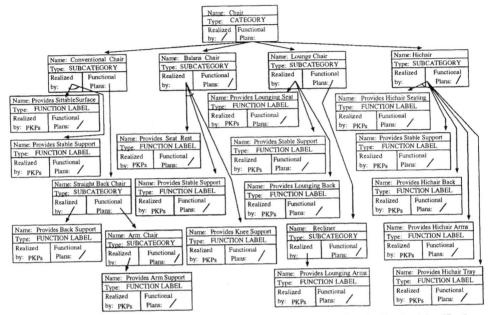

(Figure reprinted from L. Stark and K. Bowyer, Achieving Generalized Object Recognition Through Reasoning About Association of Function to Structure, IEEE Transactions on Pattern Analysis and Machine Intelligence, volume 13, pages 1097-1104. Copyright, 1991, IEEE.)

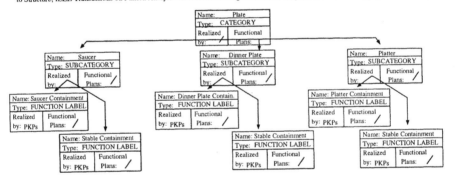

Figure 5. Representation of Basic Level Category Chair and Plate

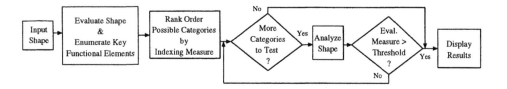

Figure 6. Flow of Control in Interpreting an Object Shape

category node, and they represent a required conjunction of additional functional properties. Thus a *conventional chair* is anything that *provides_sittable_surface* and *provides_stable_support*. A *straight back chair* adds the requirement of *provides_back_support* to the requirements for *conventional chair*. This distinguished naming of specializations of the essential functional plan may be carried more than one level deep. For example, *arm chair* adds the requirement *provides_arm_support* to the requirements of *straight back chair*.

2.4. Flow of Control in Interpretation

The input shape is defined as a set of faces, with each face defined by an ordered set of vertices. The first step in the recognition process is evaluation of the input shape to tabulate some information that will be used throughout the recognition process. This information includes the volume of the shape, center of mass of the shape, convex hull of the shape, the volume and center of mass of the convex hull, a list of convex 3-D sub-volumes of the shape, a list of groups of faces which are essentially coplanar and close to each other so that they can act as a surface for functional purposes, and all groups of faces that define enclosable concavities of the shape. (We will use the term *face* when referring specifically to one face of the input shape model and the term *surface* or *virtual surface* when referring to one or a group of faces that can act together as a functional surface.)

In the next stage, the system hypothesizes an ordered list of categories to use in evaluating the shape. This processing begins conceptually at the root of the category definition tree and flows toward the leaves. The ordering of categories is done by comparing indices computed from the shape in the first step with index ranges kept at each category node and computing an *indexing measure* for each category. The index ranges are broad, due to the varied shapes possible within each category, but help to eliminate "impossible" shapes. For example, these heuristics help eliminate evaluation of shapes the size of a conventional chair for possibly fulfilling the functional requirements of a dish or even a bed. A more detailed description of the indexing scheme is given later.

The categorization performed by the system identifies functional elements of a shape by associating them with their proper *function label* (e.g. *provides table surface*, or *provides saucer containment*). Once a category is hypothesized, that subtree of the category definition tree is used as a control structure in the analysis of the shape. If a basic level category has multiple subcategories, indexing is done again at this level.

For each hypothesized subcategory, there is a subtree that defines the *functional plan* of that subcategory. The traversal of each subtree proceeds in a depth-first manner. For example, analysis of the shape depicted in Figure 3 begins with confirming some surface(s) that could satisfy *provides king size sleeping surface*, and then analyzes whether the shape could *provide stable support* for a candidate sleeping surface. The purpose of the next section is to provide a better understanding of how evidence is gathered by the system and how an overall association measure is calculated.

2.5. Accrual of Association Measure

The system attempts to categorize an input shape as belonging to some subcategory, with some cumulative *association measure*. As noted earlier, evidence is initially gathered directly from the knowledge primitives which define each required functional property. This evidence is combined to reflect an overall measure of how well the functional properties for the category are fulfilled. The aggregation calculi used should behave so that if a required knowledge primitive invocation returns an evaluation measure lower than the current association measure, it lowers the overall evidence for that subcategory. Thus the association measure for the minimum level of membership in a category is dominated by the weakest of the set of required primitives.

When the subcategory being investigated is a refinement of its parent subcategory, in the way that straight back chair is a refinement of conventional chair, the evidence gathered at the parent subcategory node must be combined with the present evidence. In this case the behavior of the aggregation calculi should be to increase the association measure at the parent subcategory node by some factor associated to the evidence gathered at the present node.

This establishes two types of behavior for the aggregation calculi, depending upon the level of processing. An investigation of different calculi was performed using the criteria defined above [11]. It was found that a pair of T-norm and T-conorm functions provided the best performance.

2.5.1. Distinguishing Levels of Processing

An approach has been adopted that maintains an accrued measure on the interval [0,1]. Evidence gathered is combined in either a conjunctive or disjunctive manner, depending upon the *level* of processing. The first level of processing reflects how well a specific portion of the structure fulfills required functions (combining evidence at the knowledge primitive level). For example, evidence is gathered to reflect how well a surface can act as a sittable surface of a chair and whether the shape is stable when that surface is oriented parallel to the support plane. The second level combines evidence relating the interaction of different functions; for example, how well the sittable surface and back support relate to perform as a straight back chair (functional plan level).

2.5.2. Combination of Evaluation Measures from Knowledge Primitives

At the knowledge primitive level, *evaluation measures* are returned and combined to obtain an *association measure* for each functional requirement. All requirements must meet some threshold measure in order to consider the requirements of the function label as being fulfilled. Representative instances of three families of aggregation calculi [2] were evaluated. The operation that gave the best performance is the T-norm or conjunctive operation:

$$T(a, b) = ab$$

Assuming that the association measure calculated to reflect a specific functional requirement is initialized to one at the beginning of evaluation, the accrual of evidence is accomplished as:

$$new\_association\_measure = present\_association\_measure * evaluation\_measure$$

where the *evaluation_measure* is the value returned by each primitive invocation. Further processing leads us to the level at which we combine information concerning specialization properties (e.g. a straight back chair is a specialization of a conventional chair).

2.5.3. Combining Evidence across Functional Plans

Evaluation of the input shape follows a depth first traversal of the control structure (see Figure 5). For example, to establish that the shape belongs in the subcategory straight back chair it must first be established that the shape meets the functional requirements of the conventional chair. An aggregate measure is calculated by combining evidence of how well each of the functional requirements (*provides sittable surface* and *provides stable support*) was met. Processing then continues with the subcategory node straight back chair. An association measure local to the subcategory node is calculated for the single functional requirement specified for this node (*provides back support*). Now the evidence must be combined across functional plan levels. The evidence gathered in support of conventional chair is aggregated with the evidence gathered in support of straight back chair using the T-conorm operator [2]:

$$S(a, b) = a + b - ab$$

giving a final association measure to assign to the straight back chair subcategory.

Processing continues down the control structure until the functional requirements cannot be met or the graph cannot be traversed downward any further. Each subcategory node which has been visited holds an association measure on the interval [0,1]. Each subcategory node which has not been visited due to lack of evidence at a parent subcategory node holds an association measure of zero. Recognition is accomplished by reporting the nodes with the maximum association measures above a set threshold value.

3. Complexity of Shape Evaluation

In the current implementation, the input shape is assumed to be polyhedral (either because the object is truly polyhedral or because the shape acquisition mechanism produces a polyhedral approximation). A polyhedral shape with N faces is $\theta(N)$ in the number of faces, edges and vertices. An analysis of the time complexity of system execution can be broken into two main stages: pre-evaluation of the shape followed by function-based interpretation.

The first stage of the processing involves evaluation of the overall shape and formation of a list of all potential *functional elements*. Specifically, we perform the following three operations.

One, compute the volume of the shape. An algorithm due to Edelsbrunner [6] is used to divide the shape volume, and in fact all of 3-D space, into a unique set of $\theta(N^3)$ convex volumes, and then computing the sum of the volumes ($\theta(N^2)$ step) of those cells which correspond to the shape. For a polyhedral shape with N faces, computation of the volume is therefore $\theta(N^5)$.

Two, compute the center of mass of the shape, assuming that the object has uniform density. This is an $\theta(N^3)$ step.

Three, form a list of all the potential functional surfaces (individual faces of the shape and "virtual surfaces" formed by collections of faces which are "essentially coplanar"), all the enclosable concavities of the shape, and all the convex cells which make up the shape. The list of "virtual surfaces" is formed using a three-step heuristic. First, the N faces are sorted based on area (a $\theta(N \times logN)$ step). Then this list is divided into non-overlapping sublists of "essentially parallel" faces (a $\theta(N)$ step). Finally, each sublist is further divided into non-overlapping sublists of "essentially coplanar" faces (a $\theta(N \times logN)$ step). The result is a list of at most $\theta(N)$ surfaces (object faces and groups of faces that can act as a functional surface). The list of concavities is formed through a two step process. The first step involves finding all surfaces of the object that are not part of the convex hull (concavity surfaces). This is a $\theta(N^2)$ step. There are at most $\theta(N)$ concavity surfaces, each of which can be part of a single concavity. The next step is to form lists of non-overlapping surfaces, one for each concavity ($\theta(N^2)$ step).

The second stage of the processing is to interpret the shape according to the functional plan of the hypothesized categories. Each category is represented by a sequence of required functional properties, and each functional property is realized by a sequence of invocations of the knowledge primitives (KPs). The complexity of each KP is as follows.

• *relative orientation* - Finding all pairs of surfaces in a list of N surfaces which satisfy a particular relative orientation relationship is $\theta(N^2)$, there being $\theta(N^2)$ pairs in the list and the time for a surface normal evaluation and a difference between normals both being constants.

• *dimensions* - Each of the N edges of a surface is aligned with the positive **X** axis in an **X-Y** plane. In each alignment, each of the N vertices is examined and a record of the min/max **X** value and min/max **Y** kept. Thus finding all orientations of a given N-edge surface for which the extent of the surface falls within a given width and

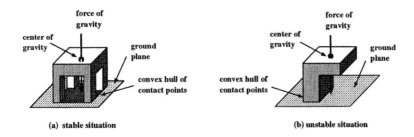

Figure 7. Testing Stability of Object Shape in Given Orientation

depth range is at worst $\theta(N^2)$.

• *stability* - Consider a given orientation of a shape to be checked for stability. Finding all the vertices at the minimum **Z** value (and so potentially in contact with the support plane) is a $\theta(N)$ step. If there are three or more non-collinear points of contact, then the convex hull of the contact points is found (a $\theta(N \times logN)$ step). Then, based on gravity resulting in a downward force, a vector is projected from the center of mass of the shape perpendicular to the supporting plane (a constant time step). If the vector projects to a point inside the convex hull, then the object is stable in that orientation and if it does not then the object is not stable in that orientation. Figure 7 depicts a stable and unstable situation. (Checking this is a $\theta(N)$ step). Thus, checking the stability of a particular hypothesized orientation of an object with N faces is a $\theta(N \times logN)$ process. In some cases, it is necessary to re-apply the stability test to an orientation already found to be stable, with forces applied to some points on the object. Any constant number of such checks still results in a $\theta(N \times logN)$ process.

If a shape is not stable in the given orientation, it may be necessary to generate all possible stable orientations. This can be done by considering all triplets of vertices (a $\theta(N^3)$ process). Hence, checking for all possible stable orientations is at worst a $\theta(N^4 \times logN)$ operation.

• *proximity* - Checking a list of N surfaces, where each surface has at most M vertices, to find all those surfaces which fall within the specified proximity to a given surface is a $\theta(N \times M)$ process.

• *clearance* - The check is made by testing each of the six faces of the clearance polyhedra to make sure that it does not intersect any of the N faces of the object, resulting in a $\theta(N)$ process.

• *enclosure* - Each concavity is defined as a set of faces, with each face defined as a sequence of vertices ($\theta(N)$ vertices). For a given orientation, a plane is introduced at the maximal height of the concavity (the level of the vertex which is farthest from the support plane). This plane is intersected with the concavity (a $\theta(N^2)$ process). It must be established that the intersection of the plane that is introduced with the concavity forms one or more closed faces and that by combining these enclosing faces to the concavity description a valid closed solid is formed. Testing for closed

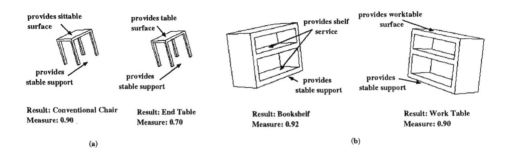

<table>
<tr><td>Result: Conventional Chair
Measure: 0.90</td><td>Result: End Table
Measure: 0.70</td><td>Result: Bookshelf
Measure: 0.92</td><td>Result: Work Table
Measure: 0.90</td></tr>
</table>

(a)   (b)

Figure 8. Examples of Shapes That Could Function in Different Ways.

faces is a $\theta(N^2)$ process. Testing for a valid closed concavity description is also $\theta(N^2)$. If the plane introduced at the *maximal* vertex level does not enclose the concavity, a new plane is introduced at the height of the next highest vertex. Therefore, the $\theta(N^2)$ process described above could be repeated up to $\theta(N)$ times giving a $\theta(N^3)$ process overall.

Evaluation of an input shape involves a sequence of operations. The maximum length of the sequence is fixed by the category definition tree and is independent of the object shape. Thus the complexity is that of the maximum complexity step, the $\theta(N^5)$ step of computing the volume.

4. Evaluation of System Competence

Appropriately evaluating the competence of a function-based recognition system is not simple. The difficulty is inherent to the nature of the approach– it is entirely reasonable (in some cases, even desirable) for a shape to belong to more than one category. For example, the shape depicted in Figure 8(a) is about equally functional as either a simple chair (a stool) or as an end table. In this case, there does not seem to be a strong preference for one interpretation over the other. Similarly, the shape depicted in Figure 8(b) can function about equally well as either a bookshelf or as a work table. However, in this case most people would have a strong preference for naming the object as a bookshelf. This preference may be driven by considerations that are not purely function-based; the presence of the regularly-spaced parallel surfaces may be considered so "non-accidental" that it must be related to what the object is "meant to be" (as opposed to what it can function as).

Thus at one level, the competence of a function-based recognition system might be measured by how correctly it can determine all of the possible functions of an object. At a more detailed level, the competence of a system might be measured by how correctly it can determine the rank ordering of the possible functions. The problem in either case is how to judge the correctness of the system's answers. The standard for comparison must be some human interpretation of the same shape. However, the human interpretation of a shape is based on a much larger domain of

Table 1
Evaluation of "Is / Is Not" Competence in Function-Based Recognition.

	HUMAN INTERPRETATION	SYSTEM INTERPRETATION	PERCENT AGREEMENT
CHAIR (is/is not)	66 / 352	56 / 331	85 / 94
TABLE (is/is not)	31 / 387	17 / 378	55 / 98
BENCH (is/is not)	31 / 387	27 / 366	87 / 95
BOOKSHELF (is/is not)	17 / 401	15 / 399	88 / 99
BED (is/is not)	27 / 391	25 / 389	93 / 99
CUP/GLASS (is/is not)	35 / 383	35 / 383	100 / 100
BOWL (is/is not)	44 / 374	39 / 360	89 / 96
PLATE (is/is not)	13 / 405	13 / 403	100 / 99
POT/PAN (is/is not)	26 / 392	25 / 392	96 / 100
PITCHER (is/is not)	2 / 416	2 / 412	100 / 99
NO KNOWN OBJECT	152	136	89

competence and possibly also on context and cues for what a shape is meant to be. Thus it is clear that the results of such a comparison must be considered carefully.

4.1. Is/Is Not Competence

We created a database of 418 shapes to use in evaluating the competence of the system. A subset of these shapes is depicted in Figures 9 - 11. One of the authors considered each shape from the point of view of "could it function as an X?". Thus each shape was assigned to one or more of the ten categories known to the system or to "no known category." If the shape was assigned to more than one category, then the categories were ordered from most appropriate to least appropriate. Each shape was also analyzed by the system. No indexing strategy was used in this analysis, so that each shape was evaluated for each category known to the system. If the evaluation measure fell below 0.4 for a given category, then the shape was considered as not capable of providing the specified function. The categories resulting in a measure of greater than 0.4 were ranked by their measure. (The value of 0.4 was chosen empirically as a threshold which gave good overall performance.)

The first level of comparison is how often the human and the system agree that a shape could provide the function of a given category, ignoring the relative ranking of categories. We refer to this as the *is / is not competence* of the system. The results of the comparison at this level are summarized in Table 1.

On the whole, the system agreed with the human interpretation approximately 85% of the time. The largest discrepancy was the 55% agreement in category table. The 14 discrepancies for this entry are depicted in Figure 9. To gain a better understanding of system performance, these discrepancies will be explained in greater detail. A more thorough breakdown of system performance will be presented later.

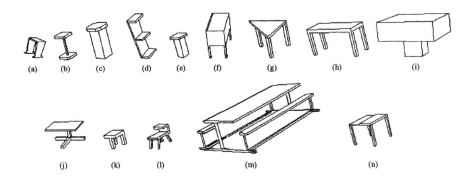

Figure 9. Disagreement between System and Human Interpretation of Tables

The shapes can be divided into subsets according to the reason for disagreement. The shape in Figure 9 (a) failed due to not being able to find a surface large enough (according to system constraints) that could function as a table. The shapes in Figure 9 (b-e) initially identified surfaces which fell in the end table size range, but later failed due to being too tall (according to system constraints) to function properly as end tables. Object (f), which has a small lip running along one side of the top surface, failed due to not being able to establish clearance. We have included in the functional properties of clearance and accessibility the requirement that the table surface be completely clear and accessible from all sides. Objects (g) and (h) did not pass because their measures (0.24 and 0.37, respectively) did not meet the threshold of 0.4. For shapes (a-h), recognition as tables could be accomplished by simply adjusting our functional property constraint values or possibly our threshold value.

Objects (i-k) were actually recognized as tables, but were flagged as potentially unstable if a force is exerted on the table surface near the edges. Objects (l-n) failed due to the lack of knowledge by the system of the concepts of offset or "tiered" table surfaces and the concept of picnic table. The system has no explicit functional plan for a picnic table, and so object (m) does not pass as a table since the table surface would not be accessible from all directions, and it does not pass as a bench since the benches are partially occluded by the table. A new functional plan would need to be added which allows such a relationship.

4.2. Preference Competence

A more subtle level of comparison is to look only at the first-ranked interpretations of each shape. We will refer to this as the *preference competence* of the system. This evaluation looks at how well the system makes appropriate rankings between categories. The results of this comparison are summarized in Table 2. The rows represent the system interpretation that resulted in the highest evalua-

Table 2
Evaluation of Preference Competence in Function-Based Recognition

System Interpretation	Human Interpretation										
	chair	table	bench	bookshelf	bed	cup/glass	bowl	plate	pot/pan	pitcher	not known
chair	50	3	1	1							6
table		12									4
bench	1	4	20								3
bookshelf				15							
bed					25						1
cup/glass						35					
bowl							38	4			1
plate							1	13			1
pot/pan									22		
pitcher										2	
not known	6	4		1	2	5					137

tion measure. Similarly, the columns represent a person's preferred interpretation. If the system's relative ranking of categories was totally independent of the human ranking of the categories, then the numbers in a given row should be evenly distributed across the columns. If the human and system interpretation were perfectly correlated, then the only non-zero entries would be on the diagonal. Over the entire database of shapes, the system's preferred interpretation agreed with the human's preferred interpretation approximately 90% of the time. Figure 10 shows a representative set of these shapes by category.

The shapes for which the system's interpretation did not agree with the person's interpretation provide more insight into the problem than do the shapes for which there was agreement. These shapes are depicted in Figure 11. Many of these are the result of the person's interpretation being driven by what the shape was "meant" to be and not easily being able to consider a shape purely on function-based grounds.

Recognition failures depicted in Figure 11 can be attributed to different factors. As explained earlier, some shapes do not pass analysis because the system cannot confirm that there is a functional element which fulfills the requirements within the specified constraint values. Objects 3-5, 8-13, 21, 36-43, and 46-49 failed due to such conditions. On the other hand, at times the system was able to confirm functional elements of the proper constraints, whereas the creator of the object believed these limits had been exceeded (objects 44-45). Another reason for recognition disagreements was due to the fact that the system was looking for stable conditions when forces were applied to specified functional elements. Objects 6, 14, 16, 17, and 19 all failed due to stability tests. (Object 14 has one leg which is shorter than the other three.)

Other shapes fail as a result of the system identifying novel orientations in which the shapes function as the designated category. For example, shapes 25-27 were labeled as unknown in the human interpretation. However, the system allows the shapes to be turned over to use the bottom of the chair as a sittable surface. Object 32, which resembles a table with a hole in the center, was found to be able to

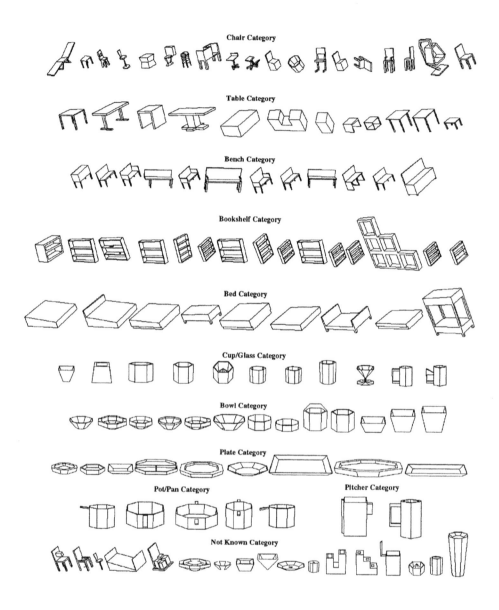

Figure 10. Shapes with Human and System Interpretation in Agreement

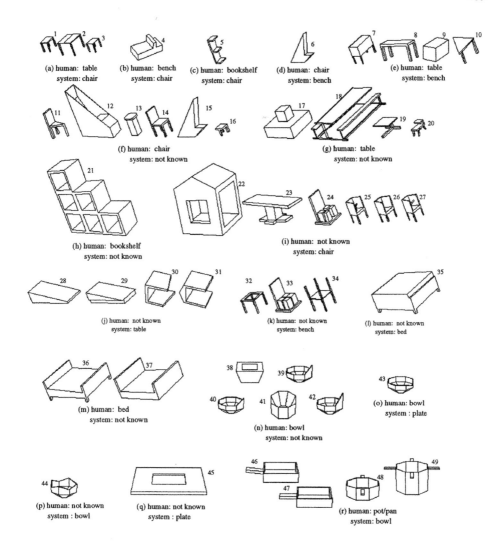

Figure 11. Shapes with Human and System Interpretation in Disagreement

function as a bench (in four different orientations) at a low measure (0.4) because the system found enough surface area of the proper dimensions at the proper height that could serve as an extended sittable surface.

Overall, we feel that the system has demonstrated great success in performing function-based recognition of shapes in its intended domain of competence. By considering the shapes in Figure 10, it is easy to verify that there are no glaring errors in the shapes that it has recognized as providing the function of one of its known categories. It takes somewhat longer to verify that the shapes classified as "no known category" could not in fact provide the function of any of the known categories, but a careful consideration of the shapes depicted in Figure 11 should be convincing.

5. Evaluation of an Indexing Strategy

We can distinguish between the *competence* and the *efficiency* of a recognition system. The system's competence is related to how frequently it correctly recognizes objects, and its efficiency is related to how much processing is required for recognition. In the evaluation of the system's competence, each shape was processed against each known category. For greater efficiency, the normal mode of system operation is to use an indexing strategy in order to avoid as much processing as possible.

The indexing strategy in the current system uses the results of the pre-evaluation of a shape to eliminate "impossible" categories from consideration and to rank order the possible (superordinate/sub)categories. The pre-evaluation yields information such as the number and size of surfaces for both the shape and its convex hull and also the volume of the shape and its convex hull. At least two quantities from the pre-evaluation are obvious candidates for use in indexing the possible categories– 1) the volume of the shape or its convex hull, and 2) the area of possible functional surfaces of the shape. The purpose of an indexing strategy is to choose a key property that best distinguishes between one (superordinate/sub)category and another. A single property could be chosen and used at all levels, or different properties could be used. The current system uses both candidates listed above.

Information used by the indexing strategy is built into the category definition tree. Each subcategory is defined in terms of one or more functional properties that are associated to functional elements of a shape. One of the functional properties may be selected as the key property. In the current version of the system, the key property for furniture is surface area, associated to a surface functional element (sittable surface of a chair, sleeping surface of a bed, ...) while the key property for dishes is volume of the convex hull. For furniture, the constraint values that are used in the knowledge primitive invocations that define the functional property result in a minimum and maximum possible area for the surface that satisfies the functional property. The constraints on minimum and maximum possible volume values for dishes were estimated from prototypical dish descriptions. Thus at each subcategory node, there is a (min, max) range for the key property for the subcategory. At each basic level category node, a similar (Min, Max) range for the key value is computed by taking the minimum of the minimums for each subcategory and the maximum of the maximums for each subcategory. At the superordinate category node, another (MIN, MAX) range is computed by taking the minimum of the minimums and the maximum of the maximums for the basic level categories.

When a new shape is encountered, the indexing strategy operates as follows. A list of the areas of its potential surface functional elements and the volume of the shape is consolidated. These values are checked against the key value range that is stored at the superordinate category node. If the object key properties do not fall within the (MIN, MAX) range for the superordinate category, then processing is discontinued with the decision that the shape cannot possibly belong to that superordinate category. If the shape meets the requirements of the key property constraints, processing continues to check against the (Min, Max) range of each of the basic level categories and an *indexing measure* is computed for each range. (The *indexing measure* is a different concept from the *evaluation measure*). At this level, area is the key property used for all categories. The indexing measure is one if the area falls directly on the midpoint of the (Min, Max) range, and falls off linearly to zero at the Min and Max values. We further incorporate the heuristic that if the surface size required for the functional property is larger, it should be weighted higher if a potential match is found.

For example, if a shape were to function as a chair, it would be less likely that the shape would have a surface large enough to provide any type of sleeping surface. On the other hand, it would be likely that a shape that could function as a bed would have surfaces within the proper size range to function as a chair, table, bench or bookshelf. The weighting factor is simply the area of the surface being tested. This is multiplied by the raw indexing measure to obtain the weighted indexing measure. Larger surfaces are therefore given a higher weighting factor.

After all of the surfaces derived from the shape have been checked, the categories are ranked according to the greatest indexing measure that they generated. A similar indexing process can then be applied to determine the order in which to consider the subordinate categories. The input shape is then considered against the possible categories in the order of their ranking, until the shape is found to fulfill the functional requirements of some category with an evaluation measure of 0.4 or better. This leaves open the possibility that the system would recognize the shape as one category when in fact there is a lower-ranked (by the indexing strategy) category which would result in a higher evaluation measure.

The ideal indexing strategy would require a "small" amount of work that was independent of the number of categories known to the system, and would always select as the first category to be considered that category for which the shape would have the highest final association measure. The amount of processing required in the pre-evaluation depends on the complexity of the shape but is independent of the number and type of categories known to the system. The number of category evaluations saved by the indexing strategy, as compared to not having an indexing strategy, is summarized in Table 3. With 418 shapes tested and ten basic level categories known to the system, a total of 4180 category evaluations would be

Table 3
Evaluation of the Effectiveness of the Indexing Strategy

	Number Of # Objects	# Of Possible Evaluations	Evaluations Actually Made	Evaluations Saved	Indexing Efficiency
No Threshold	418	4180	2607	1573	38 %
0.40 Threshold	418	4180	2207	1973	47 %

done if the system used no indexing strategy. With the simple indexing strategy described above, a total of 1573 potential category evaluations are immediately eliminated because the key value list for the shape has no matches to the key value ranges for the superordinate categories on a whole and also for some basic level categories. With recognition processing ending as soon as the shape is found to fulfill the functional requirements of some category with an evaluation measure of 0.4 or better, an additional 400 category evaluations are avoided. Thus 1973 of the 4180 total possible category evaluations are eliminated by the indexing strategy.

As mentioned earlier, there exists the possibility that, with the indexing and cutoff at the first plausible evaluation measure, the system could recognize a shape as one category when in fact there is some other category which would result in a higher evaluation measure. This occurred in 25 cases, but in 14 of those the difference in the evaluation measure for the recognized category and the highest possible evaluation measure was less than 0.125. This suggests that most recognition "errors" introduced by the indexing will be between categories which the shape fulfills almost equally well.

6. Discussion

We have evaluated a function-based recognition system that takes an uninterpreted 3-D shape as its input and reasons to determine what object categories the shape might belong to. The system has analyzed over 400 input shapes and the results largely agree with human interpretation. The greatest source of variation from human interpretation occurs with "near miss" shapes which humans label as not one of the known categories, but which have some novel orientation in which they could in fact serve the function for some category. This is a natural reflection of the fact that the system uses a purely function-based definition of the category whereas humans use other types of cues and context in their interpretation.

This work confirms that a relatively small number of knowledge primitives may be used as the basis for defining a relatively broad domain of competence. We have also shown that an indexing strategy may be devised to make the processing for function-based recognition more efficient with a relatively modest penalty in terms of accuracy.

This work does not imply that more specific representations of individual objects are not needed. If the task is to label objects within a scene as specific individual objects (e.g. "table model XYZ"), the generic representation by itself cannot provide such information. However, by first categorizing objects within a scene using a generic representation, only those objects that fall within the proper category need to be further examined using more specific representations. Matching of the individual models to the chosen object in the scene can also be completed more efficiently by using the symbolic labeling provided by the generic recognition. For example, the portion of the structure labeled as providing a sittable surface could automatically be aligned with the portion of the structural description labeled seat.

The example object shapes used to evaluate the system in this paper were all defined using a solid modeler. One of the continued research directions that we are working on is to evaluate the system's performance on models created by extracting surfaces from real image data taken with a laser range finder. Since the function-based reasoning is naturally qualitative, we expect that the system will maintain good performance with this type of image data.

REFERENCES

1　Binford, T.O. 1982. Survey of model-based image analysis systems, *Int. J. of Robotics Research*, **1**, 18-64.

2　Bonissone, P.P. and Decker, K.S. 1985. Selecting Uncertainty Calculi and Granularity: An Experiment in Trading-off Precision and Complexity, in Uncertainty in Artificial Intelligence, L. Kanal, and J. Lemmer (Editors), North-Holland Publishing Company, 217-247.

3　Brady, M., Agre, P.E., Braunegg, D.J., and Connell, J.H. 1985. The Mechanics Mate, in Advances in Artificial Intelligence, T. O'Shea (ed.), Elsevier, 79-94.

4　Connell, J.H. and Brady, M. 1987. Generating and generalizing models of visual objects, *Artificial Intelligence*, **31**, 159-183.

5　Di Manzo, M., Trucco, E., Giunchiglia, F., Ricci, F. 1989 FUR: Understanding FUnctional Reasoning, *Int. J. of Intelligent Systems*, **4**, 431-457.

6　Edelsbrunner, H., O'Rourke, J., and Seidel, R. 1986. Constructing arrangements of lines and hyperplanes with applications, *SIAM J. Computing* **15**, 341-363.

7　Kitahashi, T., Abe, N., Dan, S. Kanda, K. and Ogawa, H. A Function-based Model of an Object for Image Understanding, in *Advances in Information Modelling and Knowledge Bases*, H. Jaakkola, H. Kanassalo and S. Ohsuga (editors), IOS Press, 91-97.

8　Minsky, M. 1985. The Society of Mind, Simon and Shuster, New York.

9　Stark, L., and Bowyer, K.W. 1992. Indexing function-based categories for generic object recognition, *Computer Vision and Pattern Recognition (CVPR '92)*, 795-797.

10　Stark, L., and Bowyer, K.W. 1991. Achieving generalized object recognition through reasoning about association of function to structure, *IEEE T-PAMI*, **13**, 1097-1104.

11　Stark, L., Hall, L.O. and Bowyer, K.W. An investigation of methods of combining functional evidence for 3-D object recognition, to appear in *Int. J. of Pattern*

Recognition and Artificial Intelligence.

12 Vaina, L. and Jaulent, M. 1991. Object structure and action requirements: a compatibility model for functional recognition, *Int. J. of Intelligent Systems*, **6**, 313-336.

13 Winston, P., Binford, T., Katz, B., and Lowry, M. 1983. Learning Physical Description from Functional Definitions, Examples, and Precedents, *AAAI '83*, 433-439.